City Bountiful

The publisher gratefully acknowledges the generous contribution to this book provided by the Harriett Gold Architecture, Art, and Design Endowment Fund of the University of California Press Associates.

City Bountiful

A CENTURY OF COMMUNITY GARDENING IN AMERICA

Laura J. Lawson

UNIVERSITY OF CALIFORNIA PRESS
BERKELEY LOS ANGELES LONDON

University of California Press
Berkeley and Los Angeles, California

University of California Press, Ltd.
London, England

Library of Congress Cataloging-in-Publication Data
Lawson, Laura J., 1966–
City bountiful : a century of community gardening in America / Laura J. Lawson.
p. cm.
Includes index.
ISBN 0-520-23150-3 (cloth : alk. paper)—
ISBN 0-520-24343-9 (pbk. : alk. paper)
1. Community gardens—United States—History—20th century. I. Title.
SB457.3.L39 2005
635—dc22 2004010999

Manufactured in the United States of America
14 13 12 11 10 09 08 07 06 05
10 9 8 7 6 5 4 3 2 1
The paper used in this publication meets the minimum requirements of ANSI/NISO Z39.48–1992 (R 1997) *(Permanence of Paper).*

CONTENTS

LIST OF ILLUSTRATIONS *vii*

LIST OF TABLES *xi*

PREFACE AND ACKNOWLEDGMENTS *xiii*

Introduction. Garden Patches in American Cities *1*

PART I. EARLY URBAN GARDEN PROGRAMS, 1890S TO 1917

Introduction *17*

1. An Alternative to Charity:
The Vacant-Lot Cultivation Association *23*

2. The School Garden Movement *51*

3. The Goodness of Gardening: Gardens as Civic Improvement *93*

PART II. NATIONAL URBAN GARDEN CAMPAIGNS, 1917 TO 1945

Introduction *113*

4. Patriotic Volunteerism: The War Garden Campaign *117*

5. An Antidote for Idleness:
Garden Programs of the 1930s Depression *144*

6. Victory Gardens of World War II *170*

PART III. GARDENING FOR COMMUNITY, 1945 TO THE PRESENT

Introduction *205*

7. The Community Garden Movement of the 1970s and 1980s *213*
8. Community Greening: Urban Garden Programs from 1990 to the Present *238*
9. A Look at Gardens Today *264*

Conclusion. Sustaining a City Bountiful *287*

NOTES *303*

INDEX *347*

ILLUSTRATIONS

1. Community garden, Battery Park City, New York / 6
2. The Watts Growing Project, Los Angeles / 7
3. Philadelphia mill girls, circa 1905 / 9
4. Children at the Berkeley Youth Alternatives Community Garden Patch / 9
5. Three gardeners from a World War I war garden program / 10
6. Building the Berkeley Youth Alternatives Community Garden Patch / 10
7. Application for vacant-lot work, New York Association for Improving the Condition of the Poor / 35
8. Children in market garden, Philadelphia Vacant Lots Cultivation Association / 37
9. Philadelphia Vacant Lots Cultivation Association gardens, 1899 / 40
10. Philadelphia Vacant Lots Cultivation Association gardens, 1899–1925 / 41
11. "Ignorant of Nature: A Sociological Problem" / 56
12. "Good Gardeners" / 56

13. Planting plan, George Putnam School, Boston / 61
14. Farm school at DeWitt Clinton Park, New York / 64
15. Los Angeles State Normal School garden plan, 1905 / 70
16. East Seventh Street School garden, Los Angeles / 71
17. Garden City Bank, University of California, Berkeley / 73
18. "The Raking-Drill" / 77
19. Teacher training at the University of California, Berkeley / 79
20. Kindergarten class of African American children, circa 1899 / 81
21. Poughkeepsie, New York, school garden plan, 1910 / 87
22. Panama-Pacific International Exposition school garden plan / 88
23. Cleveland vacant lot, before / 101
24. Cleveland vacant lot, after / 101
25. Fairview Garden, Yonkers, New York / 105
26. National Cash Register Company Boys' Garden, Dayton, Ohio / 106
27. Home Gardening Association of Cleveland order form / 108
28. "Liberty Sowing the Seeds of Victory" / 123
29. Poster for the United States School Garden Army / 127
30. Children displaying a cabbage grown in their school garden / 129
31. Demonstration war garden, Bryant Park, New York City / 131
32. "Food Will Win the War" / 132
33. Cartoon: mother in her garden overalls / 136
34. Cartoon: slacker land / 137
35. B. F. Goodrich Cooperative Farm / 154
36. Subsistence Homestead garden, El Monte, California, 1936 / 161
37. FERA Airport Farm, Seattle, 1934 / 163
38. Detroit Thrift Gardens application / 167
39. Gardens for the unemployed, Jersey Meadows, 1939 / 169
40. The Food Fights for Freedom campaign's four guidelines for action / 173

41. "Plant a Victory Garden" / 176
42. Vacant-lot garden, San Francisco / 182
43. Victory gardens, Sunnydale Housing Project, San Francisco / 183
44. Victory garden, Golden Gate Park, San Francisco / 183
45. Patients in hospital victory garden / 187
46. Los Angeles woman with her first crop of radishes / 189
47. Gardening in Beverly Hills / 191
48. Young victory gardeners, Bunker Hill, Los Angeles / 192
49. Victory garden, apartment courtyard, Los Angeles / 194
50. Victory garden, Charles Schwab estate, New York City / 195
51. Mothers with children in playpens at their San Francisco victory garden / 198
52. Victory garden, Metropolitan Life Company, San Francisco / 199
53. Community garden, South End/Lower Roxbury, Boston / 214
54. Southwark/Queen Village Community Garden, Philadelphia / 222
55. Southwark/Queen Village Community Garden from street / 223
56. The original P-Patch, Seattle / 247
57. Laotian woman at community garden, Seattle / 249
58. San Francisco League of Urban Gardeners' Garden for the Environment / 250
59. St. Mary's Urban Farm, San Francisco, 1996 / 252
60. St. Mary's Urban Farm, 2001 / 252
61. Westside Community Garden, New York City / 260
62. Clinton Community Garden, New York City / 261
63. View over Wattles Farm and Neighborhood Garden / 268
64. Wattles Farm and Neighborhood Garden site plan / 269
65. Founding members of the Wattles Farm and Neighborhood Garden / 271
66. Man harvesting at the Los Angeles Regional Food Bank / 273
67. Personal altar, Urban Garden Program of the Los Angeles Regional Food Bank / 274

68. BYA Community Garden Patch site before construction / 276
69. BYA Community Garden Patch, 1996 / 277
70. BYA Community Garden Patch site plan / 278
71. BYA youth gardener preparing for market / 279
72. BYA youth selling at Berkeley Farmers' Market / 280
73. Edible Schoolyard, Berkeley / 283
74. Children in kitchen of the Edible Schoolyard / 284
75. Students and teachers at the Edible Schoolyard / 285

TABLES

1. Cities with vacant-lot cultivation associations / 28
2. Philadelphia Vacant Lots Cultivation Association results, 1897–1927 / 47
3. Chicago's war garden campaign results, 1918 / 134
4. Estimated value of war garden crops in certain cities, 1918 / 141
5. Types and number of subsistence gardens, 1934 national survey / 149
6. Expense estimates for Cleveland's garden programs, 1932 / 151
7. Garden program acreage by state, 1934 / 156
8. Results from the New York State garden program, 1932 / 158
9. Results of USDA Cooperative Extension Urban Garden Program, 1979 and 1985 / 227
10. National winners of the ACGA/Glad Bag Community Garden Contests, 1985 and 1986 / 233
11. American Community Garden Association survey results, 1990 and 1996 / 241

PREFACE AND ACKNOWLEDGMENTS

As with all research, this book represents the particular view of its author, and as such embodies a perspective that has grown out of my experiences as garden aficionado, organizer, and, ultimately, researcher. My interest in gardening began, simply enough, as a stubborn attempt to grow my own food during budget-tight college years. Discovering my love for gardening, I imagined moving to the country, acquiring a few acres, and living a more self-reliant and "simple" life. However, after a summer of training at a semi-rural organic farm, I came to realize that I loved the city, and accordingly, urban self-sufficiency became my mission. From these personal motivations grew a wider perspective that urban gardens are not only a resource for food, but also a potential catalyst for community development. Graduate studies in landscape architecture at the University of California, Berkeley, exposed me to related concepts of participatory design, environmental justice, and community planning that I applied to my understanding of gardening and urban agriculture. After reading about and visiting an array of community gardens in America and Europe, I was eager to apply the lessons I had learned.

The opportunity came in 1993 when I was approached by Niculia Williams, director of Berkeley Youth Alternatives (BYA), to design a garden for the organization. The four years that I spent as Garden Patch coordinator—and later as researcher and volunteer—represent a labor of love that required me to wear many hats—designer, teacher, community organizer, grant writer, youth supervisor, laborer, and diplomat. The rewarding experience of working with West Berkeley youth and BYA staff shed any naive faith I might have had in what gardening could accomplish in and of itself. With participants setting the agenda, the garden had to be resilient and flexible to fulfill varied priorities. Its design and program had to serve the teacher who wanted children to eat from the garden, the youth who did not care for school but found a great sense of pride in his paid employment and public role at the garden, the community gardener who regularly showed up after work to water her plot, and the volunteer who wanted to build garden furniture out of recycled materials. Sometimes conflicts had to be addressed. A minor but illustrative instance of this involved how we labeled the produce we grew for market; while many regular customers at our farmers' market stand bought our produce because it was registered organic, some BYA parents found this labeling to be alienating and preferred the term *fresh* produce. Likewise, while many onlookers praised our employment program as an opportunity for urban youth to experience nature and gardening, the youth themselves often considered the manual labor demeaning if it was not connected to leadership development, community outreach, marketing, or other skills that seemed applicable to a wider range of future jobs. But most critical was the realization that, although we received much praise for our project, funding was extremely difficult to obtain. Furthermore, even though we had invested hundreds of volunteer and youth hours in restoring the site's soil and spent thousands of dollars on materials, our land was leased from the city. While it seemed highly unlikely that anyone in Berkeley would revoke the lease, the fact that the lease stipulated that no permanent structures could be built on the site made me question why soil amending and other improvements were not considered permanent.

In 1996, I raised some of these issues to a group of garden organizers attending the American Community Gardening Association conference. Once we had put aside our pictures of happy children and inspirational stories, we found that we were all struggling with similar concerns. Everyone was committed to the idea of gardening as a resource to serve the social, environmental, and economic needs of urban, low-income communities, but everyone also felt pressured by insecure land tenure, competitive funding, staff burnout, and the need to sustain community-based leadership. We used similar language in our grant proposals—espousing multiple benefits to be achieved through our gardens—and yet we were all quite aware that gardens were a drop in the bucket compared to the unmet needs of the communities in which we worked. What I was hearing was the struggle rarely expressed in literature about urban gardens, and it prompted my query into the roots of this problem.

My investigation into the role and sustainability of urban garden programs started with interviewing garden program organizers and participants around the country. Their feedback confirmed the opportunistic approach needed to build a garden, which often meant focusing on short-term gains and losing sight of long-term goals and garden sustainability. A contemporary literature review revealed a common litany of benefits associated with gardening—nutrition, community food security, emotional restoration, education, environmental restoration, and so on. Such claims were largely anecdotal and raised the question of what aspects of the garden facilitated so many psychological and social benefits. Reaching farther afield, I investigated gardening in the context of environmental psychology, sociology, history, and cultural landscape studies. Research into the history of garden programs in America revealed patterns in organizational development, typical justifications and goals, and practical matters of construction and maintenance. It also revealed a nearly continuous presence of garden projects for over one hundred years, as well as cyclical phases of public support during times of social and economic crisis followed by obscurity once public attention had shifted.

As the gardener knows, gardens require continual care and attention. Neglecting to water for a week during dry summer weather will destroy hours of earlier work to prepare the soil and plant the seeds. Furthermore, if gardening is only about the food produced, in many cases it would be cheaper and easier to go to the store or fast-food restaurant. For the gardener, the process of growing food and flowers entails both a responsibility to nurture and an opportunity to be nurtured. Although garden programs seem to be perennial—appearing in times of crisis and disappearing in times of plenty—their constant reinvention begs the question of sustainability and the need for ongoing support. This support encompasses both the garden itself—the site and resources needed for plant growth—and the people who cultivate it. When we consider the ongoing energy that has been dedicated to communal urban gardens, we can begin to glimpse a more accommodating city, a City Bountiful, that has room enough for people to enjoy the personal and social process of urban gardening. Whereas other movements, such as the City Beautiful movement at the turn of the twentieth century, reordered the city through grand plans of physical improvements and reform, the City Bountiful calls for a subtle transformation in how we conceptualize our cities as land resources and social action. Rather than viewing a vacant lot as underutilized unless it is developed with a building, we can consider such open land an opportunity for food production and social engagement if people are there to inspire and manage its transformation into a garden. Elements of the City Bountiful already exist—the community garden, neighbors sharing harvests, school gardens, and so on. The challenge is to acknowledge the interconnectedness of its various expressions and to validate that framework by sustaining sites and participation as permanent resources to improve the quality of urban life.

This historical account of garden programs relies on a circuitous paper trail of annual reports, committee notes, pamphlets, newspaper and magazine articles, mimeographs, and some books. Annual reports and publications from garden organizations, such as the Philadelphia Vacant Lots Cultivation Association, the School Garden Association of America, the National War Garden Commission, the National Victory Garden Institute, the National Gardening Association, and the American Community Gardening Association, have provided the most in-depth information. Since the early 1900s, the United States Department of Agriculture has produced manuals and pamphlets that provide technical information for urban gardens and school gardens. I have also culled information from annual reports and publications from national organizations tangentially involved with gardening, such as the Association for Improving the Condition of the Poor, the American Park and Outdoor Art Association, the American Civic Association, the Federation of Women's Clubs, the Russell Sage Foundation, and various horticultural societies, garden clubs, and women's organizations. Many of these sources were retrieved only by returning to old bound catalogs, such as the Dictionary Catalog of the National Agricultural Library or the Department Library Subject Catalog of the U.S. Department of Health, Education, and Welfare at the Department of Education Library in Washington, D.C. Often, the authorship, publication source, and other reference information were difficult to cull from the sources. Heights of popularity for an idea or program were marked by the appearance of stories in popular magazines, such as *Country Life in America, The Craftsman, Garden Magazine, Horticulture, House and Garden, National Geographic, Reader's Digest,* and *Sunset Gardens.* Philanthropic and education journals—*Charities, Charities and the Commons, Nature-Study Review, Review of Reviews,* and *The Survey*—describe garden projects in light of the objectives of settlement workers, charity organizers, and educators. *American*

City, American Civic Annual, the *AIP Journal, Garden and Forest, Landscape Architecture, Parks and Recreation,* and *Planner's Journal* include only a few articles on urban gardens per se; however, they provide a context for understanding the activities of environmental-design professionals and city officials during various periods in which urban garden programs occur.

Books were written during the height of each phase of garden promotion. At the peak of the children's school garden movement, several books emerged, including Louise Klein Miller's *Children's Gardens for School and Home* (1906) and M. Louise Greene's *Among School Gardens* (1910). To promote the gardening campaign during World War I, Charles Lathrop Pack wrote *War Garden Victorious,* which included much patriotic encouragement as well as technical advice. Each phase was also marked by books with "how to" suggestions, whether gardening tips in M. G. Kains's *Original Victory Garden Book* or guidelines for organizing neighborhoods in Boston Urban Gardeners' *Handbook of Community Gardening.* Recent descriptions include Patricia Hyne's *A Patch of Eden* and Diane Bolmori and Margaret Morton's *Transitory Gardens, Uprooted Lives.*[1]

Although these sources provide useful insights into the way garden programs have been promoted to the public, they contain only limited information on the lifespan of various phases of urban garden programs. Most articles are "feel good" stories heavily dosed with description and anecdotal accounts. Little quantitative or logistical information, such as numbers of people involved, amount of food produced, or when the projects ended, is provided. More specific data is available on current garden programs. To supplement such information, I interviewed garden advocates and participants, conducted a mailed questionnaire survey, and visited many garden organizations and sites. The results were presented at the American Community Gardening Association annual meeting in 1999 for confirmation by an audience of urban garden organization staff and advocates.

The story of urban garden programs is enriched by examples of projects in various cities. Quite often, historical descriptions of garden campaigns list the activities in multiple cities, and it became my task to hunt down further information. Due to availability of information and matters of logistics, this book focuses primarily on larger cities on the two coasts.

This book is the product of over ten years of research, conversation, and observation. I owe considerable thanks to many people who helped along the way. The research received a jump-start from the previous scholarship of

Thomas Bassett, Sam Bass Warner, Brian Trelstad, and Mark Francis. I am indebted to the librarians at the Environmental Design Library at the University of California, Berkeley, and Annie Thacher at the Dumbarton Oaks Landscape Studies Library, as well as to many other librarians at various collections who met the challenge of helping me search for odd pamphlets and annual reports that were often omitted from current catalogs or misfiled. I hope they enjoyed the scavenger hunt as much as I did. I thank my dissertation committee: Randy Hester, Michael Southworth, and Dell Upton. For inspiration and a new eye, I am indebted to photographer Lewis Watts. Colleagues, students, and staff in the School of Landscape Architecture at Louisiana State University and the Department of Landscape Architecture at the University of Illinois Urbana-Champaign provided essential support to finish this project. Thank you to my research assistants Katherine Melcher, Angela Landry, Agus Soeriaatmadja, and Sungkyung Lee, as well as graduate students Kristofer Johnson and Chris Fellerhoff. Thank you also to the editors and staff at the University of California Press, especially Sheila Levine, Laura Harger, and Charlene Woodcock. I am sincerely indebted to Carolyn Bond for her thoughtful reading and editing during the production process.

For their inspiration and support, I thank the staff and youth at Berkeley Youth Alternatives, particularly Niculia Williams, Elizabeth Crawford, Alison Lingane, Danny Engelberg, Jason Uribe, Eric Davis, Sharon Elkayan, and the Garden Patch youth employees. Thank you to Marcia McNally for encouraging rigorous documentation of the Garden Patch throughout its evolution. Members and the board of the American Community Gardening Association opened their files and willingly shared their stories with me. Special thanks to Blaine Bonham, Alison Brown, Cory Callandra, Libby Goldstein, Tessa Huxley, Betsy Johnson, Charlotte Kahn, Lenny Librizzi, Sally McCabe, Terry Mushovic, Mohammed Nuru, Judy Tiger, and Tom Tyler. Many more names should be included, and I hope that everyone involved realizes their contribution to this effort. Thank you.

For their patient acceptance of my preoccupied state of mind, their interest in my progress, and their support, I thank my friends and colleagues: William Atwater, Shenglin Chang, Mary Edwards, Thelma Fite, Mathew Henning, Walter Hood, Nana Kirk, Anna Mehotra, Louise Mozingo, Kirstin Noreen, and Kevin Risk. Thank you to my parents, Joseph and Patricia Lawson, and my siblings and their families—Christine, Gary, Jaime, Michael, Kelsey, and Randy.

An individual grant from the National Endowment for the Arts sponsored my initial study of contemporary urban garden programs. During my studies at the University of California, Berkeley, I received financial support through the Department of Landscape Architecture's Beatrix Farrand Fellowship, a Regents Graduate Fellowship, a Humanities Diversity Dissertation Fellowship, and the Vice Chancellor for Research Fund. The bulk of the research was conducted while I was a Junior Fellow in Landscape Architecture Studies at Dumbarton Oaks during the 1998–99 academic year. A summer stipend from Louisiana State University provided support for a final summer of interviews and site visits in California and Washington.

This book is dedicated to Will Atwater, who shares my interest in community gardening, and to Kate Lawson Atwater, who we hope learns to love gardening and community as much as we do.

INTRODUCTION

Garden Patches in American Cities

CHILDREN'S DELIGHTED FACES AS THEY pull up the carrots they had planted from seeds, neighbors working together to transform a vacant lot into a garden, mouth-watering vegetables just picked from the earth—these are what come to the minds of many when they think of community gardens. Such images will likely cause experienced gardeners to nod knowingly, having enjoyed both the work and the rewards of the garden, while the general public might smile benignly, having driven past a community garden or perhaps read a story about one in the local paper or a national magazine. When we see these colorful images and hear personal accounts of positive community gardening experiences, we understand why community gardens are generally liked and considered healthy expressions of civic life.

It is not surprising, then, that people have organized to create places for people to garden in American cities since the 1890s. Looking back, we can trace a nearly continuous chain of urban communal garden efforts. In the 1890s, social reformers started the trend by promoting vacant-lot cultivation associations to provide land and technical assistance to unemployed laborers in cities including Detroit, New York, and Philadelphia. At the same time, education reformers promoted school gardens as an interactive teaching venue that correlated with school subjects and taught civics and good work habits. School gardens grew into a national movement, with an office in the federal Bureau of Education devoted to school-garden promotion. The concurrent civic beautification movement motivated women's groups, garden clubs, civic organizations, and others to support vacant-lot gardens, children's gardens, window-box gardens in tenement districts, and garden contests. During World War I, millions of Americans planted backyard and community gardens as a way to augment the domestic food supply so that more could be sent overseas. Children's efforts coalesced through the federally sponsored U.S. School Garden Army. In response to the early stages of unemployment during the Great Depression of the 1930s, families applied to private, municipal, and state agencies for subsistence garden plots and jobs in cooperative gardens and farms. In 1934, over 23 million households participated in subsistence garden programs, growing produce for home consumption that was valued at $36 million. During World War II, households participated in the victory garden campaign to grow food for personal consumption, morale, and recreation. After the war, a few remaining school and community gardens provided continuity and inspiration for a rebirth in interest that occurred in the 1970s. Once again, gardens reappeared in neighborhoods, often built as acts of resistance to urban abandonment as well as to provide resources to address inflation, express a new environmental ethic, and reconnect neighbors during a time of social unrest. Today, myriad garden programs exist, including neighborhood community gardens, children's gardens, horticultural therapy gardens, and entrepreneurial job-training gardens.

Some suggest that community gardens have an even longer history that dates back to the communal lands associated with American frontier towns. The picturesque town commons in New England that we see today started as practical communal land pasturage for sheep and cows as well as space for military drills, sports, and social events. Likewise, towns in the Southwest were laid out according to the Laws of the Indies to include a central plaza and other publicly used land. These were practical and necessary spaces for subsistence, protection, and civic functions. Allotted at the time of settlement, the commons and the plazas evolved into public spaces such as Boston Common and the plaza in Santa Fe. Community gardens, however, reflect a different type of land usage in that they tend to appear after initial development. Such spaces have rarely been planned as part of development but happen after the fact, often on deserted, derelict, or otherwise unused land. However, the need for such spaces has been both practical and appealing enough to compel citizens and institutions to initiate them. Although the contemporary city is no longer threatened with the same dangers as the frontier town was, garden advocates past and present consider urban gardens as "commons" because they are a communal resource to meet current needs associated with subsistence, protection, and civic functions.

DEFINING URBAN GARDEN PROGRAMS

The phrase *urban garden program* encapsulates various cooperative enterprises that provide space and resources for urban dwellers to cultivate vegetables and flowers. While the term *community garden* is probably more familiar to many people and dates back to at least World War I, it tends to be associated with one particular manifestation—the neighborhood garden in which individuals have their own plots yet share in the garden's overall management. The broader category *urban garden* can include more types of programs, such as relief gardens, children's gardens, neighborhood gardens, entrepreneurial job-training gardens, horticultural therapy gardens, company gardens, demonstration gardens, and more. The term *urban* broadly refers to the city, its suburbs, and the urban edge. These are the contexts for most programs that have allotted land to people with limited access to gardening space, although some programs have extended into rural and low-density areas.

The programmatic nature of many of these efforts needs to be stressed. While the idea of allotting land for gardening may seem straightforward, in fact much organization and program development is necessary. Most gardens rely on organizations and programs that coordinate gardeners, manage land, and facilitate educational or social activities. Hailed as a means to address a variety of concerns, urban garden programs have been established by individuals, philanthropic groups, educational and social reformers, civic improvement groups, governmental agencies, environmentalists, and others. Today, we generally assume that these garden projects are "grassroots" efforts, meaning that the programs develop locally with local, nonprofessional leadership. In truth, however, such ventures rely on a network of citywide, national, and even international sources for advisory, technical, financial, and political support. Quite often, the local, often voluntary leadership relies on staffed organizations and policies generated outside the community.

Although various urban garden programs may share the same impetus—to bring people and land together to garden—they can take a variety of forms. Gardens might be located on institutional grounds, public land, or private land. Some programs focus solely on vegetable gardening, while others grow flowers, native plant habitats, or culturally significant plants for the purpose of teaching. Some programs provide land where individuals garden separately, while others set up a communal garden where everyone joins together and then splits the harvest. In most projects, participation is voluntary; however, some past programs required participation as part of the economic relief package provided by a charity or local relief agency. At times, gardening has been promoted as a national effort for everyone to participate in, such as the war gardens and victory gardens of World Wars I and II, respectively. At other times, garden programs have targeted specific groups, such as children, immigrants, or the poor. Even in these cases, people who may not be directly involved as gardeners still provide support through leadership, financial contributions, and land donations. It takes many people to nurture a garden, and many types of gardens to nurture individuals and community.

NOT JUST FOOD AND FLOWERS: GARDENS THAT SERVE MULTIPLE AGENDAS

Growing food has rarely been the only agenda in urban garden programs. Unlike European allotment gardens, American urban garden programs have not had the singular goal of providing land to workers for food production. Garden programs have been established for many reasons—educational, social, economic. While the dominant rationale behind each phase of garden promotion might vary—such as food for war campaigns or to occupy the time of the unemployed—propaganda usually has included a laundry list of additional benefits that have remained surprisingly consistent over time. For example, although the World War I war gardens and the 1970s community garden movement had very different contexts, many of the same justifications were used for both, such as building morale, fighting rising food prices, and improving nutrition. Likewise, while education reform terminology may change, rationales for children's gardens remain strikingly similar. The following praise of the DeWitt Clinton Farm School in New York was written by social reformer Jacob Riis in 1911, and while it reveals the stylistic quirks of his time, similar objectives might be cited today:

> The destructive forces of the neighborhood had been harnessed by so simple a thing as a garden patch, and made constructive. And a "sense of

dignity of labor" had grown up in that of all most unlikely spots, that made the young gardeners willing and anxious to work for the general good as well as for themselves.[1]

Compare this to a 1997 description of the San Francisco League of Urban Gardeners (SLUG) youth program:

> SLUG's youth programs provide opportunities for teenagers to earn a paycheck, contribute to their communities, learn skills, and spend their free time productively and safely. Teens also discover a group to belong to, adult mentors to look up to, and a whole support network to turn to when they need help or guidance. Society has downplayed the importance of these opportunities and this support, choosing instead to pour resources into punishing youth once they have taken a wrong turn. Teens working at SLUG are building healthy futures for themselves and the city.[2]

This comparison does not mean program objectives have not changed over time; contemporary issues and local conditions shape each incarnation of the urban garden program. For instance, the paternalism that was acceptable in the early twentieth century is now replaced by goals of social justice. Tone and wording are not the only ways this is expressed. Whereas in the 1920s, middle-class women's clubs brought window boxes to tenements as a beautification gesture, today a garden organization's efforts at street beautification may include, along with window boxes and tree planting, local leadership development, lead-abatement education, and youth training programs.

Nevertheless, there are significant recurring themes. An important one is the use of gardening to reintroduce "nature" to the city. Urban garden programs provide a participatory experience that connects people living in cities, especially children, to the soil and plant and animal life. In some cases, the garden has served as a foil to the city by providing an avenue for the expression of agrarian values and ethics. At the start of the twentieth century, the urban garden was often viewed as a transitional space, an opportunity to teach immigrants and urban dwellers to love nature and therefore leave the city for the suburbs and country. In the 1930s, the garden was a testing ground for new ideas on how to integrate nature and the city, as in public housing or cooperative housing proposals that included community garden plots. This reconnection with nature is often associated with improving social and psychological health. Today, gardens are often

Figure 1. The desire for nature in the city: a community garden in Battery Park City, New York. Photograph by author, 1999.

Figure 2. The Watts Growing Project is in a residential and industrial neighborhood of Los Angeles. Many of the gardeners live in the adjacent public housing project. Inside the garden, the city seems far away. Photograph © by Lewis Watts, 2001.

described as oases of green in a concrete-dominated urban world. Thus, gardens appear in very urban places, such as Battery Park City in New York, where only a few people may have the opportunity to garden plots but many more can enjoy viewing them as passersby. In another context, a community garden plot can help recent immigrants with agricultural skills to transition into urban economies, as is the case at the Watts Growing Project in Los Angeles. Quite often, gardens include culturally significant reminders of agrarian traditions, such as scarecrows, sheds that look like red barns, or crops from gardeners' countries of origin.

A second theme is the association of urban gardens with education. Through gardening, one learns not only practical skills associated with gardening—the steps necessary to nurture seed to fruit—but also the civic-mindedness to nurture a community open space. The connection of gardening to learning has been the impetus behind various attempts to integrate gardens into school grounds and academic curricula. Interest in school gardens swelled from the 1890s to 1920s, during the war- and victory-garden campaigns, and has recently regained popularity. For older children and adults, the educational capacity of gardening translates into job training. Through gardening, one learns how to work, and why. This conviction has been put to work in the relief gardens of the 1890s, the subsistence garden programs of the 1930s, and the entrepreneurial training programs of today. A project in 1911 to teach "backward or defective boys" through market-based urban farming is the precursor of current entrepreneurial gardens targeting "at-risk" populations. In some cases, the official purpose of a program might be to teach agriculture and gardening, but more likely the goal is to teach good work ethics that translate to many forms of employment. The tangible nature of gardening allows participants to see the rewards of one's labor, the benefits of teamwork, and the importance of commitment and patience.

A third theme is the portrayal of gardens as a democratic space and gardening as an activity that brings diverse groups together in mutual self-interest. The urban garden offers an opportunity for families and neighbors to help themselves during war, depression, or civil unrest. During the world wars, the nation called upon its citizens to garden as an expression of their patriotism. The World War I campaign emphasized the dire need to produce food for households so farm-raised foods could be sent overseas, while the victory-garden campaign of World War II focused more on health, exercise, and morale than on food production for the war effort. During both wars, promotional material frequently praised community gardens as places where bosses and workers, husbands and wives, and people from varied ethnic backgrounds worked shoulder to shoulder. From the 1970s onward, the community garden movement supported a more localized camaraderie and morale. Faced with disinvestments and racial tensions, people got together to reclaim land and build food-bearing, social open spaces.

These three themes—nature, education, and self-help—interweave according to the social debates of the time. Given such broad ambitions, gardens are applied as solutions to address a range of concerns. Our gut instinct assures us that urban gardens are a means to counter urban congestion as well as depopulation, to improve education, to provide recreation, to ease labor relations, to curb social anomie, and to remedy environmental conditions. For example, school gardens appealed to turn-of-the-century reformers and educators as a way to improve education through manual training while also providing an entry point to teach immigrant families about civic duty, health and sanitation, and middle-class aesthetic values. Today,

Figures 3 and 4. *Top:* The educational capacity of gardening has justified its promotion in multiple situations and in serving a range of constituencies. This 1905 image shows a garden program for girls working in a Philadelphia mill. Reprinted from M. Louise Greene, *Among School Gardens* (New York: Russell Sage Foundation, 1910), 267. *Bottom:* Gardening continues to be a valued resource for the education, recreation, and socialization of children and youth. Children at the Berkeley Youth Alternatives Community Garden Patch in 1995. Photograph by author.

Figures 5 and 6. *Top:* The community garden expresses democratic ideals of people working together, as illustrated in this photograph from the World War I–era book *War Garden Victorious.* The three men are intended to illustrate different ethnic groups involved in the war garden campaign. Reprinted from Charles Lathrop Pack, *War Garden Victorious* (Philadelphia: J. B. Lippincott Company, 1919), facing 64. *Bottom:* Building and sustaining a garden creates many opportunities for people to work together. Volunteers helping to build the Berkeley Youth Alternatives Community Garden Patch. Photograph by author, 1993.

programs such as Philadelphia Green use gardening projects to nurture community participation, revitalize neglected neighborhood landscapes, and foster local entrepreneurship. The garden itself is rarely the end goal but rather facilitates agendas that reach beyond the scope of gardening. The many outcomes associated with gardens have also attracted support from various organizations, including beautification groups, charitable organizations, government agencies, environmental groups, and neighborhood associations. The up side of this fact is that it allows programs to draw on many interests and resources. The down side is that the high ideals associated with gardening rarely can be documented or verified. The tendency to layer multiple agendas on gardens makes achievable objectives difficult to ascertain, much less prove to a skeptical land developer or policy maker. The lack of clarity makes it possible for a garden to seem like an educational resource to one person and a regressive paternalistic institution to another.

Urban garden programs persist as a small but pervasive strategy to improve American urban conditions. The public receives garden programs favorably, particularly at times of societal change, as a satisfyingly direct and tangible means for people to improve the local manifestations of larger social, environmental, or economic crises.[3] Urban gardening has been and remains an appealing approach because it shows immediate results, is highly participatory, and is relatively cheap compared to other strategies, such as new housing, more jobs, or school reform.

Even though urban gardens have had periods of broad popular support and receive the general goodwill of the public, they rarely have been considered permanent. Most urban garden programs throughout urban gardening's hundred-year history have been designed to be temporary. Situated on borrowed land and often relying on borrowed leadership for organizational and technical support, garden programs tend to seize immediate opportunities without asking for any assurance of permanence. Once the war, economic depression, or civic crisis fades, less attention is paid to continuing the garden program. As economic and social conditions stabilize, the garden site—usually donated—becomes more valuable for development of a different kind. While garden advocates typically contend that even in stable times there are individuals and families who need gardens as a resource for food, income, or education, the communal garden does not fit into general conceptions of progress. The demise of urban garden programs is generally a result of the public's benign neglect, although some mourn the lost resource and plead for continuance. In localized cases, a garden might linger, such as some victory gardens that evolved into community gardens in the 1970s and still exist today, but this is the exception rather than the rule. With each crisis, programs have had to be reinvented, new land found, and new promotional campaigns developed to reiterate the economic, social, and personal benefits associated with gardening.

This pattern profoundly impacts current garden programs, which generally intend to be permanent. Today, most advocates assert that gardens have a long-term function as open space that promotes nutrition, education, household income subsidy, recreation, and psychological and environmental restoration. However, this goal is impeded by difficulties in retaining urban land and securing funding for educational programs, staff, and outreach. Most gardens today exist on land that is donated or leased on short terms. According to a 1996 survey conducted by the American Community Gardening Association (ACGA), only 5.3 percent of gardens in thirty-eight cities were owned or in permanent land trust.[4] Even a garden established on public land is likely to be considered temporary if the site is needed for some new development. Lest the blame be put solely on land tenure, it is important to remember that public opinion and participation are also critical factors in garden sustainability. The ACGA survey also found that the two main reasons for the loss of gardens were lack of interest by the gardeners followed by loss of land to a public agency or private owner. And yet, more gardens are built annually than are lost, so there has been a gradual increase in number even while locations and participants might have changed.

PREDICTING THE FUTURE OF URBAN GARDEN PROGRAMS BY EXPLORING THE PAST

Considering the persistent development of urban garden programs and the current interest in them, a question arises: should urban gardens be part of the permanent infrastructure of the city? The repeated re-creation of gardens suggests that there is an ongoing demand for land in cities for gardening. However, the episodic appearances of past programs raises the question of why gardens did not "make the grade": why they did not become permanent so they would be available when they were needed later. Notably, from a historical perspective, the initial phases of urban garden promotion in the 1890s coincided with the development of another new type of urban land use, the playground. School gardens, playgrounds, and vacant-lot cultivation were frequently listed together as beautification and social reform projects that philanthropic groups could initiate, with the goal that the municipality would eventually claim responsibility for the project. Whereas playgrounds eventually became part of public park and recreation systems, urban gardens rarely received the same public sanction. Early playgrounds often included vegetable garden plots and peripheral ornamental plots. Today, most people acknowledge playgrounds as public services for recreation and child socialization, but they would be hard-pressed to articulate as clear and concise a definition of the function of urban garden programs. The urban garden's multifunctionality may be at once a blessing and an obstacle to its validation as a public, permanent resource.

This book documents the history of urban garden programs and is organized around periods of national promotion. Part I describes the urban garden programs that developed in American cities during the formative years between 1890 and the First World War. Justified to serve distinct yet intertwined agendas—including income generation, education, assimilation, and beautification—garden programs of this era included vacant-lot cultivation associations, the school garden movement, and civic improvement gardening. Although these programs differed in audience and approach, they overlapped and blended together to become, in some advocates' thinking at the time, a "universal gardening movement."

Part II describes gardening programs established to help resolve local manifestations of crises that impacted the entire nation. These include the war gardens of World War I, the subsistence and work-relief gardens of the 1930s depression, and the victory gardens of World War II. Patterns established in earlier decades—such as reliance on borrowed land, leadership being drawn from civic and women's clubs, and emphasis on children's programs—continued. Differences also emerged, including a shift away from growing food for sale and toward gardening to serve household nutritional needs, and a transition from gardening as occupational training toward gardening as recreation. While still a local effort, the garden projects that emerged in this era were shaped largely through top-down directives from national organizations and federal agencies. As these efforts sought a broader participant pool than earlier programs, much more attention was given to ways of promoting gardening to the general public. Even as urban gardening was promoted, it was cast as a temporary effort that could solve an immediate crisis and then segue into recreational gardening, particularly in the suburban home yard.

Part III attends to the promotion of gardening from the 1970s to the present. Although there was a general decline in urban garden activities after

World War II, a few successful projects continued, becoming national models that inspired the renewed activism of the 1970s. Gardening became a venue for community organizing intended to counter inflation, environmental troubles, and urban decline. While many of the same impulses that had spurred urban garden programs in the past—education, nutrition, beautification, and recreation—continued to be important, the garden also came to represent community empowerment and grassroots activism. This groundswell of interest was complemented by the formation of local and national organizations devoted to "community greening" in a broader social and physical sense. While the community garden movement has continued to grow and evolve, major concerns about participation and land tenure haunt activists and organizers. Along with a general overview of the evolution of community gardening since the 1970s, these chapters include descriptions of specific citywide organizations and programs. The final chapter describes contemporary urban gardens with goals that include recreation, community food security, job training, and education.

City Bountiful explores each phase of urban communal gardening in light of its social, political, and environmental context. By examining the rationales for gardening used in promotional literature, one gets a sense of the key issues and underlying cultural assumptions surrounding gardening at the time. Highlighting organizational structure and day-to-day operations—who gardened, how they were supervised, how plots were allotted, methods of cultivation, and so on—is also instructive because such specifics reveal assumptions about participation, longevity, and outcomes. Whenever possible, quantifiable results are included, such as the number of participants or the value of food raised. However, these numbers come from promotional materials and rarely can be confirmed by other sources. The facts provided in this book are intended to illustrate broad issues that affected each locale in different ways. Cross-checking data was often impossible because of the limited availability of original sources and their tendency to emphasize positive outcomes rather than objective critique. Nevertheless, this broad overview hopefully will encourage urban historians, planners, and garden activists to dig deeper into the history of urban garden programs in specific cities around the nation.

PART I

Early Urban Garden Programs, 1890s to 1917

Introduction

STARTING IN THE 1890s, three types of urban garden programs emerged—the vacant-lot cultivation association, the children's school garden, and the civic garden campaign. Although each had its own constituency and promotion style, they shared common themes, largely due to the social, environmental, and economic context of the 1890s. Evolving at a time when ideas about city and country were in flux, the urban garden represented a merging of the best of both that might inform and inspire a changing American identity. American cities of the day were bustling industrial places, especially in New England and the mid-Atlantic states, where factories increasingly dominated economic and social conditions. While industrial interests could boast that America had surpassed Great Britain in terms of industrial output, few could say that the system was stable. The boom-bust cycles of industry rocked local and national economies. The economic depressions of 1893 to 1897, 1907 to 1908, and 1914 to 1915, for example, led to severe urban unemployment and all its related problems.

With industrial expansion came growth in the urban population. While in 1850 the total population of the United States was about 23 million, by 1890 it had grown to almost 63 million, of whom 35 percent were urban residents. Ten urban regions had populations of 50,000 or more in 1850, and by 1890 this number had grown to fifty-eight. By 1920, there were 144 urban areas with populations over 50,000 where the majority of the population—

51 percent—lived. Much of this population was foreign-born and had settled in the northeastern cities to take advantage of the region's concentrated industrial employment opportunities. By 1910, urban populations were 41 percent foreign-born, and approximately 80 percent of these immigrants had settled in the Northeast.

This urban industrial activity and population expansion influenced the form of the city. Whereas in the "walking city" of the 1850s the resources for most of one's daily needs were concentrated close together, the various functions of the city of the 1890s had become dispersed.[1] Downtown life bustled with business and civic matters during the day, but white-collar and clerical workers commuted by streetcar or train back to their homes in the suburbs and outlying towns in the evening. The garden-suburb ideal—homogenous residential communities surrounded by green and linked to the city by train or streetcar—attracted the middle class out of the city. Controlled by restrictive covenants and other means, suburban communities separated into class, racial, and religious enclaves. In the case of many middle-class and working-class suburbs, profit motives often limited their "garden" qualities so that suburbs acquired a reputation as ugly and bland, made up of cheap, uncomfortable houses. And while many who could afford to move from the city to the suburbs did so, there were still reasons to remain in the city, including proximity to work, access to multiple job opportunities, and communities of fellow immigrants. Even as many middle-class families were moving out from the city, housing options for laborers were limited, resulting in congested urban neighborhoods. Older housing was subdivided to serve more households, and tenement apartments were built to accommodate laborers and their families.

Meanwhile, industrial development, urban congestion, and lack of industrial controls exacerbated environmental problems, such as substandard water quality, inadequate sewage and sanitation, poor management of solid and hazardous waste, and inadequate ventilation. Rivers were filthy with industrial discharge. Coal smoke left its mark on buildings, laundry, and residents' lungs. Tenement apartments housed many of the urban poor, but substandard ventilation and sanitation and overcrowding led to disease, high infant mortality rates, and other social problems. The vacant land that did exist in congested cities was often held in speculation for development. In the meantime, such lots often became eyesores and health hazards when industrial and household wastes were dumped on them.

If the city and its suburbs were experiencing growing pains, the countryside was faring no better. As more people migrated from the farm to the city and the number of people farming decreased, deserted farms were becoming a common sight, especially along the East Coast. Both rural and urban reform advocates maintained that improvements in rural conditions would also benefit the cities. At an 1895 conference on the agricultural depression confronting the East Coast, held under the auspices of the New York Association for Improving the Condition of the Poor (AICP), farm advocate George T. Powell evoked a contemporary belief: "The permanent prosperity of the country is dependent on the right relations existing between rural and urban population as affecting productive and consumptive interests and the proper distribution of labor."[2] Sentimental responses in books and magazines advocated a back-to-nature ideal in which people would live on small farms and engage in home-based crafts while also working in cities for income and civic life.[3] Based on faith in scientific agriculture and entrepreneurial marketing—and a general lack of attention to financial considerations—articles in *The Craftsman* (an influential monthly magazine promoting the arts and crafts movement in America) encouraged urbanites, particularly the poor, to consider farming as a profession. Agricultural experts, however, often advised against such a move unless urbanites first obtained proper agricultural education and financial backing. Concern about what was happening in the countryside led President Theodore Roosevelt to form the Country Life Commission in 1908, which reported on rural conditions and recommended a series of interventions, including educational opportunities, improved road and mail service, social activities, and the development of scientific farming.[4]

Although the factors contributing to the congested city and the forsaking of the countryside were clearly tied to industrial and social conditions in complex ways, there was a general optimism that these problems could be remedied. Commissions and organizations made up of philanthropists, social-settlement workers, designers, business leaders, and others formed to address various health, social, and environmental problems. Maps were drawn to document the location of urban problems, such as typhoid concentrations, and to chart population density and immigrant enclaves. Giving problems site specificity showed that they were frequently concentrated in congested urban neighborhoods and supported the idea of addressing them through localized physical solutions. Meanwhile, advances in biology were changing the understanding of what caused epidemics of diseases such as typhoid and cholera. In contrast to the earlier belief that disease spread through noxious smells or miasmas that floated over poorer sections of the city, the germ theory argued that disease was bred through poor ventilation, water quality, and waste removal. Consciousness of the relationship between environmental factors and health problems strengthened concerned citizens' resolve to address conditions in the industrial city's tenement districts. For instance, in 1908 a New York group called the Committee on Congestion of the Population held an exhibition featuring maps that showed locations of disease, distribution of ethnic groups, and so on, and then proposed physical interventions ranging from model tenements to playgrounds and school gardens at these locations.[5]

To reformers at the turn of the century, solutions to urban problems required both physical and social change. Environmental determinism—the belief that changes in the physical environment produce changes in people's behavior—was the optimistic solution. Civic centers, model housing, and park systems were proposed as means to stop crime and alcoholism, "Americanize" immigrants, and spark civic-mindedness. Urban design, education reform, and social activities went hand in hand with sanitary campaigns to improve food quality, water supply, and access to light and ventilation and to address garbage removal. This was the era of progressivism, in which change was founded on a faith in benign guidance and enlightened personal actions. The optimism that infused progressivism was partially based on the assumption that an informed, properly guided public would exercise both economic individualism and voting power to advance a morally superior civil society. Most proposals were of an advisory nature and relied on rational responses by individuals to prompt change within their means. For example, reformers believed that congestion would be solved by first exposing the areas of concern, which would inspire enlightened developers to altruistically change their development patterns, which, together with proper educational guidance, would change the behavior of poor urban dwellers.

Given their faith in voluntary individual action, reformers rarely directed their efforts toward larger, systemic factors that led to exploitive hiring practices, limited environmental regulations, and discriminatory housing and employment practices. As historian Richard Hofstadter observed, the reform movement was not a battle between classes or against any particular sector but a "rather widespread and remarkably good-natured effort of the greater part of society to achieve some not very clearly specified self-reformation."[6] Yet despite the reformers' optimism, public action was limited due to both political corruption and lack of consensus regarding the role of the government and the proper use of the power of eminent domain.

Instead, reform efforts often sought to ameliorate the consequences rather than the core problems associated with the industrial economy. For example, to mitigate the conditions of congested, unsanitary neighborhoods, one solution was to demolish them and replace housing with wide boulevards and parks. Very rarely did the reformers consider where the residents of those neighborhoods would move; or if they did propose new housing districts, these were often too expensive for the people being displaced. It was simply assumed that by introducing educational, recreational, and environmental resources to the urban laboring class and new immigrants, they would be able to advance themselves to a more secure middle-class lifestyle and leave urban centers.

Urban gardens provided a small but widely appreciated approach to addressing urban congestion, immigration, economic instability, and environmental degradation. Anchored in a philosophy of environmental determinism, turn-of-the-century urban garden programs were expected to simultaneously improve both the environment and the behavior of the participants. They would link idle land with idle hands to satisfy intertwined impulses for education and beautification. They would not only provide a venue for the moral, physical, and economic development of the poor, but also result in a cosmetic improvement to the unattractive physical manifestations of land speculation and urban environmental conditions. Furthermore, in light of political corruption and the unclear role of government, urban garden programs would be of a scale to make them feasible projects for civic and socially minded groups to implement.

The period from 1890 to World War I was an important time of experimentation for urban garden programs. The first efforts to programmatically introduce gardening to urban communities began in the 1890s through school gardens and vacant-lot cultivation associations. Over the next thirty years, interest in children's school gardens grew into a national movement sponsored by both voluntary organizations and the U.S. Bureau of Education. Meanwhile, gardening was also promoted as a solution to economic problems. At the onset of the economic depression of 1893 to 1897, the City of Detroit established a vacant-lot garden program as relief for unemployed laborers. Reports of this program's success in occupying the unemployed, improving urban nutrition, and providing opportunities for the poor to earn income through their own efforts inspired emulation in many other industrial cities and influenced later garden-relief projects. In addition to these organized urban garden programs for children and the urban poor, the period of 1890 to 1917 also witnessed a volunteer-led movement to promote home gardens, children's gardens, and vacant-lot gardens as avenues for beautification and expression of civic-mindedness.

The underlying assumptions of environmental determinism and progressive reform affected how programs were implemented. Programs were structured as primarily educational and self-help resources that made do with temporary land and leadership drawn from the reform-minded wealthy and upper-middle-income volunteers. Leadership and financial support from civic groups, women's clubs, and garden clubs were often given based on the hope that gardeners would develop the skills and motivation to seek new careers and home environments, preferably farming in rural areas or at least suburban living. In the case of school gardens, the ultimate goal was to integrate gardening into the standard curricula, where it could receive public funding and support. During this period of innovation and experimentation, urban garden programs were organized along philanthropic lines, educational in intent, and interim in structure.

In the midst of the many different reform proposals, urban garden programs were a very small gesture. They were not central to the reform platform but were frequently justified as local contributions that compensated for limited municipal engagement in local problems or supplemented larger efforts. They provided a participatory outlet for reformers who were interested in making cities more livable, helping immigrants and urban children, occupying unemployed laborers until the economy picked up, and so on. These varied goals were to be accomplished by reintroducing urban dwellers to "nature" and, through this contact, filling a perceived moral vacuum in the urban living experience. Gardening was not the sole means to achieve this but was included among a laundry list of other nature-based activities, such as fresh-air outings, summer camps, and farm outings to take children and women out of cities to farms and wilderness. In particular, gardening was associated with agrarian values of self-help as well as the aesthetic ideals associated with suburban living. As social reformer Jacob Riis commented on one garden-based training program: "The children as well as the grown people were 'inspired to greater industry and self-dependence.' They faced about and looked away from the slum toward the country."[7]

ONE

An Alternative to Charity

The Vacant-Lot Cultivation Association

> But what is poverty? Poverty is the lack of wealth; wealth comes from land by labor; therefore application of labor to land should and would produce wealth enough for all. Charity organizations and other philanthropic societies are doing what they can to relieve poverty, and one means of doing it is by getting people back to the land. To help people help themselves is the only charity worthy of the name.
>
> *Bolton Hall, "Vacant Lot Garden Work for 1910,"*
> The Survey *(February 19, 1910), 2*

THE CYCLE OF EVENTS THAT led to the 1893 depression started in 1890 when the Baring Brothers banking house in England failed, which dropped prices for American goods in the world market.[1] The impact of this failure intensified as speculation resulted in a flood of bankruptcies by businesses and railroad companies. In a time when labor relations were already tense, workers took the brunt of the economic turmoil in the form of pay cuts, reduced hours, and layoffs. As more and more people applied to municipal poorhouses and charities for assistance, local leaders looked for new solutions to help the unemployed survive the temporary economic hardship while maintaining good work habits. In 1894, Detroit's mayor, Hazen Pingree, proposed gardening as an alternative to charity for unemployed laborers and their families. The return on investment cited by the Detroit project, plus the promise of other noneconomic benefits, compelled other cities, mostly in the East, to start similar vacant-lot cultivation programs.

The idea of providing land for the poor to raise food was not new, nor was it unique to America. Giving land to laborers and farm workers who did not have their own land was a tradition in France, England, Sweden, and elsewhere. In England, where the Enclosure Acts passed between 1750 and 1850 shifted communal land to private ownership, conflicts eventually resulted in a mandatory provision of allotments to the poor, enacted in 1845.[2] In 1873, there were an estimated 246,000 such allotments in Great Britain, totaling more than 60,000 acres. However, there were important differences between the English allotments and American vacant-lot cultivation associations. The English allotment system was intended to compensate for the low wages of fully employed farm workers and laborers. Land was provided on long-term leases. Further, it was assumed that the workers could maintain their plots as a supplement to their regular work. In contrast, the American programs were generally urban and intended to be temporary rather than permanent opportunities.

Vacant-lot cultivation was not considered the panacea for unemployment, yet it provided a local emergency relief measure. The immediate objectives were to produce food for families and to help industrious participants generate income through sales. The market value of the produce represented an unparalleled return on investment compared to other charity solutions. Furthermore, such programs complemented new philanthropic agendas to combine self-help, education, and the ameliorative influence of nature. Whereas other forms of charity might weaken the moral character of the recipient, vacant-lot gardeners, according to one report, "worked with a zest because they knew that they were to have the whole fruit of their labors, and they recognized that their efforts produce results, because of the careful training."[3]

THE FIRST OF ITS KIND: PINGREE'S POTATO PATCHES IN DETROIT

It was Detroit that pioneered the vacant-lot cultivation idea and provided a model for philanthropic groups in other cities interested in starting vacant-lot cultivation associations. Reports at the time pointed out that Detroit had over forty thousand Polish immigrants, most of whom worked as day laborers for a dollar or less per diem. When the economic downturn in 1893 led to unemployment, laborers' families had little recourse but to sign up with charity organizations. The demand for assistance exceeded what philanthropic and municipal services could supply. In an attempt to provide efficient relief, Mayor Pingree, a controversial social-reformist mayor, and later governor, seized upon the potential connection between the idle work force and the undeveloped land held in speculation for future development and inaugurated in 1894 a garden program called in the press

"Potato Patch Farms," "the Detroit Experiment," or "Pingree's potato patches." At the time, over six thousand acres in the Detroit area were unoccupied and held for speculative and other purposes. Organizers of the new program assumed that the poor would willingly participate and that landowners would happily donate use of their otherwise idle land.

Initially, the proposal for relief gardening faced skepticism. Some churches and other philanthropic venues ridiculed the mayor's plea for contributions to purchase plows, implements, and seeds. The pastor of a fashionable Presbyterian church reportedly asked his congregation to "give liberally and pray that potatoes might grow as had the . . . [mayor's] head and then there would not be a single hungry child left in Detroit."[4] Some newspapers included cartoons depicting the mayor as a despot named "Tubor I," with a carrot as scepter in one hand and a potato in the other. Failing to receive help from the wealthy citizens of Detroit, Pingree dramatically sold his own prize horse in a public auction to raise funds. Politics divided support for the project to such an extent that merchants reportedly refused to display prize-winning specimens grown in the gardens for fear that they would be "classed as friends and admirers of Mayor Pingree and might be boycotted by their best customers."[5]

Nevertheless, in the first year, Mayor Pingree raised $3,600 by subscription and other means, acquired 450 acres of donated land, and started his program. Most of the land was on the edge of town, on sites ranging from one to sixty acres. Based on advertisements posted in the local paper, over 3,000 applications were received, but limited funds meant that the program could accept only 975 participants. The first year's participants were described as "deserving persons and heads of families, either out of work or very poor; among them thirty widows, who, having half-grown boys, were able to properly attend to cultivation of land."[6] Individuals were assigned plots as close to their homes as possible. Each person's name and address were written on a wooden stake, and he or she was given a scheduled time to plant under the direction of a foreman. The committee provided seed potatoes, beans, and other seeds. To accommodate immigrants, planting instructions for various crops were printed in three languages. Even though the first year was dry and the plots were planted late in the season, the participants harvested a good crop of potatoes, averaging fifteen and a half bushels per plot. Gardeners used their produce for both home consumption and cash sales.

Based on returns for the first year, estimated at $14,000 worth of produce, the mayor claimed victory over his critics and earned the support of the aldermen and city council. Using this success as political leverage, in his annual message to the Common Council in 1895, the mayor announced that the project's economic and social value had clearly proven its worth:

> It seems to me the experiment has clearly demonstrated, first, that at least ninety-five percent of the people who are in destitute circumstances as a result of hard times are ready, willing, and anxious to work; second, that a large number of these people can be supported by utilizing vacant lands on the outskirts of the city; third, that a very small space of ground is sufficient to raise enough vegetables to support a family through the winter; fourth, that a majority of our citizens who own vacant land would much rather allow it to be cultivated by the poor than to pay a large tax for their support; and fifth, that the needy are therefore assisted without creating the demoralization in the habits of the people that gratuitous aid always entails.[7]

In 1895, the city council allocated $5,000 to the project, which that year produced a crop worth an estimated $44,056. With 1,546 families enrolled, approximately 25 percent of families on public relief participated. The next year, the program served 46.8 percent of families seeking public relief and the gardeners grew $30,998 worth of food, which exceeded the Poor Commission's overall budget for aid of $23,729. A report written a decade later stated that during the program's existence, it reduced the city's poor roll by 60 percent, at a cost of $3.60 per family.[8]

THE VACANT-LOT CULTIVATION IDEA GROWS

News of the "Detroit Experiment" spread through write-ups in popular magazines and charity journals.[9] The stories interested Bolton Hall, a New York lawyer from a prominent family, who decided to promote the idea for New York City. As an advocate of the back-to-the-land movement, he was impressed that the scheme not only addressed immediate sustenance needs but also provided training that could lead laborers toward more self-reliant, agrarian lifestyles. Furthermore, the program's self-help nature reduced financial strain on charities and taxpayers. In 1895, under the auspices of the New York Association for Improving the Condition of the Poor (AICP), a vacant-lot association committee formed to garner public support and funds.[10] The results in Detroit and New York, in turn, encouraged a group of prominent Philadelphians to form the Philadelphia Vacant Lots Cultivation Association. The Philadelphia project started in 1897 with 100 families on 27 acres and continued until at least 1927.

In 1898, the AICP published a summary of its own program and others titled *The Cultivation of Vacant City Lots by the Unemployed.* According to Cornelius Gardener, superintendent of the Detroit project, this report shifted vacant-lot cultivation "out of the domain of ridicule and flippancy to which it had been consigned by many."[11] The report promoted vacant-lot cultivation as a model solution to temporary unemployment because it not only warded off starvation and malnutrition but also satisfied contemporary philanthropic ideals. The AICP report predicted that "[in] future years it will be found that many philanthropists will refuse to contribute to miscellaneous charities until this plan [of vacant-lot cultivation] is in operation in their district."[12] Gardener anticipated that "in a year or two the scheme will be universally adopted in the United States, for it appeals to common sense, and besides saving taxation, it teaches people to rely upon themselves and their own efforts. Direct giving makes paupers; this method constantly reduces their number."[13]

Vacant-lot cultivation associations quickly appeared in cities across the nation. The 1898 AICP report listed projects in nineteen cities. In 1898, an article in *Charities Review,* written by three members of the Philadelphia Vacant Lots Cultivation Association's executive committee, summarized four years of vacant-lot cultivation projects in sixteen cities and towns. As shown in table 1, these first years witnessed both short-lived and expanding programs across the nation. For example, Boston had a small program of fifty-two allotments in 1895 that grew to eighty-three allotments in 1897. Seattle started with land for eighty-nine allotments and assistance to fifty-two gardeners with their own land, expanding to two hundred participants in 1897. The largest program was in Buffalo, New York, where 2,118 gardeners farmed 700 acres in 1897.

PHILANTHROPIC GARDENING FOR MORAL, HEALTH, AND ECONOMIC RETURNS

To understand why vacant-lot gardening arose as a temporary solution to the depression requires contextualizing this form of relief in the larger debate about charity in the laissez-faire economy. With growing numbers of unemployed and poor visible in cities across the nation, there was considerable debate as to who should receive charity and how it should be provided. While charity for the disabled, the elderly, and widows with children was expected, many feared that aid to able-bodied persons would thwart the work ethic and create a permanent dependency system. Philanthropic

TABLE 1

Cities with Vacant-Lot Cultivation Associations

State	*City*	*Years of Operation Mentioned in Literature*
Colorado	Denver	1895–1897
Delaware	Wilmington	1915
District of Columbia	Washington	1895
Illinois	Chicago	1897
Indiana	Indianapolis	1915
Kansas	Kansas City	1897
	Topeka	1895
Massachusetts	Boston	1895–1897
Michigan	Detroit	1894–1897
Minnesota	Duluth	1895
	Minneapolis	1895–1897
	St. Paul	1895
Missouri	St. Louis	1895
Nebraska	Omaha	1895–1897
New Jersey	East Orange	1895
New York	Brooklyn	1895–1897
	Buffalo	1895–1897
	New York City	1895–1897, 1905–1907, 1910
	Rochester	1895
	Syracuse	1895
Ohio	Cincinnati	1895
	Dayton	1897
	Toledo	1895
Pennsylvania	Philadelphia	1897–1927
	Pittsburgh	1915
	Reading	1896
	Scranton	1895
Rhode Island	Providence	1897
Tennessee	Jonesboro	1895
Washington	Seattle	1895–1897
Wisconsin	Lacrosse	1895

SOURCES: This table is based on available materials; some programs may have lasted longer. See New York Association for Improving the Condition of the Poor, "Cultivation of Vacant Lots by the Unemployed," *AICP Notes* 1, 1 (December 1898); Frederick W. Speirs, Samuel McCune Lindsay, and Franklin B. Kirkbride, "Vacant-Lot Cultivation," *Charities Review* 8, 1 (March 1898); Bolton Hall, "Vacant Lot Garden Work for 1910," parts 1 and 2, *The Survey*, February 19, March 19, 1910; Vacant Lot Gardening Association, *Vacant Lot Gardening Association Season of 1907* (New York, 1907). Also see *American City* 14, 2 (February 1916): 131–134; and short summaries interspersed in *Charities Review*.

journals, such as *Charities, The Survey,* and *The Charities Review,* provided a forum for national debate about the best approaches to improve urban conditions and distribute aid. Charity societies formed to coordinate citywide efforts, establish welfare application stations and procedures, and insure that relief was individualized to encourage self-improvement. Reformers struggled to reinterpret relief away from traditional almsgiving—a "crime against society" that promoted dependence—and toward new approaches that facilitated the moral and economic development of the needy. Soup kitchens were discouraged, since they were only a temporary measure that gathered the poor in potentially volatile masses. Gardening, on the other hand, was just the sort of self-help that not only provided access to food—if the gardener was diligent—but also kept people productive and taught them new skills.

The fact that participants stayed with the program and diligently gardened supported the stance that gardening separated the worthy from the unworthy and that most people on relief at the time were hard-working and not desirous of public assistance. According to the 1898 New York AICP report on vacant-lot cultivation, "As a 'labor test' this plan is unexcelled. Here is an opportunity for honest, moderate and not unpleasant work, and for learning a useful rural occupation; any who are able but will not work should not be helped at all."[14] Given that "in one form or another the idle man is a dangerous one, and the more intelligent and willing to exert himself he is, the more terrible he becomes," the report reasoned, "If we could but find a place for employing such men during the winter as efficiently as the Vacant City Lots could employ them in the summer, we would have gone far toward solving at once the labor question and the problem of undeserved pauperism."[15]

Besides its charitable incentives, gardening was attractive because of its health returns to urban laborers who lived in crowded tenement districts. Not only did participants enjoy better-quality food than was typical for laborers, but they also benefited from the physical exercise. Some projects, such as one in St. Louis, allowed gardeners to set up tents or shacks on their plots during the summer so they could stay overnight. Programs were lauded for taking mothers and children out of the "crowded and heated city and giving them a taste of country life so conducive to health and energy."[16] A tuberculosis specialist stated that the vacant-lot cultivation association in Philadelphia was "helping greatly the crusade against consumption."[17] Keeping people busy in their gardens diverted them from less desirable activities, such as drinking and gambling. Stories in annual reports frequently mentioned cultivators who abstained from alcohol during the growing season. The Philadelphia supervisor noted in the 1904 annual report, "During the year many incidents came to my knowledge where such habits as drinking and loafing around saloons and clubs and abusing the other members of the family have been checked on account of the gardener's time and attention being occupied in the little farm, thus enabling him to work himself out of the old ruts which have been formed during his enforced idleness."[18]

Another attractive benefit associated with gardening was its potential as agricultural training. Acknowledging the aesthetic and health concerns associated with dense urban areas, some promoters hoped that vacant-lot cultivation would entice unemployed laborers into farming and country living. They regarded gardening and farming as open fields of economic opportunity with ample room for specialization and diversification that required far less capital than other kinds of work. For instance, at an 1895 conference on the agricultural depression, held by the AICP, vacant-lot gardening was regarded as a vehicle by which "men who have no experience in agriculture can be taken from other occupations, and, by being intelligently directed, can more than support themselves on a few acres of land."[19] As a sign of success, F. B. Dickerson, Superintendent of Poor in Detroit, reported that at least one hundred families, out of approximately one thousand, moved to homes in the country as a direct result of their vacant-lot cultivation experience.[20] Even if families did not take up farming, at least they would learn to appreciate gardening and seek less congested neighborhoods where they could have their own gardens. The New York AICP vacant-lot cultivation association supervisor, J. W. Kelgaard, suggested that vacant-lot gardening provided a step toward home ownership:

> The experiment has convinced me thoroughly of one thing, and that is, if you will let the poor of our large cities get to the soil, and properly guide them for a year or two, they would become not only self-supporting, but would in a little while do much more. They could be so guided that in a few years they would own their homes, and a citizen who owns his own home is always the best kind of citizen.[21]

Furthermore, gardens were a relatively easy and inexpensive addition to neighborhoods. They appealed to many people because they replaced vacant lots that were considered health hazards, eyesores, and potential sites of criminal activity. Social reformer Jacob Riis endorsed the Philadelphia project, stating, "Your Vacant Lot Gardeners are not only utilizing soil that before went to waste, and eking out their income in the best of all ways, but you are giving them a fresh outlook upon the world that is worth all the rest, at the same time that you are helping win the children from the streets. Yours is splendid work."[22]

GARDENING AS ECONOMICAL CHARITY

Not only was vacant-lot cultivation considered an ethical form of relief, it was also attractive to charitable organizations and municipal agencies because it required little up-front money. With land and materials donated and labor usually free, little money was necessary to start a gardening program. In the philanthropic tradition, donations were collected to cover expenses for supervisory salaries, tools, materials, overhead, and in some cases the payment of wages to participants. The Philadelphia project couched its request for donations in terms of the self-help ethic, stating that "the Association appeals with confidence to the charitable people of Philadelphia for means to continue and broaden its relief work for these self-respecting unfortunates who must otherwise eat the bitter bread of charity unsweetened by the sense that it has been earned by honest toil to the limit of their impaired capacity."[23] When the project began, the press provided good publicity and the public responded liberally to appeals, with contributions that ranged from two dollars to five hundred dollars. As a 1898 *Charities Review* article noted, "The general testimony is that money for vacant-lot cultivation has been readily obtained. The idea of helping men to help themselves is attractive, and a circular of appeal showing the results achieved in early experiments usually brings the comparatively small amount necessary to provide superintendence, tools, and seed."[24]

Vacant-lot cultivation associations produced better returns on investment than other forms of charity. Donors could see the results of their investment in the form of cleaned and cultivated land, harvested produce, and gainfully occupied people. With experience, associations became more efficient and reduced their costs of operation. Each year, the expense required to plow, harrow, and furnish seeds and stakes in Detroit reportedly diminished. In 1903, the Philadelphia project reported that for every dollar donated, nine dollars' worth of produce was grown. Because one of the strongest appeals was economic value, proof of earnings was important. In order to convey the accumulated success of vacant-lot cultivation projects across the United States, the authors of the *Charities Review* article noted that "accurate accounting will do much to establish public confidence in the work, while vague estimates and extravagant claims will discredit it."[25] The article went on to urge systematic documentation of results, including both actual cash received through sales of crops and the cash value of the produce consumed by the gardener and his or her family.

The desire to be cost-effective influenced how programs ran. For example, with its goal to run for less than ten dollars per person, New York organizers envisioned their program much like a loan, with the responsibility for returns placed on the cultivator. To keep prices down, it was preferable for participants to spade their own land instead of having it plowed at the committee's expense. "When it is known that all direct aid is to be paid for by work or by part of the crop, most men will prefer to do as much of their own work as possible."[26] However, after careful documentation of their results, the AICP concluded that costs were too high to be evaluated purely in economic returns. Their analysis concluded that success also meant noneconomic benefits, such as education and moral development. "As an additional Charity, it will in the end be a failure. As an educator and a natural social development it can do and, under wide minded guidance will do, permanent and incalculable good."[27]

Like the New York project, the Philadelphia project proposed transferring costs to gardeners as a way of increasing their self-reliance and responsibility: "The gardeners do the rest and get the proceeds either by sale or through the use in their homes of the products; and since they get all that they make, the more intelligently and industriously they work the larger their income."[28] The garden plots were provided for free, but it cost—by the association's estimate—approximately five dollars per year to plow, fertilize, and provide seeds. This cost was offset through a graduated fee system. First-year gardeners paid one dollar, second-year gardeners paid two dollars, and so on, so that families who had cultivated their garden for five seasons paid the full cost of site preparation. The project's 1907 report mentioned that participants themselves donated very generously to the program, giving about one dollar for every nine dollars received in the previous three years. To address financial shortages in later years, the Philadelphia association often had to take out loans or reduce its placements. In 1922, the association joined the Welfare Federation, a central organization that solicited and distributed charitable donations.

ESTABLISHING VACANT-LOT CULTIVATION ASSOCIATIONS

Vacant-lot cultivation associations formed in many ways—as municipal agencies, as private committees, as charity organizations, or as other benevolent bodies—and they often worked in cooperation with public authorities. For example, Detroit's mayor appointed a municipal committee to establish the "potato patches." Similar to the involvement of the AICP in New York, existing charitable organizations in Brooklyn, Boston, Chicago, and Seattle stepped forth to set up programs. Independent committees created specifically for vacant-lot farming started programs in Denver, Minneapolis, and Philadelphia. In the case of the Denver project, the committee included representatives from the Charity Organization Society, the Woman's Club, and city government. Some associations, such as the one in Buffalo, were started as private associations and later adopted by public authorities.

The associations employed supervisors to provide technical information and on-site assistance. The ideal supervisor needed to have a combination of farming, business, and marketing skills; management skills; experience working with the poor; and the ability to secure support from contributors. The 1898 *Charities Review* report acknowledged the rarity of such a person and recommended paying market-rate salaries. However, in actuality many supervisors volunteered or worked for less than they would have received from other employment. For example, the Detroit superintendent, Captain Cornelius Gardener, volunteered his services, as did New York's J. W. Kelgaard, while Philadelphia's superintendent, R. F. Powell, received less money than he could have earned at another job.

Projects were promoted through local press and charity organizations. Several projects received participant referrals from charity organizations, but others stressed the need for public outreach so that those "self-respecting men and women who need help, but abhor the idea of accepting poor relief, shall not be repelled. Self-help is the central idea of vacant-lot cultivation, and if this fact is emphasized in the announcements, the people who most need the opportunity to help themselves will be attracted."[29] In some cases, the character of the applicant was checked through references or investigated by charity organization societies. In other cases, participation was mandatory. In Detroit, for instance, any family that had received aid from the city's Poor Commission for two years had to participate in the vacant-lot program or their names would be removed from the commission's list as undeserving of aid for the following winter. Mayor Pingree justified this policy as a way to assure recipients that they were helping themselves. "It is an error to suppose that because people are poor and needy, that they refuse to help themselves where an opportunity offers. From the experience of the committee, such is never the case."[30] However, some may have been prompted to participate more by the threat of not receiving aid in winter than by the opportunity for self-development.

Typically, participants signed a contract agreeing to abide by the association's regulations and to give up their garden plots if that was requested by the landowner. The gardeners were required to cultivate the land for the entire season, keep records of what was grown and how it was used, follow supervisors' instructions, and help guard the garden against trespassers. Most reports praised the smooth running of the organization, writing that the gardeners followed the rules and worked hard, with few complaining or causing troubles. In 1898, the Philadelphia supervisor noted the lack of disciplinary problems: "No complaints are made and our gardeners are a sober, hard-working set of people, who appreciate keenly the opportunity to tide over hard times and to supply their families with wholesome food, which is in large measure, the direct result of their own efforts."[31] For example, in its second year, only 11 of the 162 gardens in the Philadelphia program were forfeited, 5 because the men had found steady work and did not have time to garden and several due to construction of a new road on some of the garden land.

PARTICIPANTS

Most public-interest stories praised programs for offering large numbers of people food, occupation, and some income. These accounts usually described the participants as needy: an article in *Country Life in America* portrayed gardeners in Philadelphia's program as "the worn-out, the disabled, the incompetent, the men with large families and small salaries, widows with small children, the aged, the half-sick and schoolchildren," though later it mentioned that anyone could apply for a plot.[32] According to the few factual records available, most participants were laborers who were either unemployed or partially employed. Railroad workers, office workers, the elderly, and widows also participated. Many were immigrants of various ethnic backgrounds. The Detroit commission generalized that its participants were mostly Polish, followed by Germans, other Americans, and African Americans. New York's organization estimated that of its initial eighty-four

SCHEDULE FOR THE CULTIVATION OF VACANT LOTS BY THE UNEMPLOYED.

Application No. ________ ***Date*** ________

St. ____ No. ____ Front Rear SURNAME	RELATIONSHIP TO HEAD	NUMBER CAPABLE OF WORK	Number in family who should be at work, but unemployed	AGE	SEX	COLOR W—White C—Colored.	NATIONALITY	BIRTH PLACE	COUNTRY BORN	CITY BORN	COUNTRY BRED	CITY BRED	OCCUPATION	WORK HOURS PER DAY	STEADINESS OF WORK LAST 12 MONTHS	NUMBER OF ROOMS	RENT PER MONTH	LENGTH OF RESIDENCE IN CITY IN MONTHS	AID Yes or no	SOURCE OF AID	EXPERIENCE IN FARMING IN MONTHS	Can you get or pay for Tools, Seed or Fertilizer	REMARKS The scientific value of this data will be seriously impaired by the omission of a single answer. Please see therefore that each question is answered.
1																							W—Wife M—Mother
2																							S—Son G M—Grand Mother
3 and																							D—Daughter G F—Grand Father
so on																							L—Lodger [Abbrev'ns for relationship col.]
																							U. S. Ire. Scot. Ger. Eng. Fr. Ca. Rus. Eng. Ca. It. Scand. [Abbrev'ns for nationality col.]
																							Endorsed by

(There were 13 spaces for names.)

Reference (1) Person ________ ***or*** ***(2) Society*** ________

Name of Application Station ________

Figure 7. This application form was included in the description of vacant-lot work by the New York Association for Improving the Condition of the Poor. Reprinted from AICP, "Cultivation of Vacant Lots by the Unemployed," *AICP Notes* 1, 1 (December 1898): 4.

families, 42 percent were German, 31 percent Irish, 13 percent American-born, and the remainder from other European countries. The organization also reported that participants had lived in the city for an average of 12.3 years and had households of 3.29 persons, and that the head of household's average age was 46 years. In his 1908 annual report, Philadelphia garden superintendent James Dix described participants as "Americans, English, Irish, Italians, Germans, Scandinavians, Swiss, Scotch, French and Swedish; also Hebrews, colored people and the inhabitants of the West Indies, who compose our population and are going to become our future citizens."[33] By 1923, 60 percent of families in Philadelphia's vacant-lot cultivation association were classified as either "American" or Italian, 10 percent were African American, and the rest were of eleven other nationalities.

Vacant-lot gardening involved the entire family. Although many projects initially offered plots of land to unemployed male heads of household, it was soon realized that women—widows and laborers' wives—benefited from participation as well. In an article on vacant-lot gardening in *Garden and Forest,* the superintendent of the New York garden project was reported to have identified women as some of the most successful gardeners and said that new policies were enacted to give women plots so they could help support the family.[34] Similarly, laborers' children were encouraged to help their families through work in the garden. Vacant-lot cultivation was justified as a means for the children of poor families to earn income outside school without the negative connotations of children's wage labor in factories and industry. One story from Philadelphia described how participants' children loaded wagons, developed selling routes, and kept sales records. As the superintendent reported, "You will, I am sure, be gratified to know that these little salesmen often took home four to five dollars per week and yet never worked more than three to five hours per day. The work was done under such favorable circumstances that to them it was not work but play. It was in reality a school in which the lessons were both useful and practical."[35]

The following story, included in the 1898 annual report of the Philadelphia Vacant Lots Cultivation Association, describes the ideal participant:

> We had a mechanic, for instance, who has a fairly steady job at $7 to $9 per week, varying as work is slack or plentiful. Although not yet 40, and his wife but 33, there are 15 children, all living. The oldest child is under 15 years. It requires about one-third of this man's wages to pay rent for a poor

Figure 8. Children in one of the Philadelphia vacant-lot gardens preparing to sell the produce they raised. Reprinted from Philadelphia Vacant Lots Cultivation Association *Ninth Annual Report,* 1905.

> house. Now if from even the maximum wages, $9 per week, $3 is set aside for rent, $6 remains with which to pay for food, clothing, fuel and household expenses for 17 people. How the mother could find time after caring for the children, cooking, washing, sewing and mending to aid in support of the family, is hard to see, yet, with the help of the children she cultivated a garden quite well, although it was more than a mile from her home.[36]

LAND

An essential function of a vacant-lot cultivation association was to connect the poor with land to farm on a temporary basis. With board membership that often included influential city leaders and landowners, the associations could solicit land that the poor could not. The land was generally donated with the understanding that it would be vacated quickly upon the owner's request, with no liability for lost crops. Appealing to landowners to contribute use of their property, the superintendent of the Philadelphia project gave a variety of reasons for them to cooperate:

> To land-owners the Vacant Lots Cultivation Association in substance, says: Lend us your idle land (subject to immediate dispossession whenever required) and we will offer ample self-help to all who cannot work the usual employments. We will leave the land (on demand) in better condition than we find it. We will make of it something even better than a park or playground for the poor, or rather we will show them how to make not only parks and playgrounds for themselves but productive gardens as well out of what are now only idle lots and in many cases rubbage heaps. We will help make the city beautiful while making this idle land and these idle people useful.[37]

Once land was acquired, it was prepared for cultivation. Vacant lots often had to be cleared of rubble. If the land was not fenced, gardeners would have to devise means to protect their crops against animals and theft. Some places, such as Detroit, provided a mounted policeman or guards to keep daily watch. The land was then divided into individual gardeners' plots, the size of which varied by project and site. While New York provided one-acre sites, Minneapolis and Detroit offered one-third- to one-fourth-acre plots, and Brooklyn provided one-eighth-acre plots. Plot sizes in Philadelphia fluctuated based on the number of participants and available land. Initially, plots were approximately one-quarter acre but they shrank to one-sixth acre in 1912 and one-eighth acre in 1917, when the campaign for war gardens increased the demand.

Getting and keeping access to land was often the most difficult task the associations faced. Although vacant lots existed both within and around cities, urban development patterns and speculation limited their availability. Small lots scattered in different parts of the city were too costly to prepare and supervise. Some land ideally situated within urban neighborhoods was not offered because the owners wanted it readily available for sale and development. Often the land was at the city's edge. In New York, for instance, the organizers initially based their plan on an 1893 survey of vacant land in New York City that had revealed 17,329 vacant lots below West 145th Street and the Harlem River. However, speculation made it difficult to secure land in or near the city, and scattered sites seemed inefficient to run. In the end, the New York committee acquired three tracts totaling 138 acres (1,656 city lots) in Long Island City. The land was divided into individual one-acre plots plus a forty-acre cooperative farm. Because the site was across the river from participants' homes, the project had to bear additional transportation costs.

Some associations circumvented the unreliability of donations by renting land on more secure terms or using public land. From its beginning, the Boston vacant-lot cultivation association rented its land, a sixty-acre farm serving fifty to eighty families. In at least one case park land was used for cultivation—the New York project used 321 acres of Pelham Park in the Bronx for its 1898 season. Unfortunately, transportation costs were high there as well. Referring to the difficulty of getting land in New York City, the AICP report stated, "It is easier to get the people back to the land than the land back to the people."[38]

The shifting of garden sites over time during the Philadelphia project's long tenure illustrates the impact of land development on vacant-lot cultivation opportunities. When Philadelphia began its program, the organizers used information from the assessor to develop a list of available plots of land in each ward. In the first year, about forty persons offered the use of their land in lots that varied from a building lot to a sixty-acre tract. Twenty-seven acres were divided into 96 plots of about one-quarter acre each, serving 528 participants. Most of the land was located in the western and northwestern part of the city, about five miles from City Hall. In 1898, most of the land was in West Philadelphia, in the neighborhood bordered by Fiftieth and Haverford Streets and Seventy-Fifth and Masters Streets. All land was donated with the stipulation that it would be returned to the owners within ten days of notice. While sites were indeed reclaimed from time to time, such a severe turnaround was never demanded. Six years later, the association developed an arrangement with the Pennsylvania Railroad Company to manage railroad properties for employee gardens, with the understanding that extra land could be used by nonemployees.

As the economy improved, pressure to use garden sites for other purposes increased. In more prosperous times, the Philadelphia Vacant Lots Cultivation Association lost two or three sites annually. The 1923 annual report mentioned a record amount of construction in Philadelphia that limited the availability of land, resulting in the loss of 197 garden plots that were not replaced. Addressing the record low number of applicants and garden assignments in 1927, supervisor James Dix noted that poor weather conditions alone were not the problem: "Operators and builders in Philadelphia

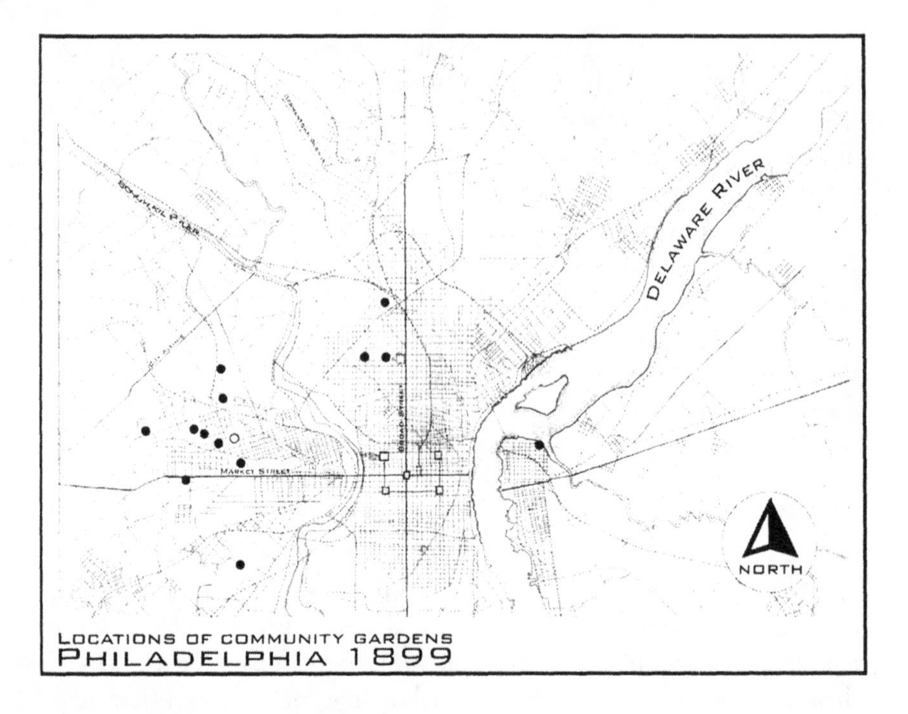

Figure 9. Philadelphia Vacant Lots Cultivation Association gardens, 1899. Drawn by Angela Landry and Agus Soeriaatmadja.

have kept our gardeners in a constant state of uncertainty for the last ten years."[39] Land insecurity discouraged participants from investing a lot of time and energy in their plots. James Dix could only empathize with reluctant participants: "It would be unjust to criticize them too severely for they do not apply for gardens for the mere sake of riding a hobby. A garden is really a business proposition with them."

Making land permanently available to the industrious poor was bandied about but never directly addressed. After the second year of the Detroit gardens, the committee recommended that the city purchase a farm on the west side and another on the east side of Detroit for permanent relief farms. No information exists to suggest that the land was purchased. There were fears associated with the possibility of land permanently accessible for the poor to garden. One reader wrote a letter, printed in *Garden and Forest,* to protest the vacant-lot cultivation experiment as a wedge against property rights:

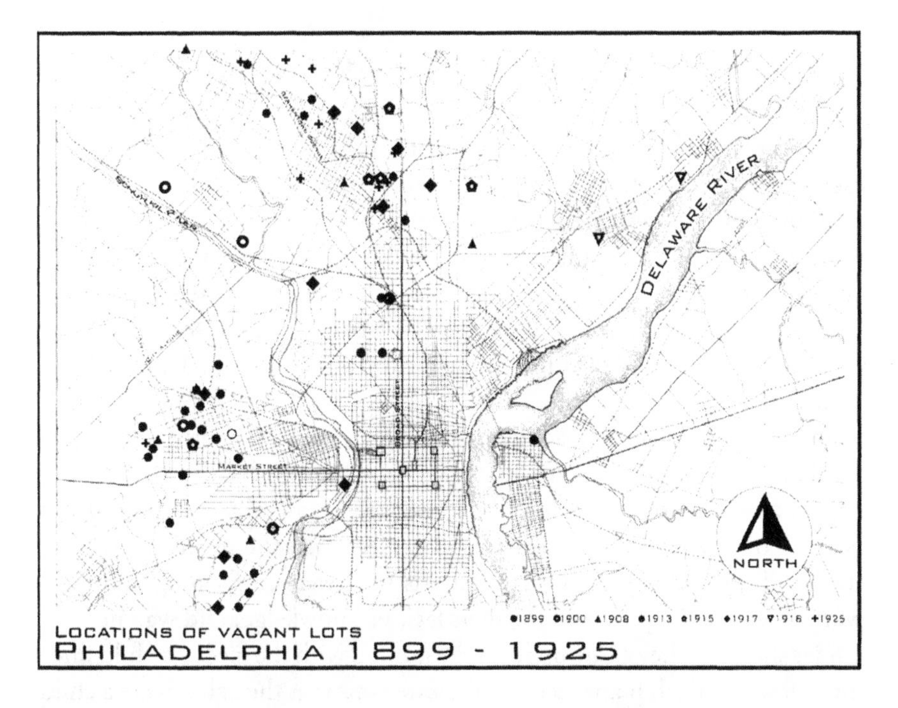

Figure 10. Composite map of all Philadelphia Vacant Lots Cultivation Association gardens from 1899 to 1925. Drawn by Angela Landry and Agus Soeriaatmadja.

> Let the masses once realize that with free access to land they can make their own living; give them that free access in the name of charity, and it will not be long before the philanthropic features of the case will be forgotten, and access to land will be demanded as a matter of right, and questions of title between the state and the individual will be subjected to the same remorseless scrutiny and review as similar questions between individuals are decided now.[40]

This person went on to suggest that participants in such projects would maintain their self-respect only if the land were rented at fair valuation: "No other plan will meet the exigencies of the case and remove all fear in the future of a dangerous spirit of independence among the classes whose necessities are now the object of philanthropic solicitude." However, renting the land raised concerns too. Landowners who worried that the gardens would make their land less attractive to possible purchasers were just as hesitant about renting the land as they were about loaning it.

The vacant-lot gardeners grew a wide variety of vegetables, including beans, peas, cabbage, tomatoes, carrots, turnips, onions, lettuce, and radishes, and some fruit as well. Potatoes, being a storable crop, figured prominently in both suggested and mandatory planting plans. For instance, the Detroit program required that approximately half the plot be used to cultivate potatoes. Some associations provided the gardeners with exact planting plans, while others allowed more latitude. Often, seeds were free to participants, provided either through donations or by the United States Department of Agriculture.

Some participants had prior gardening experience, while others had to be taught. Over half the participants in the AICP New York City program had some knowledge of agriculture, but most were unfamiliar with intensive farming methods appropriate for small gardens.[41] As the superintendent of the AICP project reported, "Ignorant as they were to farming and farm work, they made up by zeal for their lack of knowledge, and watched the tender shoots as they came up from the ground with the glee of a child finding a new toy. Each plant received the care almost as though it were a child itself."[42] Supervisors provided advice through site visits, experiments, demonstrations, and written materials. Generally, supervision and demonstrations were kept informal so they would not interfere with the gardeners' possible employment opportunities. Some programs limited technical support to basic gardening instruction, while others used the garden to teach state-of-the-art cultivation practices in hopes of shifting laborers into the agricultural economy. One of the most intensive training programs was in Philadelphia, where superintendent R. F. Powell considered his work as not only charity but also an agricultural training school "in which those most in need of instruction are taught the best methods of truck gardening in the most practical way without loss of time or expense to themselves and at a very slight cost to the community."[43] Under his direction, gardeners experimented with specialty crops, standardized cultivation practices and winter gardening under glass, and also conducted a less successful experiment in raising Belgian hares for food and fur, with rabbit hutches placed in participants' backyards, cellars, and garrets.

Most of the garden produce was consumed by family and friends, and the rest was sold at good prices. Unlike the poor-quality food available in the markets, vacant-lot cultivated vegetables were fresh and nutritious. As the Philadelphia program's 1898 annual report observed, gardening families enjoyed better vegetables at their dinner tables than did their nongardening neighbors, whose "scanty means would force them to take quantity rather than quality into consideration." For those who wanted to sell their produce, there was a ready urban market. Because of its freshness, produce grown by vacant-lot gardeners reportedly earned the highest price possible. "Gardeners do all their own marketing and invariably get as high prices as hucksters, and often much higher, as the quality of their vegetables, and the fact that they are always fresh, make them much more desirable than the stale ones usually sold from the wagons."[44] According to the New York report, the immediacy of the gardens meant that "many a planter was able to peddle his bunch of radishes, picked only a few hours previously . . . for five or six cents, whereas the wholesale price at market was only two and a half or three cents."[45]

This market advantage translated into good earnings for participants. The superintendent of the New York project reported that the gardeners earned the equivalent of skilled mechanics' wages, or four dollars per day—considerably more than the standard farm wages of seventy-five cents per day (with lodging). While some contemporaries argued that vacant-lot cultivation was unfair competition to truck farmers because the gardeners were given free land and seed, most participants did not undersell the independent producer. Rather, the gardeners commanded better prices and more profit than regular truck farms due to the freshness of their produce.

COOPERATIVE FARMS

Many vacant-lot cultivation associations also developed cooperative farms where workers were paid hourly or daily or received a percentage of the proceeds based on time worked. Rather than cash, the association in Rochester, New York, paid workers in fuel or provisions from the Poor Store. Where the gardeners were paid, they typically received a below-standard wage, justified on the grounds that it allowed more people to be employed while discouraging them from relying on the program as a permanent job.

As early as 1895, the New York AICP program included a forty-acre cooperative farm to employ those most in need of money. The farm was intended to be self-supporting—covering all its expenses and dividing the profits among the workers. The program initially considered paying the workers with meal tickets in lieu of cash: "The outlay is small, since men living so cheaply can afford to work cheap."[46] In the end, forty men were hired at ten cents per hour plus half interest in produce sales. Given that the men who secured employment elsewhere left, the majority of workers were elderly, infirm, or disabled, which in turn meant that productivity was lower than expected from a business perspective. The supervisor of the New York cooperative farm summed up the project: "Notwithstanding the fact that this class of labor was unable to do in three days what a good man would do in one, the Co-operative Farm, I am glad to say, was a success."[47] From an expenditure of $967, the farm produced crops valued at $1,068. Staying within the charitable sector, most of the produce (potatoes, cabbage, tomatoes, turnips) was sold to the relief department, while other customers included the Convent Good Shepherd, St. Joseph's Asylum, Mt. Sinai Hospital, and Governor's Island. The executive committee felt the cooperative farm played a major role in the program because it "stimulated the men to friendly rivalry, created a public spirit against loafing, and whilst affording immediate wages to the cultivators, can be made to contribute largely toward the other expenses of the Committee."[48]

The Philadelphia association also ran a cooperative farm, from 1901 to 1909. In its first year, the farm was a three-acre site and the workers were paid $1.25 per day. They grew $364 worth of produce, and the net profit of $180 was divided among workers in proportion to the number of hours each one worked. In 1903, this venture was expanded to a seventy-seven-acre farm near Trenton, New Jersey, on land donated by Bolton Hall, who believed that "if these people were given profitable and pleasant employment and their freedom, this would, in a large measure, stop their tramping."[49] In 1904, over fifty men, women, and children worked on the cooperative farm. The 1909 annual report urged securing a permanent location for a cooperative farm to facilitate emergency employment needs and compensate for lost garden sites, but by indications in later reports, 1909 was the cooperative farm's last year.

EFFORTS AFTER THE DEPRESSION

Reports of vacant-lot cultivation associations in charity, horticultural, and popular journals diminished after 1898. The last mention of Pingree's "potato patch" program was in 1896. The AICP-sponsored New York vacant-lot cultivation project probably ended in the late 1890s as well, as we have no reports from after that period. However, some programs may have lingered on, finding new rationales for support. One such program was the Philadelphia Vacant Lots Cultivation Association, which continued until at least 1927. During its tenure, the program continued to serve the underemployed, although this required some changes to garden locations and programs. In fact, the project described itself as doing charitable work by providing food, job training, and moral instruction for the chronically unemployed. By continuing for all those years, the project was in place to serve during subsequent depressions as well as in the World War I war garden campaign. The association justified its presence as a needed resource in good and bad times:

> The demand for gardens has not abated the slightest so far as we are able to observe, notwithstanding the great demand for labor that has prevailed in the city throughout the season. Many of the gardeners are fast coming to regard the work as a permanently established institution; they rely upon it and make their plans from one season to the next, just as men in other avocations do. Thus they feel secure from want, and confident of their ability to keep the wolf from their door, notwithstanding the fact that in nearly every case they are debarred from regular employment in shops and factories because of their age or physical disabilities of some kind or other.[50]

Some of the same gardeners who had participated during the depression continued to utilize the program. The 1912 annual report listed 442 families, of which 225 were first-time participants, 92 had gardened for two years, 55 for three years, 35 for four years, 18 for five years, and 17 for over six years. Most of the gardeners who had participated for five or more seasons were senior citizens. In fact, the proportion of participants over the age of fifty increased from 30 percent in 1897 to 62 percent in 1902. The superintendent noted that the younger gardeners had moved to homes outside the town, where they could garden "in a larger and even more successful way."[51] During the general enthusiasm for gardening during World War I, the 1918 annual report noted, only 43 percent of the 1,156 families were new to the program.

Not only did the intended type of participant change over time, but the program activities and garden locations shifted as well. The association continued to provide land to families for independent gardening. In addition, they established a cooperative farm and a job-placement program, supported the entrepreneurial ventures of participants, and developed a corps of workers to do other landscape projects, such as preparing playgrounds and school gardens for the Civic Club and other institutions. Over time, the locations and availability of gardens changed.[52] In general, garden sites were located on the outskirts of the city. While some sites were used for quite a while—a few for fifteen years or longer—most were loaned for three to five years. As a result, the association was constantly looking for new land.

Noting the continued growth of the Philadelphia Vacant Lots Cultivation Association, the 1901 annual report announced that "Philadelphia was not the first to adopt this excellent system, but the Philadelphia Association has developed the plan more fully and has put its relief work upon a broader, more permanent basis than any other organization in the country."[53] Whereas most of the other vacant-lot cultivation associations had ended once the economy had picked up, the association reported in 1904 that its success even in times of economic prosperity "proves conclusively that this plan of relief adopted in the dark days of industrial depression is a thoroughly efficient and greatly needed system even under conditions of general prosperity such as those we have enjoyed during the past few years."[54] Table 2 shows the changing participation, number of acres cultivated, and financial results of the program from 1897 to 1927.

Even though most of the vacant-lot cultivation associations ended by the turn of the century, the idea of vacant-lot cultivation as a form of economic relief occasionally resurfaced during economic downturns. New York City's AICP program ended in the late 1890s, but the concept was revived in the city first by the Salvation Army in 1904, and again from 1905 to 1907 as the Vacant Lot Gardening Association.[55] During this same period, the Philadelphia vacant-lot cultivation supervisor was asked to advise other American cities, including Buffalo and Rochester, New York; Worcester, Massachusetts; Bridgeport, Connecticut; and Coatesville, Pennsylvania. Unfortunately, there are no records that show whether these cities did in fact start programs. Although the idea of vacant-lot cultivation intrigued many, emulation of the programs of the late 1890s was most likely limited. In the Russell Sage Foundation's 1936 report on relief programs during American depressions up to the 1920s, vacant-lot cultivation was not mentioned.[56]

That said, the idea did influence other philanthropic ventures. For instance, many advocates saw the potential of establishing gardens at institutions, such as sanitariums, hospitals, and prisons. As advocate Bolton Hall noted, such institutions had not only available land, but also available labor. Gardening was justified as offsetting the costs of inmate upkeep as well as providing healthy recreation and agricultural training so that inmates would not eventually resettle in the slums. For example, the superintendent of the New York State Reformatory for Women felt that gardening was beneficial to two types of women—"those whose nervous systems

TABLE 2

Results from the Philadelphia Vacant Lots Cultivation Association, 1897–1927

Year	*Number of Acres*	*Participant Total*	*Crop Value*	*Earnings per Garden*	*Total Cost*	*Cost per Garden*
1897	27	528 persons or 96 families	$6,000	$60.00	$1,825.33	$18.25
1898	40.5	770 persons or 140 families	9,700	59.87	2,266.76	14.00
1899	73	1,495 persons or 249 families	4,810.80	49.35	2,650.30	9.07
1900	130	2,386 persons or 480 families	24,600	47.30	3,962.48	7.62
1901	158	2,946 persons	30,000	47.46	4,480.94	7.09
1902	198.5	3,775 persons	50,000	62.80	5,556.80	7.00
1903	275	3,609 persons	36,000	47.00	4,837.00	6.16
1906	—	800 families or 4,000 persons	52,000	65.00	6,616.34	—
1907	—	800 families	54,000	—	7,200.00	—
1908	10 farms	412 families	—	55.00–70.00	9,918.38	—
1909	124.5	452 families	—	—	8,903.83	—
1910	82.5	355 families	—	—	7,229.38	—
1911	72	388 families	—	—	8,211.44	—
1912	—	422 families	25,000	—	7,813.98	5.00
1913	20 farms	548 families	28,000	154.00–160.00	8,105.17	—
1914	20 farms	603 families	32,000	—	8,427.98	—
1915	23 farms	671 families	—	—	9,458.89	—
1916	100 acres (22 farms)	611 families	30,000	—	9,807.61	—
1917	160 acres	1,145 families	70,000	—	17,310.74	—
1918	—	1,156 families	90,000	—	25,561.49	—
1921	—	3,145 persons or 664 families	55,000	80.00	13,846.00	9.00
1922	—	632 families	53,000	83.00	4,830.32	7.64
1923	50	2,216 persons or 488 families	—	—	—	—
1924	—	2,238 persons	42,000	90.00	—	—
1925	—	467 families	38,000	80.00	—	—

(continued)

TABLE 2 *(continued)*

Year	Number of Acres	Participant Total	Crop Value	Earnings per Garden	Total Cost	Cost per Garden
1926	—	2,000 persons or 476 families	$38,000	$80.00	—	—
1927	—	260 families	—	—	—	—

SOURCE: Statistics culled from Philadelphia Vacant Lots Cultivation Association Annual Reports, 1898-1927.

NOTE: For the years 1904, 1905, 1919, and 1920, no data are available.

are somewhat out of order" and who needed fresh air to better their appetites, sleep, and quiet; and those women diagnosed with "a superabundance of energy."[57] Occasionally, articles in *The Craftsman* reiterated the benefits and pleaded for philanthropic and organizational support to train and facilitate urban dwellers in agricultural and rural occupations. The magazine frequently included pieces about farms at reform schools and institutions, as well as resettlement efforts:

> We cannot expect our poor, our sick, our unfit, our hungry in the city to get together and say how fine a thing it would be to live in the country, to train their children to be contented farmers,—this is quite beyond them; we have only to realize how far it is beyond ourselves even as thinking people. It is our business today if we know how to think, to go among these people with the message, to find out just what openings there are throughout the country, just what can be done with the city's hungry surplus, to form a connection between them and the new rural life and to see to it that not only is it made possible for them to become a part of this life, but to help them see the truth so that they want to get there, and that after they reach the promised land, it shall in truth make good to them.[58]

The vacant-lot cultivation associations may have had international influence as well. The Philadelphia Vacant Lots Cultivation Association promoted its program through lectures and exhibits around the country and in Europe, including Liege, Belgium; London; and Milan. They advised organizations in England and Scotland wishing to start similar projects. Indeed, in 1904 the association took credit for advancing the allotment-garden movement in Europe:

> But it is in European countries, France and England especially, that the helpful features of the vacant lots cultivation plan have come to be most appreciated. In 1899, through a lecture delivered for the Musee Sociale of Paris by the director of the Philadelphia Vacant Lots Association, and through the distributions of copies of the annual report of the then current year, the work was first taken up in France. The idea spread rapidly; even the great railways put it into operation along the lines of their various systems; and so great has been its growth and so important are the results that in 1904 a Congress of Cultivators of Workingmen's Gardens, held in Paris, was attended by more than seven hundred delegates. More than one hundred and thirty villages, parishes, and cities in the Republic now have the work underway, and during last season (1904) six thousand four hundred gardens were under cultivation.[59]

This claim may be exaggerated, since Europe had a tradition of allotment gardens that began much earlier than American vacant-lot cultivation associations did. However, combining food production with education and moral development may have been a particularly American approach.

OUTLOOK: SUCCESS OR FAILURE?

Vacant-lot cultivation associations were praised as a new kind of relief program based on the ideal of helping oneself, yet how successful were they at meeting their objectives? Most contemporary accounts were optimistic and not very critical. One of the few analyses of the period, the AICP report, acknowledged that some programs were not successful, usually due to organizational factors. Three reasons were cited: limited participation due to distrust of charity, active speculation in suburban land that made access to land difficult, and lack of managerial ability. In addition, financial returns were often limited by lack of seasonal preparedness. Another frequently mentioned problem was the lag between the time when public interest was raised for the project and the first planting, which resulted in shorter growing periods and smaller yields. Further, time was lost due to the participants' inexperience and the time needed to train them. However, as the 1898 *Charities Review* article commented, "Notwithstanding all the drawbacks and the ignorance of the workers concerning farming methods in general, the financial return has been satisfactory to both the workers and to those in charge of the several movements. Where the work has been discontinued it has usually been for reasons other than lack of encouraging financial results."[60]

As a temporary measure, vacant-lot cultivation was generally considered worth the effort. However, the few voices that rose to suggest permanency were drowned by the returning economic prosperity. To serve the immediate needs of the poor, the programs relied on donated land and volunteer leadership. For the sake of efficiency and to meet demand, donations of larger tracts at the city's edge were desirable; however, these sites were also attractive to developers once the economy improved. The vacant-lot cultivation associations had not attended to the long-term possibilities of land as a resource for the urban poor to raise food and provide other social benefits. While gardening was touted as benefiting participants in terms of health, income, and possible job opportunities, the provision of land for gardening was not institutionalized for permanency and the pleas of garden supervisors for permanent farm sites went unheeded. Instead of land access, it was the job-training and moral aspects of gardening that continued to draw public support. Garden projects were intended to influence individual moral uplift and impel families to leave the city for the country or suburbs, where, through their own resources, they would acquire their own land.

TWO

The School Garden Movement

> As a means of education, gardening under intelligent and sympathetic direction is very valuable. For children to work outdoors, with their feet in the soil, their heads in the sunshine and their lungs filled with good fresh air, is good for them physically. To learn something that may contribute toward their support is good for them morally. Direct contact with the soil, with the fundamental forces of nature and with plant and animal life and the constant adjustments which must be made among these, is highly educative in itself and furnishes raw material of the very best kind for the work of the school. I look forward to the time when school gardening and home gardening under the direction of the school may become an integral part of the work of all the schools.
>
> *Philander P. Claxton, U.S. Bureau of Education Commissioner,*
> School Garden Association of America Annual Report
> *(New York: School Garden Association of America, 1916), 2*

GARDENING WITH CHILDREN IS A recurring theme in the evolution of urban garden programs. Every phase has included some intention to involve children, whether as students, volunteers, or members of a family. Perhaps the most impressive era for children's gardening was the promotion of school gardens that started around 1890, with the establishment of the Putnam School in Boston, and continued for about thirty years. Before this time, children's garden programs had been individual efforts with a local focus. For example, late-nineteenth-century garden promoters traced the history of school gardens back to 1691, when George Fox willed a tract of land near Philadelphia "for a playground for the children of the town to play on, and for a garden to plant with physical [medicinal] plants, for lads and lassies to know simples and to learn how to make oils and ointments."[1]

However, what was different in the 1890s was the movement's national scope and the enthusiasm for gardening as part of public school education.

Concerns during this period about urban and rural conditions, child development, and civic improvement found a shared solution in school gardens. Because gardening was considered a means to address a range of educational, social, moral, recreational, and environmental agendas, children's garden programs enjoyed a broad base of support from teachers, government agencies, institutions, garden clubs, social reformers, and civic groups. Prominent figures such as social reformer Jacob Riis, landscape architect Frederick Law Olmsted Jr., and President Woodrow Wilson praised gardening for its contributions to youngsters' education, health, industrial training, and general civic-mindedness. Exhibits on children's gardens appeared at such influential events as the 1908 Congestion Exhibit in New York, the 1908 International Tuberculosis Exhibit in Washington, D.C., and the 1915 Panama-Pacific International Exposition in San Francisco. Special-interest stories in local newspapers and national magazines, such as *The Craftsman, Garden Magazine,* and *Country Living,* as well as articles in philanthropic and professional journals, spurred general goodwill and interest in starting programs. Children's gardens began to appear in cities across the nation and its territories, so that by 1906 the United States Department of Agriculture was estimating over 75,000 school gardens.[2] When the federal Bureau of Education established the Division of Home and School Gardening in 1914, gardening received official endorsement as an educational resource in school curricula.

These gardens were called by various names—school gardens, school home gardens, children's gardens, school farms, farm schools, garden cities, and others. Whether part of a school curriculum or an extracurricular, nonschool activity, advocates considered all children's gardens to be in the movement. In her 1910 book, *Among School Gardens,* M. Louise Greene, one of the leaders of the movement in the early twentieth century, defined the school garden as "any garden where children are taught to care for flowers, or vegetables, or both, by one who can, while teaching the life history of the plants and of their friends and enemies, instill in the children a love for outdoor work and such knowledge of natural forces and their laws as shall develop character and efficiency."[3] However, for many of the advocates, the underlying agenda was to integrate gardens into the public school curricula. Gathering support for this and getting schools to accept the responsibility of school gardening led to the development of national organizations and an abundance of promotional materials.

PRECEDENTS: EUROPEAN AND CANADIAN AGRICULTURAL EDUCATION AND SCHOOL GARDENS

The legitimacy of school gardens was sometimes argued on the basis that a lineage of educational gardening dated as far back as ancient King Cyrus of Persia[4]—a long history reassuring advocates that school gardens were more than a fad. However, what piqued American interest the most was the recent expansion of garden and agricultural education in Europe and what this might represent in terms of American educational and economic competitiveness. Advocates of school gardens in America frequently cited European successes with school gardens and agricultural education, and promotional materials often described school garden programs as the culmination of progressive school reform that had reached a more advanced stage in Europe. The kindergarten concept of pre-school training to enhance children's physical and mental development, developed by Austrian educator Frederick Froebel in the early 1880s, had already gained acceptance in both Europe and America.[5] As even the name suggests, kindergartens frequently included gardening and other forms of nature study as educative and developmental resources. Other European educational leaders, such as Erasmus Schwab in late-nineteenth-century Vienna, advocated school gardens as a means to teach agriculture and vocational sciences to older children. In several European countries school gardens and agricultural curricula were available and sometimes mandatory. American promoters cited an 1869 Austrian imperial school law requiring a garden and a place for agricultural experiments at every rural school. Similarly, they noted that France established obligatory agricultural instruction as early as 1882, and later decreed that every school should have a garden attached to it. Reports of school gardens in the German states, Sweden, Belgium, France, England, Russia, and other countries not only inspired American educators but also raised concerns that America was falling behind in an essential educational development. Some advocates suggested that Europe's advanced agricultural training threatened American market competitiveness. For example, in 1904, the state superintendent of public instruction for Michigan went so far as to suggest that Europe's agricultural training meant immigrants would be more skillful and thrifty farmers and would quickly outstrip the average American farmer.[6]

Closer to home, Americans looked to Canada's recent promotion of school gardens as both model and competition. School gardening in rural Canada had floundered from 1872 until 1903, when the effort received renewed support from the commissioner of agriculture as well as financial support from philanthropist Sir William Macdonald. The program was mainly focused on rural populations as a means to teach them scientific agriculture as well as gardening as a domestic science. The effort led to the establishment of Macdonald schools throughout the eastern provinces of Canada. These schools were part of the school consolidation effort, in which five or six small rural schools were replaced with one central, graded school that could support school gardens and manual training for other rural-based trades. In addition, traveling instructors taught agriculture and gardening to more remote rural schools. American education reformers and agricultural experts both cited the Canadian program as a good model for American school gardening, particularly in rural areas.[7]

GARDENS TO ALLEVIATE URBAN CONDITIONS AND INSTILL A LOVE OF NATURE

Those who promoted school gardens often justified gardening as an antidote to poor urban conditions. Promotional literature often prefaced its discussion of school gardens with dire descriptions of the American city and its impact on children. The congested and dangerous streets, rubbish-filled vacant lots, overcrowded homes, and dearth of parks created environmental conditions detrimental to children's health and development. Concerned citizens blamed the city for producing poorly developed children susceptible to crime and delinquency. For example, Fannie Griscom Parsons described the urban conditions that inspired her to found the DeWitt Clinton Farm School in New York City:

> City children are alienated from their human birth-right of trees, fields and flowers. Encased amid bricks, stone, concrete, trolleys, trucks and automobiles, the crowds of people in our streets are as giants to them, and the blue sky overhead is seldom seen. These conditions are making our children hard and unfeeling. Deprived of their natural lives, impelled by the restless energy of youth, they find mischief the only diversion possible, and they become easy victims of vice and crime.[8]

Many people believed that nature-based experiences could provide the necessary alternative. To an adult audience, many of whom had grown up in the country or suburbs, it was unthinkable that urban children would not enjoy the outdoor experiences of their own youth. To expose the contrast between these experiences and those of urban children, Manhattan principal Margaret Knox described the urban child's conception of the rites of spring as a justification for school gardening:

> When the signs of spring asked for by the teacher who expected to get in reply something about the swelling buds, the flowing sap, the softened earth, brings only the answer, "yes ma'am, I know when spring is here because the saloons put on their swinging doors," is it not worth while to lead such a child to notice other signs of spring? To me this is what [a] school garden means in a crowded city district.[9]

Urban children who had little experience with plants and animals were at risk of losing their "natural heritage," the lessons learned from nature-based experiences that supported civic and cultural mores. Basing her argument on an evolutionary perspective, Ellen Eddy Shaw of the Brooklyn Botanical Garden justified gardens as "a racial instinct": "It is as old as civilization, for it was one of those influences which caused primitive, nomadic man to settle in one spot, because of his start at soil culture. It behooves us first to train a child along the easiest avenues, those of inheritance, to give him the benefit of the wholesome influence of the soil and to foster love of the country and its activities."[10]

Changes in child labor laws, mandatory school attendance, and working parents meant that children had unsupervised free time after school and in the summer. Convinced that a "bad boy" was only a case of misdirected energy, reformers sought legitimate activities that would positively engage youth. To avoid the street's bad influence, children needed programmed, productive activities when they were not in school. School gardens not only provided an alternative environment, they also extended the opportunity to control children's time and activities, and provided a social outlet and an activity that would encourage the children's proper behavior and development. According to James Jewell, writing in a 1907 Bureau of Education bulletin, "School gardens in the slums of a number of cities have taught more civic righteousness than all the police courts or college settlements have been able to do."[11] Anecdotal accounts repeatedly reported that children were better behaved as a result of participation in garden programs. The school garden taught lessons of cooperation, patience, politeness, responsibility, and kindness. In a 1911 article titled "What Ails Our Boys," social reformer Jacob Riis praised one school garden for its success in curbing youth gang activity: "The destructive forces of the neighborhood had been harnessed by so simple a thing as a garden patch, and made constructive.

Figures 11 and 12. *Left:* This photograph, with the caption "Ignorant of Nature: A Sociological Problem," is included in Louise Klein Miller, *Children's Gardens for School and Home* (New York: D. Appleton and Company, 1904), 3. *Right:* This photograph, with the caption "Good Gardeners," accompanies a story in Miller's book about the Boys' Garden in Groton, Massachusetts, where boys not only learned to garden but also participated in other civic improvement projects. Reprinted from Miller, *Children's Gardens for School and Home,* 57.

And a 'sense of the dignity of labor' had grown up in that of all most unlikely spots, that made the young gardeners willing and anxious to work for the general good as well as for themselves."[12]

EDUCATIONAL REFORM AND SCHOOL GARDENS

Looking for antidotes to the negative influence of the urban environment, many who turned to the schools for solutions found them inadequately prepared to address the changing social and industrial conditions. For one thing, schools were unable to keep children enrolled through grade school, let alone high school. Some of the factors leading to poor retention rates included parents who did not consider school to be worth the expense and effort, the pull toward on-the-job training rather than abstract classroom education, and the need for older children to earn money to help support the household. Education reformers struggled to find new ways for schools to engage students and the community.[13]

To counter criticisms that school curricula were oriented to the few who would go to college instead of the many who were bound for trade and manual careers, school reformers began advocating manual and industrial training. By including such classes as woodshop, cooking, sewing, and gardening, schools not only could provide hands-on learning experiences but could also compensate for the on-the-job training that had recently been limited by changes in child labor laws. The school garden was included as part of a rational sequence of industrial training. For example, in Washington, D.C., in 1905, schools added gardening as the sixth-grade boys' manual training. Similarly, Philadelphia introduced gardening as manual education for lower-grade boys while the girls learned sewing. Appeasing concerns about the limited opportunities for children to earn money, the garden provided entrepreneurial training under the auspices of education without the risk of exploitation. The garden taught the value of work in relation to earnings and such attributes as efficiency, practicality, economy, and thrift. Youth could earn some money from the sales of produce and be spared the dangers of factory work.

Child development theories of the day criticized the classroom format as inhibiting children's natural inquisitiveness by making them sit all day in the artificial environment of a classroom. Education reformers sought to remedy this situation by providing active learning through playgrounds, recreation programs, and nature study.[14] In contrast to book-based learning, nature study relied on children's hands-on experiences and their manipulation of natural forms and phenomena. The school garden was a logical component of a nature-study curriculum because it provided a place for children to see firsthand such natural processes as growth and decay, nutrient cycles, and the interrelationship of plants and animals. Through gardening, "[children] very early get the idea of the great interdependencies of animals, vegetables and minerals and soon realize that plants are the connecting link between the mineral and the animal worlds."[15] Furthermore, the child could see tangible results, whether the disappointing results of neglect or the bounty of a well-tended garden.

Gardening was not considered a "panacea for all educational ills." Rather, it was generally considered part of a larger movement for change in school structure and curricula. The garden complemented the classroom by providing an alternative context in which to engage children. In his report on school gardening, Otis Caldwell, botany instructor at Eastern Illinois State

Normal School, measured the success of school gardening by its ability to engage the pupil in self-learning:

> We can put things in his way to help him develop properly, and keep from him some of the things that fail so to help him; but we cannot do his development for him. If he is to have knowledge of living things, of the elementary principles of life, of industry, of economy, of beauty, and justice, he must grow into these things by means of first hand experience with them. To obtain this growth and to eliminate some undesirable things now in school, the school garden should certainly prove efficient.[16]

A 1907 report to the Bureau of Education justified activities such as gardening by stating that children who participated in school garden programs achieved greater development in a given time and accomplished more in their regular studies than did children who did not participate.[17] Various reports stated that garden participation increased overall school attendance and preparedness to learn.

AGRICULTURAL EDUCATION AND SCHOOL GARDENS

Since the urban school garden was meant to reintroduce nature into the city, the movement borrowed heavily from the political and technological resources engaged in improving rural education through agricultural training as well as other educational reform. As early as 1862, with the passage of the first Morrill Act, the federal government was advancing agricultural education through land-grant universities and agricultural extension services, while also encouraging agricultural training in secondary schools, and gardens in schools generally. Education emerged as a unanimous concern in the 1909 Country Life Commission hearings. "Everywhere there is a demand that education have relation to living, that the schools should express the daily life, and that in the rural districts they should educate by means of agriculture and country life subjects. It is recognized that all difficulties resolve themselves in the end into a question of education."[18] Proposed solutions included rural school consolidation, specialized courses of study, agricultural training, and school gardens. Government agencies provided technical support through bulletins and circulars that included agricultural lesson plans and advice. These resources, while designed primarily for rural schools, also advanced urban programs. In the face of rural depopulation, some advocates proposed school gardens as a way to encourage urban youth to leave the city to pursue truck farming and other agricultural careers in rural areas, rather than seeking "overcrowded occupations in the city."[19]

The broad desire to expand agricultural education in primary and secondary schools was only partly related to a need for practical training in farming; it also drew considerable appeal from the wish to maintain the cultural influence of agrarianism and country living. Late-nineteenth-century changes in agricultural technology, shifting agricultural economies, and industrial opportunities elsewhere meant that fewer people—even those who grew up on farms—would take up farming as their profession. However, while the economic changes may have discouraged farming as a profession, there was still a strong cultural allegiance to rural lifestyles. For the farmer and nonfarmer alike, agricultural training taught much more than cultivation. "Agriculture has a far bigger content than the term farming," wrote Ernest Babcock and Cyril Stebbins in the introduction to their *Elementary School Agriculture,* continuing, "It is, in fact, a great composite of the fundamental sciences and, during the progress of civilization, has come to include a long list of elementary arts and technical industries.[20]

In addition, agriculture had a social perspective, conveying important social skills associated with agrarianism. Agrarianism was not viewed as solely a profession; it was considered the bedrock of American citizenship.[21] The characteristics associated with farming—independence, hard work, honesty, and so on—were pertinent not only to the farmer but to the businessperson as well. In his manual for teachers and pupils, H. D. Hemenway, Director of the School of Horticulture at Hartford, estimated that 90 percent of successful businessmen had grown up on farms and learned productive industry at an early age through farm life. He concluded, "There is no kind of training that squares itself for all-round development like agriculture."[22] According to Henry Griscom Parsons of the Children's Farm League in New York, the constant object lessons associated with gardening taught children "the dignity of labor, and the love of doing productive work; the virtues of thrift, honesty, courtesy, and obedience to law."[23] Thus, agricultural education was not just about technical skills associated with state-of-the-art farming techniques but also about ethics and moral development. There was a stated need to elevate appreciation for the farmers' role in society and promote agriculture as a worthy profession. This attitude was optimistically described by Hemenway:

> While we may not be able to make many farmers and gardeners, we may help to make much better men and women. It is hoped that we may check

the flow of people to the city and turn some back again to the country. The school-garden creates a love for industry, a love for country, for nature and things beautiful, and makes boys and girls stronger, more intelligent, nobler, truer men and women.[24]

Many educators and advocates saw the possibility of combining agricultural training and nature study in a hierarchy of agricultural training that included children's gardens. At a minimum, gardening would impress upon rural and urban children alike the value of scientific knowledge and appreciation of the farmer's role in society and in relation to nature. Younger children in primary schools would receive nature-study lessons, and each grade thereafter would advance with more technological understanding of plant propagation, cultivation, and other skills. Youth who wanted to pursue agricultural careers could advance to specialized agricultural courses in normal schools (high schools) and colleges, since it was generally acknowledged that urban schools could not adequately teach agriculture. In her 1905 report to the Department of Agriculture, Beverly Galloway, chief of the Department of Agriculture's Bureau of Plant Industry, proposed that the garden in rural contexts should focus specifically on practical agricultural training while urban gardens should serve as aesthetic, moral, and physical training. Similarly, in the Farmers' Bulletin issue titled *The School Garden,* L. C. Corbett outlined two classes of work—one for rural teachers, who should focus on laboratory experiments to demonstrate more sophisticated and applied lessons in plant nutrition, propagation, and variety yields, and one for urban teachers to teach simple operations in the maintenance of vegetable and flower gardens.[25]

FROM THE GEORGE PUTNAM GRAMMAR SCHOOL TO GARDENS ACROSS THE NATION

Credited as the catalyst for the school garden movement, the George Putnam Grammar School in the Roxbury district of Boston, Massachusetts, established its first garden in 1891. This venture was the result of one teacher's efforts, with the support of the Massachusetts Horticultural Society. Inspired by a tour of school gardens in Europe, Henry Lincoln Clapp returned to his position as Master of the George Putnam School and established a children's wildflower and fern garden on the grounds. In 1900, he added a vegetable garden that was tended by thirty seventh-graders. The garden eventually included eighty-two vegetable garden plots, each twenty-

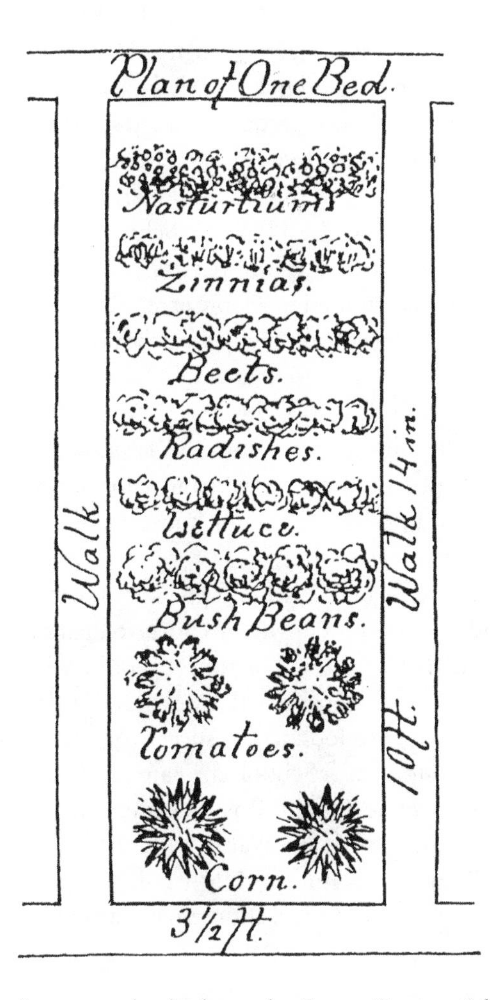

Figure 13. A plan for one student's plot at the George Putnam School Garden. Reprinted from *Report of the Committee on School Gardens and Children's Herbariums of the Massachusetts Horticultural Society* (Boston: Massachusetts Horticultural Society, 1902), 9.

eight square feet. The Massachusetts Horticultural Society continued to support this venture and encouraged replication in other Massachusetts schools by initiating a school garden contest. Putnam School won every year from at least 1891 to 1902, and the $15 prize money was used to buy equipment and expand the garden program.[26]

School gardens began to appear around Boston and then throughout Massachusetts. In 1901, the Boston Normal School and the South End Settlement House joined forces to start a garden at Dartmouth Street and Warren Avenue. This garden, located in a crowded part of the city, served eighty seventh-grade students. School garden advocates formed the School Garden Committee in 1901 to promote more gardens. As a result of their efforts, a school garden in the North End was established on the site of a torn-down tenement adjacent to the Hancock School. With time, the school board accepted school gardening on a systemwide scale. However, until 1904, these projects were funded to a large extent by the Twentieth Century Club, with a small appropriation from the schools themselves. According to a report produced for the state board of education, by 1906 there were 210 school gardens in fifty cities and towns in Massachusetts.[27] Assistance, both financial and technical, came from the Twentieth Century Club, the Women's Auxiliary of the American Park and Outdoor Art Association, the Massachusetts Civic League, settlement houses, and other women's clubs and village improvement societies.

In a very short time, school gardens appeared in other East Coast cities and then spread westward. Glowing reports of gardens in New York, Philadelphia, Chicago, Cleveland, Dayton, St. Louis, Minneapolis, and elsewhere were intended to encourage others to start gardens.[28] Teachers and civic groups were invited to look to the successful programs, such as Manhattan's DeWitt Clinton Farm School, the Fairview Garden in Yonkers, the National Cash Register Company Boys' Garden, the Hartford School of Horticulture, and the garden programs affiliated with southern African American schools such as Hampton Normal and Agricultural Institute in Virginia and Tuskegee Normal and Industrial Institute in Alabama.

THE DEWITT CLINTON FARM SCHOOL

One of the most frequently mentioned children's gardens was the DeWitt Clinton Farm School in New York City, organized in 1902 by Fannie Griscom Parsons, who was a member of New York's school board. Although the garden was a cooperative effort that involved the parks and school departments, "it is to the cheery, energetic personality of the volunteer originator that the scheme owes its success. Any afternoon—challenging the hottest suns beneath a wide straw hat—may be found this mother of seven who has adopted the hundred."[29] The site—at Eleventh Avenue and Fifty-Third Street—had been a dumping ground and truck storage facility but was slated for a new park. The farm school was intended as an interim activity; however, it quickly gained a dedicated constituency and progressed as though permanent. In its first year, a 114-foot-by-84-foot plot was prepared, requiring heavy street-breaking plows to break the ground and imported loam. That year, twenty-five children from the surrounding tenement district worked under the supervision of gardeners assigned by the Parks Department. These city-employed gardeners were reported to be mostly Swedish immigrants who had experience in school gardening from their homeland. The next year, the garden grew to one hundred feet by two hundred feet and served 277 children. As the time drew near for the park's construction, Parsons lobbied for the farm's preservation. Citing the popular appeal of the farm in the community, Parsons noted, "And so a whole neighborhood is reached. Could any beautifully laid out park with Keep-off-the-grass signs bring such joy as this, and must the garden go when the new park materializes?"[30] Eventually, a smaller version of the farm school was included in the park's final design.

The DeWitt Clinton Farm School was a model for how to garden in a congested urban area. It consisted of a grid of beds that could be efficiently assigned to individuals or classes. The site also included a piggery and a small building to serve as a farmhouse—a "dear little country seat" where children could pretend to be host and hostess. The garden was surrounded by a three-foot fence on which adults were reported to lean to watch the events of the garden. Parsons described the park's garden in 1905:

> Down "Broadway" as the center path has been named, amid groups of busy little farmers each tending a claim 4 feet by 8 feet, containing a stalk of corn in the center of a row of beets, to the right and left of which grow carrots, peas, lettuce, radishes, and onions, here and there a teacher holds the attention of an interested group as he explains the wonders of growth, soil, etc.[31]

ADVOCATING FOR SCHOOL GARDENS

As was the case at the George Putnam School and the DeWitt Clinton Farm School, the school garden was often the work of a single teacher or other individual. Much of the promotional literature encouraged teachers to start projects with their classes, in hopes that the school system would eventually support the effort. In many cases, school gardens were started by or relied heavily on the support of women's clubs, mothers' associations, horticultural clubs, and other civic groups. These groups helped to raise

Figure 14. The DeWitt Clinton Farm School, built in 1902, was originally intended to be temporary until a park was built on the site. When the park was finally designed, it incorporated a smaller version of the farm school. In 1910, the farm school won a *Garden Magazine* award for its general appearance. Reprinted from Henry Parson, *Children's Gardens for Pleasure, Health and Education* (New York: Sturgis and Walton Company, 1904), frontispiece.

funds, secure land, and provide incentives to students and teachers in the way of prizes, contests, and exhibits. City agencies, universities, institutions, civic clubs, settlement houses, newspaper agencies, private institutions, companies, and individuals sponsored garden programs as well.

As the various efforts around the nation gained recognition, it became clear that school gardening was a national movement. Two national voluntary organizations were founded in 1910: the School Garden Association of America and the International Children's School Farm League.[32] The purpose of the former was to promote school gardens as an integral part of the educational system. Its objectives were to arouse public interest in the educational value of school gardens, to discuss the evolution of the movement, to distribute useful school garden literature, to conduct experiments, and to promote the establishment of school gardens. The association's standing committees addressed topics such as garden clubs, local garden associations, city school gardens, and rural school gardens. In 1912, its membership included nearly two thousand principals, teachers, and citizens. Similarly, the International Children's School Farm League, which grew out of the demand for information received by the DeWitt Clinton Farm School—formed "to promote and unify a world-wide interest in children's gardens."[33] The league's first effort was a demonstration children's garden for the 1907 Jamestown Exhibit, for which the league worked with local children to develop a garden and farmhouse modeled after DeWitt Clinton. Members of the league also gave lectures and provided displays at conferences, including—according to their reports—the first national city planning conference in 1909.

While most school gardens relied on the financial and personnel support of philanthropic sources, there was still a strong desire among advocates to integrate gardens into the public school system. Firmly convinced that gardening was an important educational resource, advocates such as School Garden Association of America president Van Evrie Kilpatrick urged the continued effort to promote public school gardens. In his 1910 annual report, Kilpatrick argued that since children were already obligated to attend schools, working through the school system avoided multiplying the number of institutions that had command over children's time. He maintained that "all children's gardens, community gardens, home gardens, or school gardens, are to be encouraged, but the development should be fostered by the regularly organized schools."[34] M. Louise Greene foresaw the problems of ambiguous leadership and uncertain access to land for gardening programs not under the auspices of the school system: "So long as the educational value of school gardens is not fully recognized by the local school boards, just so long will they be dependent for their support upon philanthropic societies or upon the good will of private individuals, and be subject to the discouragement of loose tenure and shift of locality as land values rise."[35]

In 1914, almost twenty-five years after the founding of the Putnam School garden, the federal government set up the Bureau of Education's Office of School and Home Gardening. The office started with three staff members under the leadership of C. D. Jarvis. Because it had limited authority over local school boards, the office primarily assisted school gardens through publications and public promotion. In 1916, the Bureau of Education published *Gardening in Elementary City Schools,* which Commissioner of Education P. P. Claxton introduced as a "comprehensive statement of the best means of organizing and directing this work, to the end that the largest possible educational and economic results may be obtained."[36] Written by Jarvis and based on a survey sent to school superintendents, the bulletin reiterated many of the reasons to garden, including vocational training, teaching thrift and industry, and providing a means for youth to earn income, along with the benefits of recreation, health, improved morality, and joyful living. Acknowledging that funding for school gardens had passed "through the same troublous vicissitudes as music, manual training, drawing, and home economics," Jarvis suggested that programs start small and use the support of local voluntary organizations to garner success and public support.[37] Garden programs needed to demonstrate their popular appeal and educational worth if school administrators were to accept gardening into the curriculum and budget. Citing the relatively inexpensive nature of gardening—estimating that for every $1 invested in the employment of a gardening instructor, $20 worth of produce was grown—the bulletin urged financial support from local school boards.

To varying degrees, many schools and school boards took on the responsibility. Some schools assigned a single teacher to be in charge of all garden and nature-study activities. Others gave extra pay to a teacher who supervised gardens as an after-school, weekend, and vacation activity. The Office of School and Home Gardening recommended one garden teacher for every two hundred children of gardening age, which was a typical population for an eight-room school. The teacher, assumed to be female, would devote her mornings to teaching agriculture, nature study, elementary science, homemaking, or other special subjects. To compensate for her time spent after school, on weekends, and in the summer, the office recommended that the teacher receive $200 to $300 over the average teacher's salary.[38] Other school districts employed a districtwide garden curator or supervisor. In 1905, Cleveland, Ohio, formalized a systemwide gardening program and employed a curator of gardens, Louise Klein Miller. Her duties included supervising school gardens, giving lectures in the schools, inspecting flower shows, arranging for fall festivals, and improving school grounds.[39]

Whether or not gardening was accepted as part of the public school curricula, garden programs relied on a network of support. Collaboration between schools and civic or philanthropic groups was a common and often necessary strategy to get projects started and/or to sustain them. Philadelphia's school garden program epitomized the kind of cooperation necessary between civic groups and the school board. It took the combined efforts of the Philadelphia Vacant Lots Cultivation Association, the board of education, the Public Education Association, and the Civic Club to initiate the first two school gardens in 1904—while in their various reports, each of the involved parties claimed to have initiated the idea. The Vacant Lots Cultivation Association secured and prepared suitable land for school gardens, while the other agencies supported teaching expenses. A 1905 report praised Philadelphia's foresight in including school gardens in the public school system:

> The history of school gardens in Philadelphia already well shows that it *matters not so much whether the ideas of one* or another *group of individuals are carried out as that the work shall be sanctioned and supported by the people.* It is because school-gardens have from the start been conducted as a legitimate part of free public education in Philadelphia that the pride and interest in them have been so widespread and immediate.[40]

How well the various supporting agencies collaborated was a critical factor in a program's success or failure. One of the few reports that addressed the difficulty of balancing public and private support was a 1918 report by the Detroit Bureau of Governmental Research.[41] Detroit's school garden program had started under the leadership of the Twentieth Century Club, which until 1913 bore all expenses connected with promoting children's and community gardens. In 1913 the board of education began to support the program, and in 1915 responsibility was shifted to the recreation department. The report was very critical of the lack of organization and supervision given by the recreation commission. Money that had been allocated for training garden recreation leaders and supervision was not spent. The report recommended that public agencies lead the programs, and it concluded that since gardening had proven itself an educational and civic resource, similar to kindergartens, school gardens should be the responsibility of the board of education.

SCHOOL GARDENS IN THE NATION'S CAPITAL

> What might have been playgrounds for the children or little gardens to dig and be happy in, have been desecrated by being used as ashpiles; where the eye might have delighted to rest upon God's green earth, fragrant with flowers, it has been offended by unsightly heaps of broken crockery and all such familiar articles as go to make up the regulation "dump pile." So that while our city, with its wide streets and beautiful parks, is capable of being what

President Roosevelt says it should be, a model city, it can never attain that distinction until it shall have availed itself of the advantages [with] which it is so richly endowed for rich and poor alike.[42]

Depicting Washington, D.C., thus in a 1903 *Charities* article, Elizabeth Rafter went on to describe the successes of school gardening in the District through the combined efforts of civic, educational, and governmental sources. In 1902, the Civic Center of Washington and the Washington branch of the National Plant, Flower, and Fruit Guild started a home garden program. The Department of Agriculture provided support through free seeds, which were distributed through two settlement houses to about 150 applicants. Two experts from the Department of Agriculture demonstrated how to plant gardens and window boxes. After the planting, volunteers visited the children's home gardens to see their work and provide advice, and at the end of the season a picnic was thrown and prizes were given. The following year, the same institutions formed five children's garden clubs, each with ten to twenty children who collectively received instructions and shared a garden. In locating plots of land for this work, the main goal was "to show the children what could be done with just such materials as they already had in their own yards" while also not minimizing the need to secure and protect a shared plot. In 1903, this project also included five hundred home gardens visited by volunteers who advised the children.

About the same time, the Bureau of Plant Industry received a request from Washington's Normal School Number Two for assistance in introducing gardening into the school curriculum. L. C. Corbett, horticulturist for the bureau, presented five lectures to the students on soils, germination, cuttings, grafting and budding, and the adornment of school grounds. In 1904, the Department of Agriculture provided a one-sixteenth-acre site on their grounds, a small greenhouse, and a workroom. A group of sixth-grade boys used the garden for manual training while the girls received training in sewing. The boys raised $55 worth of produce, which they donated to hospitals. Beverly Galloway from the USDA praised the project: "Experienced farmers sometimes fail to do as well."[43] The next year, the Department of Agriculture provided a greenhouse for African American pupils at Normal School Number Two. These youth, trained in the greenhouse, then taught children in other schools, so that by 1905, every African American school in the District had a garden unless it lacked the grounds for one.

In 1910, Washington had four large school gardens on vacant lots as well as school-ground improvements at one hundred White and fifty African

American schools. Congress appropriated $1,200 for school gardens but did not allow money to go toward instructors' salaries. Susan Sipe, botany teacher at Normal School Number One, was hired as supervisor of nature study and school gardens of the District of Columbia on a nominal salary that was paid by outside sources. She collaborated with the Department of Agriculture, producing various articles and pamphlets that informed the national movement.

CALIFORNIA'S CORNUCOPIA OF SCHOOL GARDENS

In his 1905 bulletin, *School Gardens for California Schools,* B. M. Davis noted that although the state had no comprehensive school program or mandate for gardens, school gardens were starting to appear throughout the state through local initiative.[44] In Southern California, a school garden program had been developing at the Los Angeles State Normal School since 1898. Situated on a plot 64 feet by 122 feet, the garden provided nature study and botany education to elementary school children. The garden was laid out in a utilitarian manner but included flowers, shrubs, and water features for aesthetic effect. A similar but larger garden was established at Chico State Normal School in Northern California in 1902. In 1910, the Los Angeles School District established a districtwide school garden program.[45] Garden teachers were hired to assist classroom teachers with garden-related activities, while civic and women's groups provided additional help.

One of the first school district gardens was built on a vacant lot adjacent to the East Seventh Street School. Here, agricultural teacher Mrs. Marie Aloysius Larkey, assisted by Ms. Merle Smith, developed a small garden in which each class, kindergarten through sixth grade, had its own plot. Later, the garden was laid out for individual gardens. In 1913, the school won a prize as the most improved urban school garden in the *Garden Magazine* children's garden contest.[46] In addition, several high schools in the Los Angeles area established agricultural training and garden programs, including Hollywood High School and the new Gardena Agricultural High School. In 1912, the district organized the primary and secondary agricultural training and school gardens into a district-wide agriculture department for the county and city. The department included a supervisor with four assistants, each responsible for garden programs in a section of the city. In the first year, there were over one hundred school gardens, sixty-five of which were on borrowed land. Thirteen thousand students received

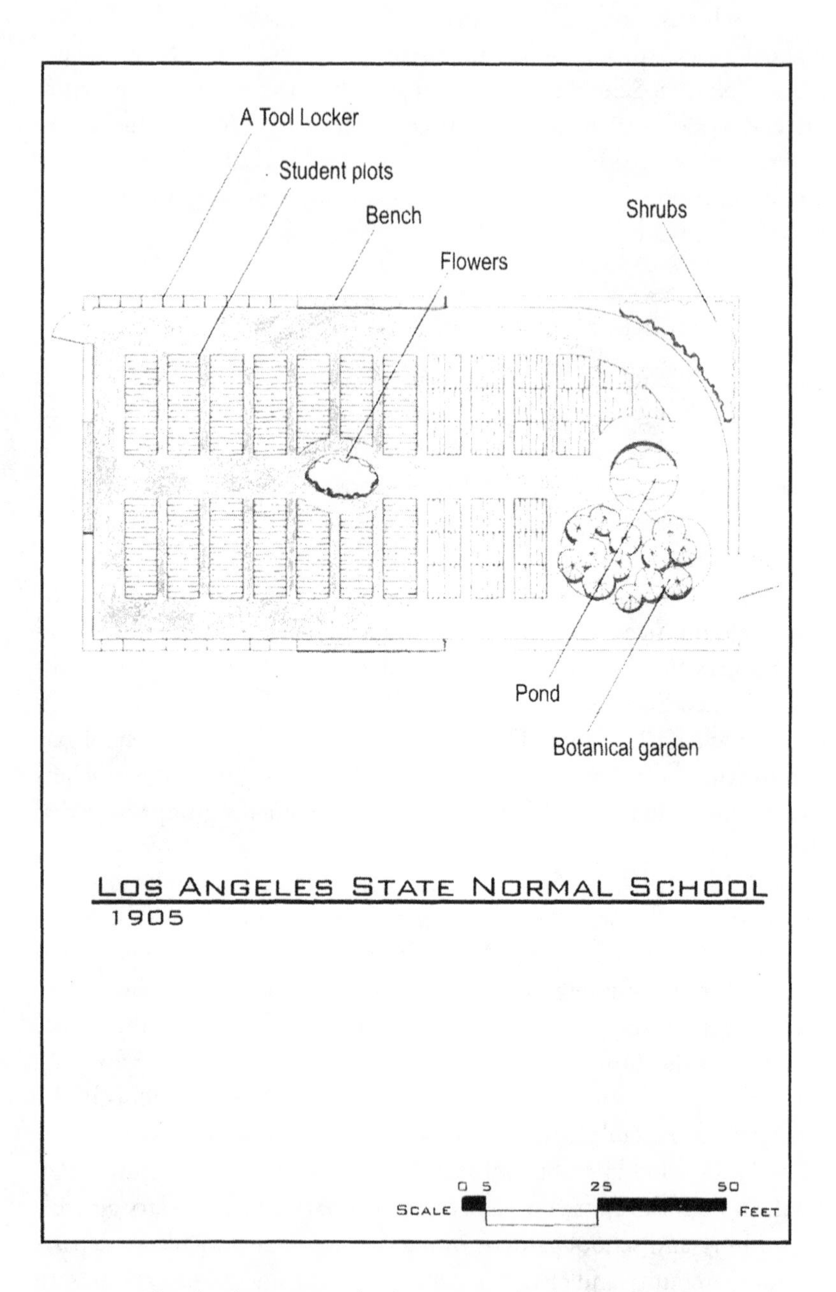

Figure 15. Plan for the Los Angeles State Normal School garden. Based on a plan in B. M. Davis, *School Gardens for California Schools* (Sacramento: State Printing Office, 1905); redrawn by Agus Soeriaatmadja.

Figure 16. The East Seventh Street School garden in Los Angeles, 1913. The garden won a *Garden Magazine* award for most improved school garden. Courtesy of the Security Pacific Collection/Los Angeles Public Library.

regular instruction. The program evolved into a graduated agricultural education program, starting with the fourth grade and continuing through high school.

Civic interests soon joined the effort. In 1914, the chamber of commerce began to promote clean-up campaigns and home gardening through its "Beautify California for 1915" campaign. Identifying children's participation as a means to get the word out to many families and neighborhoods, the chamber worked with the school board to encourage children's home gardens, vacant-lot gardens, and school gardens. Over 90,000 children, described as City Garden Soldiers, were reportedly involved in school gardening as a result of this effort.[47] The campaign also included a designated planting day in September, organized donations by local nurseries, and a contest in which over eight thousand schools, women's clubs, and municipalities competed.

Of the various garden projects developed in California, the Garden City at the University of California, Berkeley, illustrates a unique contribution by the state's public university and the agricultural extension network.[48] The program began as a way to meet a state-determined goal of promoting the ideals of country life through nature study, gardening, and agriculture for elementary school pupils, and expanding rural vocational agricultural training in the upper grammar grades. As an initial step, the university published *Suggestions for Garden Work in California Schools* in 1909.[49] In 1911, the Division of Agricultural Education of the University of California offered a one-acre site on the Berkeley campus, along with water, tools, and seeds, to start a garden program for fifth- and sixth-grade children from Whittier School in Berkeley. The garden was located on the northern border of the campus and within view of the president's house. Called the Garden City, the garden program was an extracurricular activity in which children participated one hour per week, supervised by university faculty and students. Within a year, over two hundred children, aged six to sixteen years, had individual garden plots of six feet by nine feet. There were also communal plots where children could watch experiments in fertilization and irrigation. To promote an entrepreneurial spirit and a sense of responsibility, the program included a Garden City Market where children could sell their produce, retaining 90 percent of their earnings for themselves and giving 10 percent to the Garden City Bank toward lease of the land, maintenance of shared tools, and other group needs. Over fifty mothers and fathers visited and purchased the first harvest of produce—rutabagas, lettuce, greens, and radishes.

The expressed goal was to provide agricultural education but also to use gardening to teach civic and moral duty. "We grow flowers, vegetables and children in California gardens," stated Cyril Stebbins of the University of California in a 1912 *Nature-Study Review* article. The Garden City was based on a philosophy that "the school garden is a miniature world patterned after the universe."[50] Indeed, the program was organized like a miniature city. "Through the garden, the school may repeat this civic history and the children may be brought in contact with community factors."[51] The Garden City included an elected government: a mayor, city council, garden commissioner, street commissioner, tool commissioner, city clerk, and two police officers. "Citizens" were required to work at least one hour two days per week.[52]

The university's College of Agriculture expanded this program, sending university students to rural and urban schools throughout California to establish clubs and Garden Cities. To reach a larger audience, in 1911 the college began publishing a monthly newsletter, *The Junior Agriculturist,* along with teacher supplements. The four-page pamphlet was intended for children around the state who were studying agriculture through the university's extension programs. Circulation rose to five hundred children within

Figure 17. Children learned recordkeeping, banking, the electoral process, and more as part of their involvement at the Garden City, University of California, Berkeley. Courtesy of the Bancroft Library, University of California, Berkeley, UC Presidents Archive CU-5, ser. 1, 53:72.

six months. The pamphlet included technical information, suggestions for experiments, contest rules, short articles by children, and general information on school garden programs around the state. When Stebbins, its editor, left the University of California to teach at Chico State Normal School, already well known for its school garden program, the publication moved with him.

Besides the Garden City, the Berkeley campus also had gardens where teachers could study agricultural education. Summer courses were provided for elementary and high school teachers in nature study, school gardening, and agriculture. Along with working on the Berkeley campus, the teachers went on organized excursions to the university's farm in Davis, now the University of California, Davis campus. An exhibit on agriculture in secondary schools was added to the extension's "demonstration train" that traveled around the state with exhibits on agriculture, nutrition, and sanitation. During the 1909–10 season, over seventy thousand people visited the train during its trips, including a small percentage of schoolchildren.

THE GARDEN IN THE CURRICULUM

The ability of gardens to facilitate many different educational agendas was a two-edged sword. Debate arose about the dominant purpose of school gardens—was it scientific training or fostering an emotional connection to nature and country living? Was it beautification or science? Concerned that gardening was being taught in a sentimental fashion that lacked rigor, the federal Bureau of Education's C. D. Jarvis questioned whether the intent of gardening had been misinterpreted as "a means to an end rather than as a subject having real value in itself."[53] He felt that the tendency to focus on aesthetics and emotional results obscured vital economic and industrial lessons. Some garden programs developed standardized instruction; however, as Roland Guss, director of school gardening in Cincinnati, stated, too much gardening was "hit or miss" or "rule of thumb" and degenerated into sentimentality.[54] Instead, gardening needed to fit into the larger school curriculum, with a strong foundation in science and connections to other subjects both in and out of the classroom. Arthur Dean, chief of vocational schools in New York state, emphasized the need for rigor: "While the Progressive believes that high school should teach scientific agriculture, vocational handwork, household arts, he does not believe in wasting the time of young people of secondary age in plain sewing, mere mixing of ingredients, making tabourets, and digging in the school garden unless these subjects have a scientific treatment in the school and are supplemented by home practice."[55]

There were two primary approaches to the development of school garden curricula: "practical work" and correlation to other studies. For schools intent on agricultural training, the garden was the laboratory for a wide range of lessons. This was particularly relevant for rural schools in the Midwest and West. According to Fred Bolster, who supervised the teacher training program at the University of California, lessons should include soil science, fertilization, plant propagation, botanical knowledge, seed ordering and testing, good cultivation, use and care of tools, irrigation, use of animals in garden work, plant enemies and friends, and plant budding and hybridizing, as well as landscape art, the experimental method, and recordkeeping.[56] According to a syllabus produced by the U.S. Bureau of Education in 1915, school gardens offered applied lessons in hotbeds and cold frames, soil preparation, planting, care of growing crops, marketing, and garden equipment.[57]

Besides the "practical lessons" of agriculture and gardening, many advocates sought opportunities to link garden activities to other academic subjects, such as art, writing, math, language, and geography. Concerned that gardening was being pigeonholed as manual training, a 1905 *Nature-Study Review* editorial note emphasized the capacity of gardens to serve the larger curriculum: "Gardens are such splendid concentrations of natural objects, especially the living, that they surely have the possibilities of great educational value in disciplines other than manual and in information which has practical, intellectual, aesthetic, and moral bearings. It is here, rather than in the practiced management, that we see the present problem concerning the school garden movement."[58]

In 1904, W. A. Baldwin, principal of the State Normal School in Hyannis, Massachusetts, facilitated this goal of correlation by developing a chart that correlated garden activities with reading, oral and written language, spelling, arithmetic, drawing, manual training, geography, and history. It showed, for example, that by ordering seeds, children learned how to write orders, calculate the amount of seed to order, and keep records of bills. Instruction about soil could include oral, written, and drawing lessons. This chart was displayed at the Educational Exhibit in St. Louis and reprinted in many sources. To aid children's learning, teachers were advised to have students keep notebooks where they could record weather patterns, germination rates, expenditures, and earnings, as well as write poetry and essays or draw sketches inspired by the garden. From his experience with school gardening at Tuskegee Normal and Industrial School, George Washington Carver described gardening as a source of material for instruction in other subjects:

> Nature study as it comes from the child's enthusiastic endeavor to make a success in the garden furnishes abundance of subject matter for use in the composition, spelling, reading, arithmetic, geography, and history classes. A real bug found eating on the child's cabbage plant in his little garden will be taken up with a vengeance in the composition class. He would much prefer to spell the real, living radish in the garden than the lifeless radish in the book. He would much prefer to figure on the profit of the onions sold from his garden than those sold by some John Jones of Philadelphia.[59]

School garden instruction was developed for children of all ages in accordance with child-development theories of the period. For example, it was appropriate for kindergartners to share a small garden that they visited daily for lessons in nature study and cooperation, while older children learned responsibility and competitiveness through cultivating individual plots. Most reports, including one by the Bureau of Education, considered the fourth through sixth grades particularly important times for children to garden on individual plots. For those schools that included gardening in regular class work for older children, one to two hours per week sufficed. In addition to class periods given to nature study and garden work, gardens were also tended during recess and before and after school. For most programs, whether school-supported or not, the majority of time dedicated to the garden was after school and on Saturdays.

Given the great number of children participating in garden programs, discipline was essential. The goal was to balance discipline with freedom. Garden exercises often focused on efficiency and uniformity, such as the correct way to carry water or group exercises in synchronized raking. Children were assigned tools and expected to return them clean. Discipline was often described as militaristic. Commenting on the Fairview Garden in Yonkers, Mary Leland Butler stated that the boys did better in school because of the training they received in the garden. "Discipline almost military is imposed on them—and they like it."[60] M. Louise Greene also stated, "It may be well to insist that such discipline as necessary should be almost military. The children like it better, provided the spirit is not that of the martinet, but one of mutual helpfulness, expressed in firm, gentle, unyielding yet sympathetic manner."[61]

PREPARING TEACHERS

The expectations placed upon instructors of school gardening reveal the multiple, enmeshed objectives concerning the garden. In addition to technical competence, the ability to establish good rapport with children and their families was required. The teacher had to balance freedom and discipline, keeping strict control but also encouraging the development of individualism. Teacher responsibilities included maintaining the garden and keeping good records of weather, work, harvest, visitors, student attendance, and various educational outcomes.

The widespread concern that gardening should teach children modern

Figure 18. The original caption reads: "The Raking-Drill, Carroll Garden, Philadelphia, PA." Reprinted from M. Louise Greene, *Among School Gardens* (New York: Russell Sage Foundation, 1910), 89.

cultivation techniques and scientific methods meant that teachers needed technical skills in gardening. A *Garden and Forest* article that praised the virtues of teaching agriculture in schools also raised the concern that teachers, who were usually low-paid and had taken up teaching as a temporary occupation, were not equipped to teach modern cultivation methods.[62] As more schools established programs, normal schools, teachers' schools, and colleges began to provide teacher instruction in school gardening. One of the first was the Hartford School of Horticulture, a private school established in 1900 that provided an array of courses for children and teachers.[63] In addition, many books and bulletins were available to help the teacher or civic group establish a school garden, such as Hemenway's *How to Make School Gardens* and Greene's *Among School Gardens.* The Bureau of Education produced a range of circulars containing technical information on planting as well as suggestions for activities and lesson plans.[64] Various universities and normal schools produced informational materials, such as the Hampton teacher's leaflets and Cornell leaflets. The *Nature-Study Review* provided descriptions of model programs as well as suggestions for activities. In 1909, *Garden Magazine* included a regular section on children's gardens written by Ellen Eddy Shaw of the Brooklyn Botanical Garden. Stories of projects, cultivation information, children's correspondence, and garden contest information were included to enthuse readers as well as increase the rigor of school garden programs.

The U.S. Department of Agriculture was also an important resource for teachers. Through its expert staff, facilities, and national communication channels, the department provided both theoretical advice and practical services to the school garden movement. As a clearinghouse for information, the department's Office of Experimental Stations catalogued the courses of study at all agricultural institutions, provided photographs and slides for teachers' use, sent representatives to visit and report on school gardens, and requested information from teachers for government bulletins. In 1905, the agency produced two bulletins—*School Gardens,* which provided a description of the school garden movement in Washington, D.C., as well as short summaries of projects around the country; and *The School Garden,* which suggested laboratory exercises related to soils, plant physiology, grafting, insects, diseases, and other subjects. The 1912 bulletin *Some Types of Children's Garden Work* described projects in the Midwest and West visited by Susan Sipe, who was working for the agency.

However, while knowing how to garden was important in a teacher, some advocates considered the teacher's ability to influence children more critical than the transmission of technical skills. Commenting that "nowhere does character count more than in the intimacy between children and teachers which the garden fosters," Greene said that "[tact], good judgment, justice, firmness, gentleness, directness, sympathetic understanding of child nature, normal sensibilities, a wholesome sense of humor, tolerance, patience, ready forgiveness and large hopefulness are fundamental qualities for a teacher."[65] Greene went on to describe the appropriate dress of a teacher, assumed to be female. Wearing old torn clothing considered "good enough for gardening" would be construed not as economy but as "slouchiness and disorder and lack of thrift." Instead "a practical shirt-waist suit or wash dress, or clothes of a color that does not show the stain of dirt and soil, are needed."[66]

REACHING CHILDREN

Given the broad interest in school gardening, the involvement of the schools, and the support of parks departments, settlement houses, and civic groups, garden programs could reach out to a wide range of children and communities. Many regarded the school garden as a resource for encouraging social unity, since children worked side by side and had to balance

Figure 19. The University of California provided garden training for teachers. The garden was just west of the president's house on the Berkeley campus. Reprinted from Fred Harvey Bolster, "The High School Garden," M.S. thesis, University of California, 1913, 64. Courtesy of the University of California.

teamwork with individualism. It also served as a meeting ground for rich and poor as well as for different ethnic groups. A study of school gardens in Michigan emphasized the democratic flavor of the work:

> While the children of the rich and of the poor sit side by side in the same schoolroom and receive instruction from the same teacher, they do not necessarily learn the great lesson of social equality, but when they stand side by side and engage in tilling the soil they learn that "in work there is no shame" and that the real man is the same, whether clad in overalls or in broadcloth. These are the lessons in my judgment that the children of America cannot learn too early.[67]

While they were a valuable experience for all children, gardens were considered particularly so for the poor, disabled, minorities, and young delinquents. Greene stated that children of the poor "find working in a garden preferable to sorting at the public dump, hunting greens or minding babies at home. They prefer to bring the little ones to the garden and interest them in big brother's crops even to the point where little brother helps."[68]

The successful school garden programs at southern schools for African Americans and at Native American schools received praise and were frequently cited as examples of how gardening could educate immigrants and minorities.[69] Both Tuskegee Normal and Industrial Institute and Hampton Normal and Agricultural Institute had garden programs. Prior to 1902, Hampton expanded its traditionally strong agricultural training program for African Americans and Native Americans by establishing a school garden program at its practice school, Whittier School. A 1902 report in the *Southern Workman* described the two-acre garden as having over two hundred separate plots, each gardened by two children. Some children were initially reluctant to garden, but this was overcome once the children were involved:

> While the little ones were pleased, the older girls thought it a disgrace to work in the fields, and it was necessary to use compulsion. After two years there is no pupil in the school who does not look forward with eagerness to the work. If it is necessary to be absent from school, he thinks it must not be on "gardening day."[70]

In the case of Native American children, gardening was conceived as a means to both provide vocational training and foster cultural assimilation. Having learned scientific agricultural practices, young Native Americans—who often had access to land and were assumed to take up agricultural careers—could obtain more lucrative results. Furthermore, the children's exposure to gardening was considered a means to encourage parents to "enjoy a vegetable diet and thus realize the necessity for conducting a home garden."[71]

Gardening also fulfilled the educational needs of handicapped and mentally disabled children when it was part of the manual training offered by state institutions, asylums, and hospitals.[72] Chester Tether from the State Normal and Training School, Oswego, New York, described the adjustments needed in a school garden for children with mental or physical disabilities, such as working in smaller groups, shorter work periods to accommodate short attention spans, and selection of plants that are likely to sprout quickly. Several garden programs worked with invalid children, such as the collaboration of Cleveland's Home Garden Association with the Society for the Prevention of Tuberculosis to start a garden for sick children who could not be sent to the country.[73]

Related to school garden programs, farm schools for delinquents and juvenile offenders were preferred over prisons because they provided oppor-

Figure 20. African American kindergarten children in their garden at the Hampton Normal and Agricultural Institute, circa 1899. Courtesy of the Frances Benjamin Johnston Collection, Prints and Photographs Collection, Library of Congress, photo no. LOT 11051-4 P&P.

tunities to learn industrial skills and good habits. Because the children and young people grew their own food, and could earn money through sales to support their upkeep, such ventures were considered financially expedient as well. For example, a 1909 *Craftsman* article praised a proposed parental school—a live-in school for truant boys of New York City—in part because it would include a one-hundred-acre farm in Queens to teach boys agriculture, horticulture, and manual training.[74] A similar institution was the Chicago Parental School, established to serve habitual truants and incorrigibles "even if they haven't committed a crime yet."[75] A farmer directed the cultivation, with the "boys being used where their physical ability allows." Several such programs were set up on a cottage or family system, whereby each group had its own playground, chicken yard, and small garden.

One of the most interesting examples of targeted garden programs was the one set up for the low-income refugees of the April 18, 1906, San Francisco earthquake who were temporarily housed in Golden Gate Park.[76] As part of the tent school established in the park, a small area at Ninth Avenue and Lincoln Way was assigned by Superintendent John McLaren for a school garden. Under the direction of a supervisor, children received seeds and a ten-foot-square allotment. According to reports by Bertha Chapman, nature study director in Oakland, California, the program was not intended to be strictly educational but rather was an "attempt to show these children of the city streets how many good things to eat and how many attractive flowers can be raised on a tiny patch of ground they can care for themselves and with all how much fun there really is in digging and planting in the soil."[77] One of the agendas was that children would inspire parental interest in gardening. She commented:

> Good pure air and vigorous exercise under influences that make for cleanliness is doing much to help these people to learn how to live. After seeing these happy children learning out in the open air by doing things for themselves with their own hands and heads, we must wonder if perhaps for many of them the earthquake and fire may not in after years prove a blessing rather than a calamity.[78]

The garden also brought ethnically diverse children into a single school program. Chapman's description of the children involved in the post-earthquake gardening program focused on the diversity of the participants:

> When all were provided with seeds, these tots marched in glee to plant their four rows. Round faced and almond-eyed baby Sing worked beside a tall dusky Portuguese boy, while two little people fell naturally in their moment of deep interest to chattering in their native Yiddish, and there beyond glowed the merry face of Pat with red hair and freckles. Some of the children showed signs of having careful mothers, but many more, poor children of the street, let one read the sorry story of their birth in their sad pinched faces and dull eyes.[79]

THE APPROPRIATE PLACE FOR A SCHOOL GARDEN

School gardens took three forms: landscaped school grounds, plots of land with student gardens, and home gardens. The impetus for school-ground improvements, such as planting trees on Arbor Day and planting vines to cover blank walls, combined the agendas of nature study and civic beautification. By introducing the principles of landscape gardening into the neighborhood, the school grounds became an "object lesson for the residents of the community in which the garden is located."[80] Yet these activities, while drawing support from various civic improvement groups, tended to have only limited educational value. When it came to teaching children through growing vegetables and flowers, schools and other groups procured pieces of land at or near the school for that special purpose. If space was limited or most children had backyards at home, some schools provided "model" or demonstration gardens in which children learned lessons to be replicated in their own home gardens.

As school gardening gained national recognition, there was debate as to the relative merits of gardens located at schools versus at home. On the one hand, school gardens were accessible and easily supervised, and served as school laboratories. Their disadvantages included limited space, expense to the school, and difficulties related to summer vacation, theft, and vandalism. On the other hand, home gardens provided access to otherwise underused land, could be maintained throughout the summer when school was out of session, facilitated civic improvement agendas, and left security concerns to the household. While school gardens encouraged personal responsibility for public property, home gardens sparked a sense of ownership and private property while also reducing the household's cost of living. Furthermore, by having school gardens at home, teachers had a legitimate yet informal reason to interact with parents and visit homes, in a manner that fit with the "friendly visitor" strategy of charitable aid during this period.

In the minds of some, the distinction between home and school gardens was unnecessary. The lines blurred when children brought family members to visit their school garden plot, or when a teacher visited a home garden. School Garden Association of America president Van Evrie Kilpatrick voiced his frustration over what he considered to be an artificial distinction, stating that "a school garden is any garden conducted anywhere by the school. It maybe at the school, in a vacant lot, in a park, along the sidewalk, or at the homes of the pupils. . . . The garden at the school is a model of teaching and sets an ideal of work. The garden at the home of the child is the living result which is bound to follow the garden at the school."[81] C. D. Jarvis of the Office of School and Home Gardening considered the best arrangement to be a small school demonstration garden in conjunction with home gardens. Not only was this approach more economical, it addressed problems of supervision, security, and summer vacation and also provided a connection between the school and home and encouraged family involvement and individual entrepreneurship.

A school's demonstration garden or shared garden was usually located on school property, or if no land was available at the school proper, vacant land could be acquired, usually either donated or leased at a minimal rent of one dollar a year to establish a legal contract. Such vacant lots were frequently described as trash- and can-strewn lots prior to the introduction of the garden. If a vacant lot was not available, Jarvis recommended acquiring suburban tracts of land and arranging with trolley car services for reduced fares for the young gardeners visiting their garden patches. Where a plot of earth was not obtainable teachers were encouraged to start roof gardens, window boxes, or room gardens. Urban schools, especially in the industrial cities of the East Coast, had to be creative and opportunistic in acquiring land. A 1917 study by the City of New York reported locations of school gardens as follows: 76.23 acres on school grounds, 7.8 acres in parks, and 13.16 acres in vacant lots, as well as over 1,033 home gardens of varying sizes.[82]

In some cases, children's gardens were located in parks. After the success of the DeWitt Clinton Farm School, New York began developing school gardens in other parks. In 1910, the School Garden Association of New York noted an agreement with the City of New York to set aside six spaces for school gardens in three parks—St. Nicholas, High Bridge, and Colonial—for the use of schools in the vicinity. In 1911, the New York Parks Department built their second school garden at Thomas Jefferson Park, which accommodated approximately 1,400 children. St. Mary's Park School Garden, on a quarter acre of unused parkland, was granted by Parks Commissioner Thomas J. Higgins, who had once been on the board of education. Park Commissioner Charles B. Stover is quoted as stating, "I look forward to the school garden work as giving promise to greatly reducing the lawlessness and vandalism which is shown in our city parks."[83] School gardens were located in parks in other cities as well, such as Hartford, Connecticut; Pittsburgh, Pennsylvania; and Poughkeepsie, New York.[84]

Playgrounds, which were gaining recognition as an important form of children's recreation at this time, occasionally included gardens. Writing for *Garden Magazine,* Ellen Eddy Shaw promoted children's gardens and playgrounds as complementary activities that provided a good balance of play and work.[85] Playground advocates debated the appropriateness of this practice. Joseph Lee, president of the Playground and Recreation Association of America, suggested that placing garden plots along the perimeter fences of parks and playgrounds optimized children's educational opportunities. With proper teacher supervision, the gardens would not hurt the "aesthetic effect" desired along the borders.[86] Another playground advocate, Henry Curtis, discouraged this practice, pointing out that gardens did not share the same goals as playgrounds—one being manual training while the other was recreation.[87]

No matter where children received instruction in gardening, the home garden remained a desired end product. Whether in combination with a school garden or by itself, the child's home garden provided additional opportunities for the educational system to affect home life. Louise Klein Miller's influential book *Children's Gardens for School and Home* offered this advice: "All plans and instructions must take into consideration the home of every child; he should in the garden find something which he can cultivate in a small way at his home; it may be a flower-garden on a fire escape, a strawberry vine in a tin can. He will have received an impetus and a love for 'green things growing.' "[88] Through gardening with his teachers and schoolmates, the child would gain both an appreciation of gardening as pleasurable activity and also an aesthetic appreciation of how a garden improves the home and neighborhood. Alongside the curricular motivations, the U.S. Bureau of Education supported school-sponsored home gardens for their civic benefits. Describing the ultimate goal of the school garden as the creation of "strong-bodied, efficient, and contented citizens," Dr. Jarvis was convinced that the home garden would teach important social values, particularly those associated with property-owning and middle-class lifestyles:

> With a common knowledge of the principles and possibilities of crop production, the wage earner of the future will not need to measure his income solely by the size of his pay envelope. He will consider, also, the productive capacities of his garden plat and the extent to which it will reduce the cost of living. He will see the advantages of a suburban home, contrasted with the crowded and unwholesome tenement.[89]

SCHOOL GARDEN DESIGN: THE MORAL FORCE OF A STRAIGHT LINE

The layout of the garden was determined by both site conditions and utility. Most diagrams and "how to" books laid out children's gardens in simple geometric patterns for reasons of effective teaching, supervision, and allotment. Manuals suggested staking out beds as rectangles separated by walking paths. These beds varied in size based on the children's age and the demand for space. The rectilinear layout not only reflected efficient land allocation but also correlated with educational goals of teaching discipline and order. As Miller advised, "The greatest care is necessary in planting, which must be done in a systematic, orderly manner. A garden-line is a necessity. There is great moral force in a straight line."[90] Aesthetics tended to be minimized. Describing the typical functional layout of rectilinear beds for classroom and individual plots, Dick Crosby noted that generally the school garden "does not lend itself to the realization of landscape effects, but it is a system that best fixes personal responsibility and awakens the greatest personal interest."[91] However, some gardens did include flower borders and lawn areas to create more parklike settings.

A combination of communal and individual areas was ideal. Most gardens included a central area where demonstrations and group activities were held and the whole garden could be observed. While the communal plots aided in instruction and team building, the individual plots were considered essential to building individual character. Explaining that one of the benefits of individual plots was the promotion of a sense of ownership, L. C. Corbett stated, "The idea that 'that what's mine is my own' becomes very strongly developed, with the natural sequence that such possessions must be properly protected and all rights concerned respected."[92] The following description of the Poughkeepsie garden describes a customary layout:

> Three of the four corner sections of the garden are divided into individual vegetable gardens of 4 feet by 8 feet. There is a flagpole in each of these sections, and the flag waves proudly over the section which has been approved by a committee of inspection as showing the greatest improvement for the past week. The remaining section is devoted to observation plots of flowers and vegetables, which are cared for by all the children as common property, and which furnish a pleasing variety. A shed near the garden was made over two years ago and fitted with tools in racks and with all the necessary implements.[93]

To encourage school gardens and reward attention to their aesthetic achievements, *Garden Magazine* started a children's garden contest in 1910.[94] Over five hundred entries, which included photographs, personal accounts, and reference letters, were received the first year. Prizes were awarded for both individual gardens and groups. The winners received a series of nature books. The first year's winners included Fairview Garden in Yonkers, New York, for best display at fall exhibit; DeWitt Clinton Farm School, New York City, for the "finest looking garden," and Tracy School in Lynn, Massachusetts, for best improvement of school grounds. The School Garden Association of America also promoted garden design principles by presenting a demonstration garden at the 1915 Panama-Pacific International Exposition in San Francisco. This garden included student plots and demonstration areas as well as ornamental plantings, a sun dial, a pond, and a covered pergola and platform for teaching.

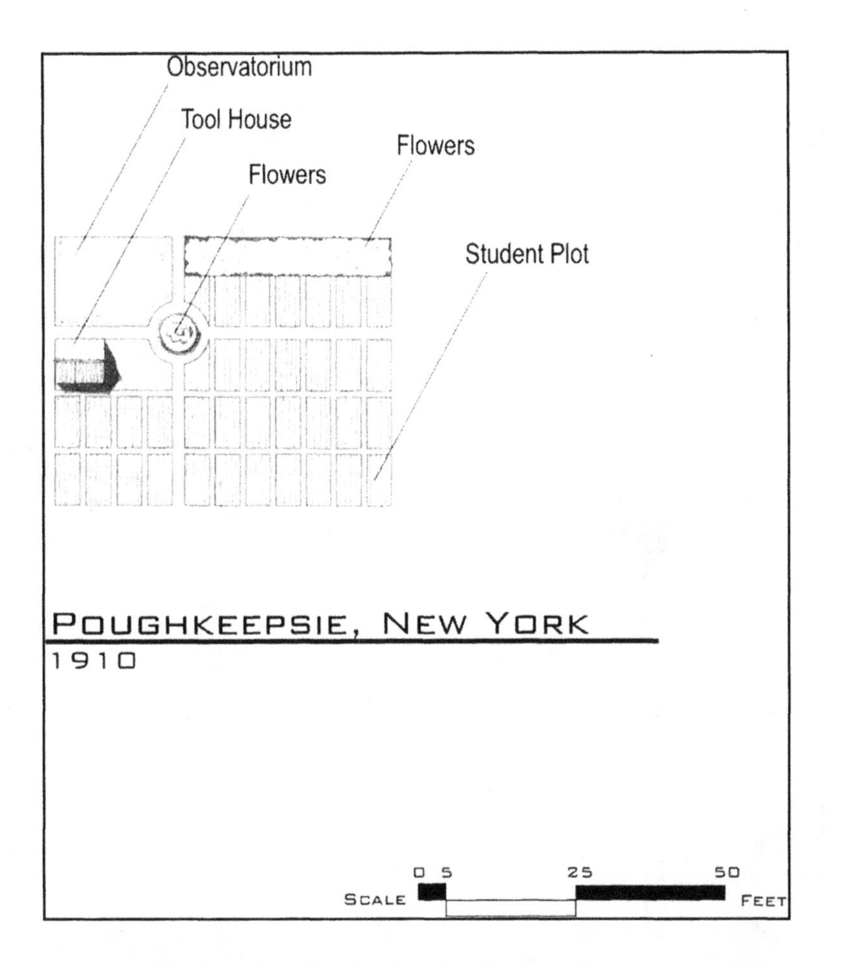

Figure 21. The Poughkeepsie school garden illustrated how a geometrical layout facilitated efficient land allotment while also providing opportunities for aesthetic features. Based on the plan in *Garden Magazine,* March 1910; redrawn by Agus Soeriaatmadja.

THE COSTS OF RUNNING A SCHOOL GARDEN

Advocates often promoted gardening because start-up was inexpensive and because it yielded a good return on investment in the form of nutrition. However, initial efforts often required pro-bono work on the part of the teacher and/or donations of time and money from other sources. Tools, seeds, and instructional materials could be acquired cheaply or were donated. Materials to amend the soil, such as manure and street sweepings, were often donated by city streets departments.

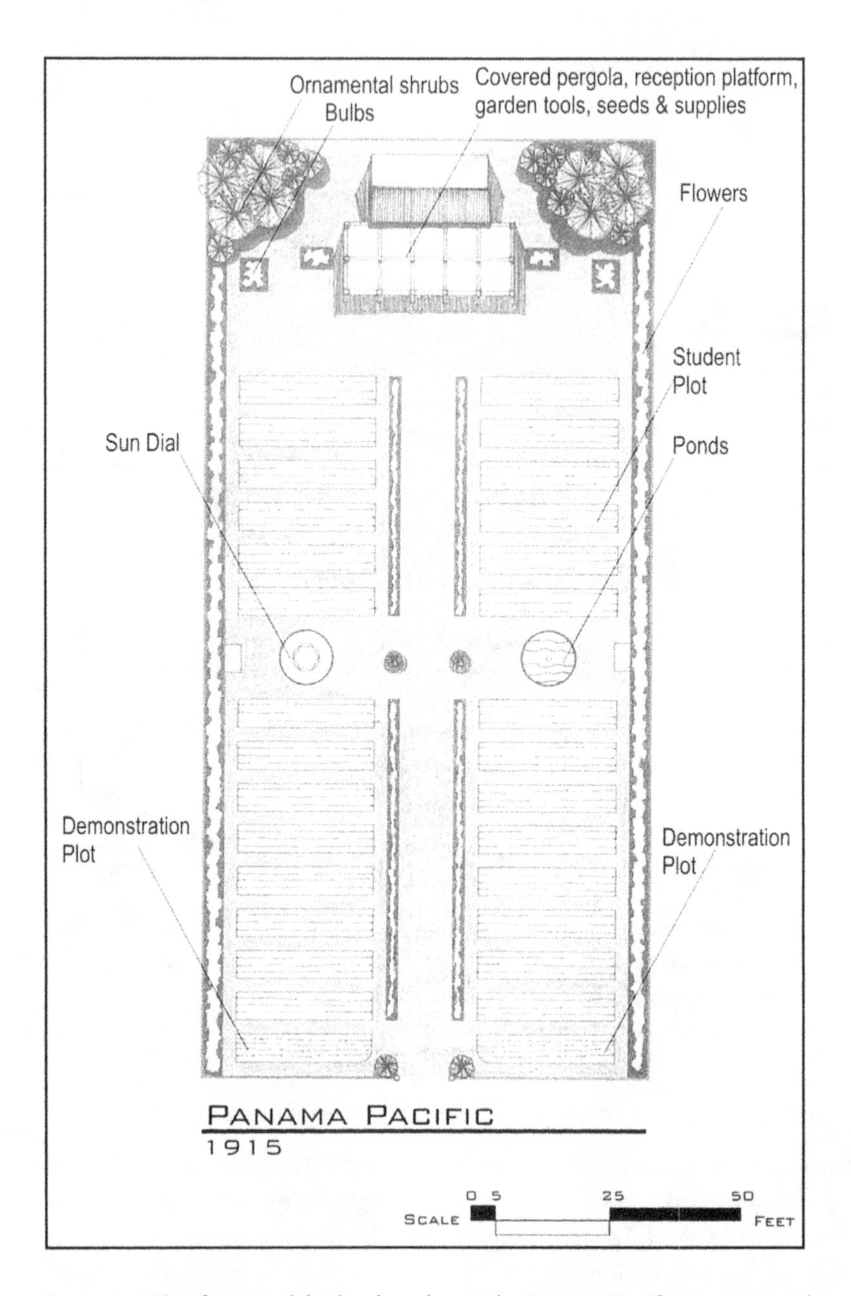

Figure 22. Plan for a model school garden at the Panama-Pacific International Exposition, San Francisco, 1915. Based on the plan in School Garden Association of America, *Annual Report,* 1915; redrawn by Agus Soeriaatmadja.

In *Among School Gardens,* Greene provided a detailed description of expenses ranging from the cost of twine to salaries. According to her 1910 estimates, a garden with five hundred individual plots at ten feet by sixteen feet and some demonstration plots would cost five dollars per child for a five-month season. It was hoped that the good returns on such a small investment would convince the local school board to support the program and expand it. Some programs, particularly the ones not affiliated with public schools, charged students a nominal fee for their plots, to be recouped through sales of vegetables. This approach was also considered a way to encourage the child's sense of responsibility. For example, the Hartford School of Horticulture had a graduated tuition, ranging from five dollars for the first year to twelve dollars for the fourth year; students who were unable to pay could volunteer one hundred hours of labor.

GARDENING AND HARVEST

The selection of crops was based on utility, ease of cultivation, and the relevance of certain plants to academic lessons, such as nature study and science. In *Among School Gardens,* Greene recommended a base of seven vegetables—radishes, beans, beets, carrots, onions, and two lettuce varieties. These seven basic plants provided an array of lessons in plant families, leaf structure, and life cycle. For instance, lettuce not only is edible, but its delicacy makes it readily responsive to poor gardening practices such as overcrowding, lack of light, or overwatering. Other crops suggested, if space was available, included corn, chard, and flowers. To illustrate geography and American cultural history, some programs grew items such as peanuts, cotton, and sweet potatoes. Age and gender also determined what was planted. Fast-growing crops, such as radishes, were especially important for smaller children, while older children might conduct longer-termed experiments. Gardens such as the one at the Whittier School at Hampton Institute included larger fields where the older boys could grow grain and field crops and learn how to use farm machinery. A few advocates suggested that boys should grow vegetables while girls grew flowers; however, this was by no means a universal policy. Hartford's director, Hemenway, described the yield of one third-year garden of ten feet by sixty feet:

> [Thirteen] and one-half quarts shell beans; ten quarts wax beans; six quarts lima beans; fifty beets; six cabbages; forty-four ears of corn; eighteen roots of celery; forty-two heads of lettuce; ten onions; fifty-eight quarts Swiss

> chard; six quarts peas; one peck potatoes; sixteen five-cent bunches of parsley; three hundred and fifty-nine radishes; nine quarts spinach; forty-three summer squash; one hundred and thirty-five tomatoes; thirty-eight turnips; eleven quarts of Valentine beans; and for flowers: three hundred and twenty-five nasturtiums; one hundred and seventy-four pansies; thirty-five snap-dragons; one hundred and thirty-five stocks; and six hundred and ninety verbenas. At the regular market price the vegetables were worth over fifteen dollars, without taking into consideration the flowers at all. The crops are so arranged that after the fourth week in the garden there is something to take home after every lesson.[95]

Starting in 1900, the Bureau of Plant Industry prepared and distributed four different seed packets to serve distinct needs: individual vegetable gardens, individual flower gardens, school grounds decoration, and a packet to illustrate agricultural crops in several sections of the United States. For instance, the vegetable seed packet was based on a plan for a 5-foot-by-16 ½-foot garden that would include radishes, lettuce, bush beans, beets, and tomatoes. According to a 1903 article in *Charities,* the USDA had sent out over ten thousand seed packets to teachers in twenty-four states and Puerto Rico.[96] While free seeds helped defray expenses, the Bureau of Education and other promoters advised teachers not to use free seeds as this could thwart the ethic of self-reliance.[97]

The tangible product of these efforts—the food—was eaten at the school, taken home, used in exhibits, donated to the needy, or sold. The Bureau of Education and other advocates emphasized that products, both fresh and canned, should go toward family consumption. After citing a contemporary study that reported that 60 to 80 percent of the laboring man's wages went towards food, Henrietta Livermore justified the Fairview Garden in Yonkers, New York because it enabled boys to provide their families with vegetables for all or part of the year.[98]

Others promoted market gardening as an educational opportunity. Many advocates, including Jarvis, felt that garden programs were best justified if they had an economic incentive. Countering criticism that youth market gardens commercialized the children's efforts, Jarvis noted that many children needed to earn a living. However, "to give boys and girls experience and training that may be used as an immediate means to a livelihood is not the primary object of gardening in school. It is a very valuable corollary, however, in the turning out of strong-bodied, self-reliant, and useful citizens."[99] Programs for high school boys often included marketing as a way of teaching them accounting, personal responsibility, scheduling, and interpersonal relationship skills—all of which were considered necessary for success in the business world. As Jarvis stated, "Children should be taught the value of money as measured by work."[100] The Bureau of Education recommended that parents support their children's entrepreneurial training by making business arrangements with their children to pay for their products. Many programs encouraged older children to market their produce to neighbors, boarding houses, restaurants, hotels, and hospitals. Some schools sold at public markets and so expanded gardening instruction to include lessons on marketing preparation and goods display. Based on her national tour of school gardens, Sipe observed that projects east of the Rockies tended to have individual plots in which the gardener owned and sold all that was raised, while west of the Rockies projects emphasized commercial sales along with cultivation, with proceeds helping to fund the program.[101]

SCHOOL GARDENS AFTER WORLD WAR I

As we will see in chapter 4, school gardens continued to gain momentum through World War I, and many advocates were convinced that they would eventually become a permanent part of school curricula. However, advocacy diminished noticeably in the 1920s. While gardens continued to operate in various localities, the federal Bureau of Education closed the Office of School and Home Gardens in 1920. A 1923 survey of 725 cities found that only thirty included school garden activities as part of their nature-study courses and a mere five cities offered courses particular to school gardens.[102] *Nature-Study Review,* a major source for information on school gardens as well as other nature-based learning projects, published its last issue in 1924. And in 1939, the School Garden Association of America was absorbed into the Department of Garden Education of the National Education Association. While the rationale for gardens as part of education remained justifiable in the minds of many people, the movement to legitimize gardens as part of the school—on a par with music class or the playground—withered.

The main reason cited at the time for the decline of support for the gardens was the readjustments after the war. As I describe in chapter 4, during World War I children's gardening efforts were called into the general food production campaign. Some proponents, such as Van Evrie Kilpatrick, believe the shift away from the educational imperative toward simple food production led to neglect after the food crisis ended. However, this explanation ignores certain definitional and organizational troubles that had existed before the war. School gardens were opportunistic endeavors that needed the support of a wide range of interests, in the hopes that one day they would be part of the public school system. In order to sustain these interests, the garden was rarely considered simply a place for children to grow vegetables but instead was tied to economic, moral, ethical, and educational agendas.[103] In order to keep the interest of philanthropic groups, women's clubs, civic groups, agricultural supports, and social reformers, the school garden had to remain loosely defined. The pressure to legitimize it in the school curricula begged the question of its purpose, something that many supporters did not want to strictly define for fear of losing support. The school garden, while a useful resource for a range of educational objectives, was promoted for accomplishing many other goals, with little evaluation as to its successes.

That said, while the school garden movement of the 1890s to 1920s failed in its national mission to integrate gardens into the schools, the idea of the school garden has withstood the passage of time. Children's gardens persisted from this period on, with or without a national campaign. Articles on school gardens continued to appear in educational proceedings, such as the National Education Association annual reports, though they tended to highlight a particular school or city program rather than a national trend. The reemergence of children's gardens with each later phase of urban gardening affirms the presence of a desire to engage children in gardening—a scheme that has often relied on communal land and organized programming.

THREE

The Goodness of Gardening

Gardens as Civic Improvement

A FULL PICTURE OF URBAN garden programs at the turn of the twentieth century must include special attention to the myriad civic and beautification programs that were started by improvement societies and included gardening promotion. As the preceding chapters have revealed, civic agendas were infused into the promotion of garden programs, which motivated federal, municipal, and voluntary organizations' involvement. For instance, while a vacant-lot cultivation association may have had economic relief as its primary agenda, the by-products of improved morale and cleaner neighborhoods encouraged civic-minded citizens to donate land and funds. Likewise, while educational goals may have spurred advocacy for school gardens, primarily in school settings, the capacity of children's gardens to awake aesthetic consciousness endeared these efforts to promoters other than educators. To many people, any opportunity to promote gardening in any form was a good thing, yielding positive aesthetic and social results, and it mattered little if the program was promoted as economic, aesthetic, educational, or all three. As a result, groups in cities, towns, and villages across the nation established campaigns for school gardens, cultivating vacant lots, window-box planting, and home gardening.

To understand why support for urban gardens came from such a range of clubs and associations and why garden programs took so many forms, one must contextualize gardening within the general civic-improvement movement at the turn of the century. The period's faith in environmental determinism—the belief that an improved physical environment would influence social behavior—led designers, politicians, social reformers, and civic-minded individuals to propose new order in the American city as a way to address urban congestion, rural depopulation, immigration, and other problems confronting the nation. The reshaping of the American city and countryside would not only treat immediate environmental problems; society also would benefit from the improved citizenry who had been inspired by beauty to higher moral and ethical standards. Transformed American cities that rivaled the grandeur of European capitals were touted as the proper physical settings in which to achieve America's democratic ideals. Progressives believed in a new American city where the heterogeneous mix of cultural groups would meld into one society with shared values and similar lifestyles. This ideal required a change in the city's physical form, and given the laissez-faire economy of the day, such change meant the involvement of not just government but also business, voluntary associations, and individuals. The principles of environmental determinism influenced the activities of philanthropic and civic associations as well as professions such as planning, landscape architecture, and social work.

Swept up in the City Beautiful movement, environmental designers—architects, landscape architects, and planners—developed plans to redesign the American city and its surroundings.[1] This quest for orderliness led to grand schemes in which separate functions were designated for various parts of the city, with different aesthetic standards for the downtown civic centers and residential communities. Urban cultural and civic cores were proposed as neoclassical campuses. Existing industrial cities were to be transformed according to plans featuring radial street patterns, the clustering of civic and governmental buildings, classical architectural details, uniform tree planting, and civic parks and plazas. Park systems shaped the larger city, with elegant pastoral parks in the civic area, connecting parkways, and more intensively programmed neighborhood parks and playgrounds.[2] To promote family life and pastoral values, the garden suburb and planned industrial village were proposed as the ideal residential contexts. The garden suburb was made up of mostly single-family dwellings surrounded by trees and gardens and linked to the city by train, streetcar, and, eventually, the automobile. For the laborer who could not afford the garden suburb, reformers and designers proposed the planned industrial village as the preferred alternative to congested urban neighborhoods.[3] The planners' strategy to decongest the city was to move factories and worker housing to the urban periphery, which could be rationally planned to include residences, stores and shops, parks, playgrounds, and schools, as well as a complete infrastructure and sewerage system. Often, the "solution" to congested tenement districts was demolition, to make room for new civic centers, axial boulevards, or pastoral parks. Promoters naively assumed that urban dwellers displaced by such projects would find their way to more agreeable housing in peripheral garden suburbs, industrial villages, or the country.

Model developments, such as the City and Suburban Homes of New York in the 1890s, were meant to inspire developers and working-class families to consider new housing alternatives that integrated aesthetic, hygienic, and civic components. However, the costs associated with these comprehensively planned communities made them prohibitively expensive for the families they were intended to serve. The much-publicized Forest Hills Gardens on Long Island were developed by the Russell Sage Foundation to illustrate ideal working-class housing. However, to make the requisite 15 percent rate of return on investment, rents were set too high for laborers' families. Working with the formula that a family could only put one-fifth to one-quarter of its income toward rent, most private developers and even philanthropic housing corporations could not build the housing planned for the laboring class and make the expected minimum investment return.[4]

Rather than propose new developments, other reformers sought to improve urban neighborhoods by mitigating the environmental and health hazards as well as the social and economic injustices that underlay urban problems. The City Functional or City Social movement focused its attention on water supply, sewerage, public transportation, and tenement reform.[5] Leadership for such projects frequently came not from environmental design professionals but from settlement house staff, women's organizations, and other voluntary improvement groups. For instance, the only voice in favor of urban neighborhoods at the first national city planning conference was social settlement worker Mary Simkhovitch, who reminded the audience that "the reason the poor like to live in New York is because it is interesting, convenient, and meets their social needs. They live there for the same reason I do; I like it."[6]

Together, the appeal of newly planned cities and the urgency to improve existing conditions in urban neighborhoods catalyzed a more general civic-improvement movement. Although improvement associations had existed as early as 1853, the turn of the century witnessed an increase in civic organizations involved in improving the visual and sanitary conditions of cities. Commissions, women's clubs, municipal art societies, gardening clubs, civic improvement organizations, and village improvement societies actively pursued beautification projects in American cities and towns. Comprehensive plans drawn up by planners and designers enticed volunteer groups to propose projects in their cities and towns as interim efforts until larger visions could be implemented. Similarly, City Social advocates raised awareness of a range of social concerns that could be addressed locally, such as the need for improved infrastructure and paved streets. As described by Jessie M. Good, founder of the National League of Improvement Associations, "The aim of all improvement associations work is practical education in civics and the cultivation of a love of everything beautiful in nature and art."[7] Stating that "civics is the ounce of prevention which is worth more than a pound of cure of philanthropy," Imogen Oakley of the General Federation of Women's Clubs defined civic work as "all work that benefits the city as a whole, and that helps every person in the city, high and low, rich and poor, fortunate and unfortunate."[8] With goals of increasing civic pride, elevating respect for property, and encouraging civic participation, improvement societies initiated good-road campaigns, street-tree planting, conservation efforts and forestry, and the development of libraries, schools, public baths, parks, open-air theaters, and public restrooms. They also facilitated social events, boys' and girls' clubs, and school gardens. Through membership fees and fundraising activities, groups raised funds to hire designers, build projects, and employ supervisors of garden projects and playgrounds.

To guide these efforts, professional designers, planners, and national civic-improvement organizations published books and articles that provided illustrations, examples, and guidelines. For instance, Charles Mulford Robinson's influential book *The Improvement of Towns and Cities* promoted the larger City Beautiful intention while also suggesting directions for smaller, incremental projects, such as plantings along thoroughfares and in nature strips along streets, landscape gardening for transit facilities and private residences, and gardens in vacant lots and school grounds. In a 1904 *Craftsman* article, landscape architect Warren Manning distinguished between the role of city officials making improvements such as street lights, tree planting, and road repair and the role of improvement societies to "inaugurate activities of which little is known in their community; such as the beautification of school and home grounds, and the establishment of school gardens and playgrounds."[9] Magazines such as *The Craftsman, Country Life in America,* and *Garden Magazine* included glowing reports of improvement groups' projects in hopes that they would be emulated in other cities and towns.

THE URBAN GARDEN AS CIVIC PROJECT

While rarely considered central solutions to the sanitary, aesthetic, and social concerns of the city, urban garden programs were attractive improvement projects because they were immediately feasible and produced tangible results. In an era so influenced by notions of environmental determinism, gardening also seemed an obvious avenue for social change. Besides improving the visual character of neighborhoods, a properly directed garden program was also considered an avocation that could cultivate good taste and potentially change people's character, habits, and social behavior. Louise Klein Miller argued for the civic importance of gardens: "Men, women and children cannot be well as long as their home conditions are exposed to disease breeding and fly breeding streets, alleys and back yards." She went on to suggest that "hundreds of acres of waste and unproductive land in vacant lots, back yards—a civic blemish—could be made to provide occupation, food, health, and moral and intellectual growth for hundreds who demand and require help for existence."[10]

One eyesore that regularly caught the attention of civic groups was the urban vacant lot, which was not only visually offensive but also potentially hazardous to health and safety. Land speculation that held land idle until development pressure raised prices was blamed for the presence of vacant lots in both the suburbs and the city. The solution to this blight was to replace trashy, weedy lots with food and flowers. Civic groups lent support by encouraging clubs and families to adopt vacant lots, assisting with materials such as seeds and soil amendments, and sponsoring contests. The vacant-lot beautification agenda of civic groups differed from the work of the relief-inspired vacant-lot cultivation associations, although the results of the latter were also recognized as a means to beautify and to awaken the civic consciousness of the poor. "We hear much in these times of clean-up movements and city beautifying campaigns," noted the superintendent of the Philadelphia Vacant Lots Cultivation Association in his 1913 annual report. "While our work was not organized, nor has it been conducted, with that as its aim, yet the results it has accomplished along this line place us in a position to claim the greatest praise."[11]

In denser areas, dooryard and window-box gardens were promoted as a way to add greenery to tenement districts. For instance, the New York Municipal Art Society had a Committee on Flowers, Vines, and Area Plantings that donated both seeds and instructions on window-box gardening to children in tenement areas. In 1908, the New York branch of the National Plant, Fruit, and Flower Guild facilitated five hundred window-box gardens by selling affordable boxes and providing plant materials.[12] In an address to the American Park and Outdoor Art Association, Dick Crosby of the U.S. Department of Agriculture described one effort to establish window gardens on one side of a block: "None of the efforts were very pretentious, but there was a striking contrast between these homes and those across the street and in adjacent streets. I had never before realized how great a change could be produced in a gloomy downtown street by a little tidying up of the front yards and the planting of a few flowers and vines."[13]

Beautification of public institutions, particularly school grounds, married civic beautification with the school garden movement. As we saw in chapter 2, the school garden movement relied on the support of civic groups. And while educators often focused on the school garden's educational and social justifications, it was often the program's capacity to promote beautification and civic leadership that inspired support from women's organizations, garden clubs, and civic associations. In their eyes, a school garden was a physical improvement to a neighborhood that encouraged aesthetic appreciation and civic-mindedness, which might lead to more neighborhood improvements. Moreover, through the children, the parents and larger community could be reached with the message of gardening. As a foil to the untidiness of many urban neighborhoods, school gardens provided "radiating centers for Civic improvement."[14] Although writing about the educative capacity of gardening for *Nature-Study Review,* teacher Alice Patterson also noted the civic outcomes: "Wherever there are children's gardens rubbish disappears, the unsightly objects are removed or transformed into something beautiful by the use of vines and tall annuals."[15] Through the process of building school gardens—both at home and in the school—children "look beyond their premises to the streets, parks, and other property and lend a hand toward making these more attractive." According to Patrick Geddes, an influential Scottish planner, children's gardens were an important part of civic improvement, "for it will be for this coming generation of child-gardeners themselves to make the Garden City."[16]

But probably the most promising aspect of gardening as civic improvement was the home garden campaign. Many lamented that the "garden" component of the garden suburb was lacking. Writing a series of articles on home gardens for *The Craftsman,* Mary Rankin Cranston noted that "backyards and beauty have not usually been considered synonymous terms" because of their utilitarian use and aesthetic neglect.[17] If a poor environment reflected and produced inferior people, the solution was to get people to fix up their environment and thus themselves. In another article in the series, Cranston wrote, "Just as the interior of a house discloses the inner life of the family, so do the home grounds reflect the family's ideal of the larger, or civic, life. A well-kept, orderly garden indicates a responsible personality, a neglected dooryard is a sign of shiftlessness."[18] In the annual report of the American Park and Outdoor Art Association, Edwin Shuey of Dayton, Ohio stated, "It is no reflection upon the value of parks or public gardens to say that the surest and best method of improving conditions of the masses is to encourage home planting and cultivation of vines and flowers even in the smallest dooryards of the crowded streets."[19]

Gardens were already a fundamental component of the ideal residential developments being promoted by designers and reported in the popular press. Model garden suburbs and industrial villages regularly included landscaped grounds, home gardens, and allotments. For example, Arthur Comey and Warren Manning's design for the Billerica Garden Suburb, twenty-one miles outside Boston, included vegetable gardens and an education program to teach gardening to residents.[20] The imagery of the private garden figured into debates regarding whether to regulate and retool tenement districts or raze them and relocate the poor to new, comprehensively planned housing developments.[21]

As civic groups jumped on the bandwagon to promote urban gardening, their rationale for support often included personal incentives along with social and aesthetic benefits. As reasoned by Edward Hartman, secretary of the Massachusetts Civic League, gardens provided a healthy hobby, nutritious food, household savings, and family recreation. In contrast, the man without a garden was likely to waste money on hobbies with no social relevance:

> The man without a garden has generally no avocation or, if he has, it may be as useless as postage stamps, brass buttons, clips; or if it is golf or any one of fifty often followed it keeps him away from his home, he is usually a poor man, and the barroom is his club where he wastes his money and his health, his family has stale vegetables or none, he loses time from his work

and is often unemployed, he has no pride in his home or its appearance, he is a wanderer. Such men do not help to build villages, do not become useful citizens, are a tax on the community because they fill the hospitals, the courts, the jails, the wayfarers' lodge.[22]

As a hobby, gardening held promise as a common interest that could be shared by people of different classes and ethnic groups. M. Louise Greene maintained that gardening created "a bond of sympathy and pleasure among the poor, the well to do and the wealthy. There is no hobby that may be so inexpensive; no subject of conversation less likely to become disagreeably personal; no topic offering better opportunity of give and take in the matter of experience than flowers."[23]

Some civic groups focused on one project, such as the promotion of local school gardens. Others worked on an array of efforts, including school gardens, backyard garden campaigns, and contests. These different kinds of projects were grouped together under umbrella terms like *cooperative gardening, extension gardening,* and *universal garden movement,* which conveyed a broadly held faith in gardening as a means of social betterment.[24]

THE AMERICAN PARK AND OUTDOOR ART ASSOCIATION'S NATIONAL CAMPAIGN

To illustrate the niche that urban gardens occupied in civic work, we can look to the activities of one organization, the American Park and Outdoor Arts Association. This organization was founded in 1897 as an information resource for architects, landscape architects, parks commissioners, and other officials responsible for physical design.[25] However, by 1903, it had shifted its outreach to the general public. The main purpose of the organization was to inspire outdoor art and recreation, thus promoting physical improvements to cities that would complement social reform efforts. As stated by the organization's president, Clinton Woodruff, in a 1903 *Charities* article, "The association stands pre-eminently for a 'more beautiful America' and for a democratic art, for a utilization of all the influences represented under these designations in building up a more intelligent, efficient and, in the best sense of the word, patriotic citizenship."[26] To achieve this, the association advocated for increased recreational opportunities and an eight-hour workday and supported the development of parks, social centers, libraries, and school gardens and the use of school buildings and grounds as communal centers.

Figures 23 and 24. *Top:* A group of children in Cleveland are about to transform this vacant lot into a garden. Reprinted from M. Louise Greene, *Among School Gardens* (New York: Russell Sage Foundation, 1910), facing 96. *Bottom:* The same Cleveland lot transformed through the children's work. Before-and-after images such as these were popular in promotional materials. Reprinted from Greene, *Among School Gardens,* facing 96.

As early as 1898, the association's annual reports were depicting school gardens as town improvement projects. Alongside articles on home improvement, park systems, roadside beautification, and vacant-lot cultivation associations were articles on model school gardens, such as the DeWitt Clinton Farm School. A special session at the association's 1902 annual meeting focused on school gardens and led to the formation of a standing committee on the topic. In 1903, the organization produced a special report on school gardens and playgrounds that highlighted garden programs in Boston, New York, and other cities and provided a state-by-state report on the status of nature-study, school-ground improvements, and school gardens. When the organization merged with the American League for Civic Improvement to form the American Civic Association in 1904, a Department of Children's Gardens was formed to furnish information and help promote the movement. Dick Crosby, vice president of the new department and also a Department of Agriculture employee, justified the involvement in school gardening: "There is no more potent influence for better civic conditions in America than an educated youth in whom is developed a 'critical discernment of beauty and excellence' in nature and in art, an abiding love of these things, and a feeling of personal responsibility for better civic conditions."[27] Specific proposals for action included offering educational programs for teachers in school gardening, lectures, leaflets, and placement services to link school districts with qualified teachers. Even if they were not directly involved in setting up school gardens, association members were encouraged to assist in securing suitable locations, supporting training programs, and persuading school officials, neighbors, and friends of the benefits of school garden programs.

WOMEN'S ORGANIZATIONS AND THE PROMOTION OF GARDENING

Women's organizations played a vital role in local civic improvement, and particularly urban garden promotion, at the turn of the century. Women's participation in civic improvements generally was justified as not only Christian duty but also as good civic housekeeping.[28] In his book *Constructive and Preventive Philanthropy* Joseph Lee legitimized women's involvement by arguing that "[a] woman has a feeling about dirt which men only pretend to have. The reaction which the sight of dirty streets produces in her, when once she has come to look on the matter as being within her sphere, is something of which every head of family has learned to stand in awe."[29] Furthermore, the participation of women was often considered a precursor to getting men involved to accomplish bigger projects. Women's efforts were assumed to be smaller in scale yet able to raise awareness and eventually attract men. Richard Watrous, secretary of the American Civic Association, commented that women "have been leaders in organized effort and have enlisted the sympathy and actual cooperation of men and associations of men in their laudable undertakings. Hundreds of cities that have distinguished themselves for notable achievements can point to some society or several societies of women that have been the first inspiration to do things."[30]

In cities and suburbs alike, middle-class women organized clubs and associations around such interests as literature, gardening, civic improvement, and specific local projects. The various improvements initiated or supported by women's groups in various communities coalesced into the National Federation of Women's Clubs, which was initially founded in 1890 as a literary club but quickly expanded to include civic improvements. The federation hosted biennial conventions and sent out circulars suggesting projects for civic clubs. Other national women's organizations included the National Plant, Fruit and Flower Guild and the Woman's National Farm and Garden Association, as well as women's auxiliaries to men's clubs.

In brochures, circulars, annual reports, and articles, gardening was often mentioned as an appropriate civic project for women's groups. Suggesting civic projects for women's clubs, the 1910 circular of the National Federation of Women's Clubs advocated the promotion of home gardens and the care of vacant lots and school grounds, along with other street and park improvements. Illustrations of projects already under way provided both inspiration and how-to advice. The accomplishments of women's groups were publicized in the annual reports of women's clubs and garden clubs as well as in national magazines like *The Craftsman, Garden Magazine,* and *American City.* For instance, to encourage respect for women's garden clubs, Francis King wrote an article published in *Garden Magazine* highlighting the activities of women's garden clubs in Philadelphia, Ann Arbor, Cleveland, Chicago, Newport, and New York. An article on civic gardening in *The Craftsman* described how the Baltimore Women's Civic League and Garden Committee had instituted school gardens, sponsored a backyard and window-box contest, and initiated vacant-lot gardens.[31]

THE YONKERS FAIRVIEW GARDEN

One project initiated by a women's organization that received national attention was the Fairview Garden in Yonkers, New York. In 1903, the Civic

League of the Women's Institution, led by its president Mary Leland Butler, established an extracurricular children's garden. From the beginning, this project was not part of the school system but instead "a garden where practical gardening is taught to the children of the public and parochial schools, who would otherwise be only too apt to spend the vacation idly, if not worse than idly."[32] Its promoters also cited the garden's positive influence on the vicinity. The mayor, who lived in the immediate neighborhood, praised the garden for "the benefit it has been to the children, particularly in the congested sixth ward."[33]

In the garden's first year, thirty-six boys gardened on two vacant lots. By 1911, the garden had expanded to 3.5 acres and served six hundred boys and girls. The cost of running the program was approximately $1,500 per year, which was covered by donations as well as weekly dues of two cents—considered less a contribution to the budget than a way to promote responsibility and teach a "sense of proprietorship." A trained gardener provided instruction and management with the assistance of a clerk, a manager, and volunteers. In addition, twelve boys were elected as monitors to police the grounds and give tours.

The project was intended not as a training program but rather as a social and recreational outlet. It eventually expanded into home-garden promotion and included a seed-ordering program for public schools, overseen by a volunteer committee of women. In 1909, they distributed 33,000 penny-packets of seeds, ordered by 3,300 children in Yonkers for home gardens. A home-garden competition and an exhibit helped spark interest. The association extended its programming to provide activities for children in the off-season as well. They leveled the garden for a temporary basketball court and two football fields, and flooded part of the garden to make a skating pond in the winter. They also acquired a nearby building to serve as a clubhouse where sewing, reading, and dance classes and more were offered.

A MODEL OF CIVIC GARDENING: THE NATIONAL CASH REGISTER COMPANY'S BOYS' GARDEN

When National Cash Register Company president John H. Patterson asked, in the mid-1890s, why his foreman and workers did not live in the neighborhood adjacent to his factory in Dayton, Ohio, he was told that the neighborhood—Slidertown—had a bad reputation. Determined to change the neighborhood, he used his own factory premises as a starting point. He hired the landscape architecture firm Olmsted Brothers to redesign the factory grounds and adjacent cottage gardens as models for residents to emu-

Figure 25. The Fairview Garden, Yonkers, New York. This garden won a *Garden Magazine* award in 1910. Reprinted from Greene, *Among School Gardens,* frontispiece.

late. The work continued with civic improvements in the neighborhood, such as rubbish and fence removal and instructions in home gardening, complete with illustrations of "the right and wrong way of planting a lawn."[34] Prizes were awarded to residents with the best-kept homes, best window boxes, and best vine plantings. Slide lectures to illustrate good and bad landscape design were presented to Sunday schools, clubs, and various improvement associations. Patterson encouraged women in the neighborhood to form a neighborhood improvement society to continue the beautification efforts. In expectation of positive results, the neighborhood even changed its name from Slidertown to South Park.

Along with these improvements, Patterson established the National Cash Register Company's Boys' Garden in 1897 as a means to thwart vandalism and disruption by boys in the neighborhood. Believing that the boys were led into mischief by idleness, he saw a connection between the lessons of responsibility learned from garden chores and the boys' ultimate success as good citizens and workers. "[Mr. Patterson] realized that it is possible for a boy to go to school too long; to have too much poured in, and not enough drawn out; to become bookish and lose touch with the active world. If boys are to achieve success, they must be given the opportunity not only to receive impressions, but to put these impressions into practical use."[35]

The vice president of National Cash Register Company, F. J. Patterson,

Figure 26. The National Cash Register Company Boys' Garden was started by the company in 1897. The original caption reads, "Boys should be formed, not reformed." Reprinted from Greene, *Among School Gardens,* facing 21.

donated a three-acre site, which he had previously envisioned as a kindergarten playground. In 1897, the city prepared approximately two acres of the site, and forty boys from the community, ages eight to sixteen, were given seed packets and individual plots of 10 feet by 130 feet. A full-time gardener helped the boys before and after school. The first year, the company paid out $3,500 to provide the garden plots, the teacher, seeds, tools, and site preparation. The company also offered fifty dollars' worth of prizes for the six best gardens, plus dinner at the Officer's Club. By 1904, the garden had expanded to seventy-two plots and was inaugurating a two-year training program from which the youths received diplomas. The annual contest had enlarged to offer ten prizes that included bronze medals, cash, and a trip to study park systems in various cities. As the promotional materials emphasized, the program was meant to "form, not reform," the boys by cultivating in them an interest in nature as the way to good character.[36]

THE HOME GARDEN ASSOCIATION OF CLEVELAND

The Home Garden Association of Cleveland, also called the Children's Flower Mission, was an extracurricular garden program with educational and civic goals. In 1899, workers of the Goodrich House Settlement sought to increase adult and child participation in civic matters by promoting home gardens. They sold seed packets with instructions to students and sponsored a contest with prizes for the best gardens exhibited in the fall. City children who grew the most vegetables and plants in their gardens would receive gold watches as prizes. According to a 1904 report by the organization, "The work of the association is based on the theory that by individual effort much valuable work can be done in the way of beautifying home surroundings, and that if each household performs its part in the work the beauty of orderliness and cleanliness will soon assert its supremacy over disorder, dirt and debris."[37]

The program continued to expand, so that in 1900, 48,000 seed packets were sold and the organization established itself as a national provider of penny seed packets of flowers and vegetables used in children's gardens. Its 1915 promotional brochure boasted that as of 1914, it had received orders from 34,140 schools requesting almost five million packets of seed, 75 percent of which were for home gardens. The brochure included 160 varieties of vegetable and flower seeds as well as roses, shrubs, vines, and bulbs. In addition to seed sales, the Home Garden Association promoted school grounds improvements, vacant-lot gardens, and community gardens through lectures, flower shows, and contests, and established an "exchange garden" where gardeners could swap plants. The organization sent its *School Garden Manual* to women's clubs involved in gardening. It also initiated a training garden to instruct older boys in truck farming.

In its seed brochures, the organization encouraged children to form a Junior Civic Improvement League, whose goal was to promote beautification and civic duty, largely through gardening. Members signed a pledge card that stated:

> I want to help make our town a better place to live in, and to this end I promise to comply with the following rules to the best of my ability[:]
>
> 1. I will help to clean up yards, streets and alleys.
> 2. I will plant flower seeds, bulbs, vines, shrubbery, etc.
> 3. I will help to make a garden, and keep lawn in good condition.
> 4. I promise not to deface fences or buildings, neither will I scatter paper or rubbish in public places.
> 5. I will not spit upon the floor of any building or on sidewalk.
> 6. I will try to influence others to help keep our town clean.

The Home Gardening Association.

SEEDS FOR 1904.

Price One Cent a Packet.

Mark opposite the variety the number of packets wanted.

Separate Colors Cannot be Ordered.

Variety	Packets	Variety	Packets
Aster, mixed, Scarlet, White, Blue & Rose, 15 inches high		**Nasturtium,** a climber, Yellow, Orange and Red, 6 ft. high	
Bachelor's Button or Cornflower, Blue, Pink and White, 2 ft. high.		**Nasturtium,** bush, Yellow, Orange and Red, 1 ft. high.	
Balsam or Lady Slipper, Mixed Colors, 2 ft. high.		**China Pinks,** mixed, Pink, Scarlet, White and Lilac, 6 inches high.	
Calliopsis or Coreopsis, Yellow and Brown, 2 ft. high.		**Phlox,** mixed, Scarlet, Pink and White, 1 ft. high	
***Cosmos—Mixed,** White, Pink and Red, 5 ft. high.		**Scarlet Runner,** A climber, Scarlet, 7 ft. high.	
Four-O'clock, Yellow, White and Crimson, 2 ft. high.		**Verbena,** mixed, White, Scarlet, Purple, 6 inches high.	
Marigold, Yellow, 1 ft. high.		**Zinnia,** Scarlet, 2 ft. high.	
Morning Glory, a climber. Mixed Colors, 12 ft. high.		****Gladioli Bulbs,** Red, Yellow and Pink, ONE CENT EACH	

Return this envelope to the teacher, with your money. Do not put money in this envelope.

No. of packets.................. Amount ....cents.

Write your name here..

Address..

Grade.................................... School..

Your seeds will be delivered to you in THIS ENVELOPE about May 1st. Prepare your garden in April. Select the sunniest part of your yard, but avoid a place where the dripping from the roof will fall on the bed. Dig deep—a full foot. Soil with well-rotted manure dug in, will give better results than poor soil.

Four-O'Clock, Bachelor's Button, Marigold, Calliopsis, Zinnia, Morning Glory and Nasturtium are the easiest to grow successfully.

*Cosmos is not recommended for smokiest districts. Blooms in October.

**Gladioli Bulbs should be planted right side up, in a good, rich soil, in a sunny situation, six inches deep and six inches apart. Will send up one stalk of bloom three months after planting. Flower stalk may need support by tying to a stick. The bulbs should be taken up in October and planted next spring. Store where they will not be frozen. Will make a fine display in school yard.

Many Window Boxes should be planted. TRY ONE.

Figure 27. Seed packet order form for the Home Gardening Association of Cleveland. Reprinted from facsimile in Miller, *Children's Gardens for School and Home,* 33.

7. I will always protect birds and animals, and all property belonging to others.
8. I promise to be a true loyal citizen.

I may not be able to do all these things, but I will do as much as I can to help our town and community.[38]

THE MINNEAPOLIS GARDEN CLUB AND ITS VACANT-LOT CAMPAIGN

Transforming vacant lots into thriving gardens was the impetus behind a sweeping civic improvement effort engineered by the Minneapolis Garden Club. Motivated by an upcoming civic celebration, the garden club started its campaign in 1911 by planting 325 vacant lots with vegetables and flowers and planting flowers along major arterial streets. It helped to establish ten school gardens on vacant lots and distributed twenty-two thousand packets of nasturtiums to children. The garden club later supported a greenhouse and 1½-acre garden for gardening classes at the Central High School. It also encouraged seven hundred nonmembers to plant vacant-lot gardens and eighteen thousand others, mostly office workers and prosperous workingmen, to start home gardens. The city—with five thousand acres of vacant land in 1911—came to be called "The City of Lakes and Gardens" by 1915.

The club provided seeds and lectures and held a contest. It produced a booklet of gardening instructions specific to Minneapolis's climate and planting conditions. For a fee—$1 in 1911 and $2.50 in 1913—the club prepared each participant's vacant lot for cultivation and supplied seeds, instruction, and supervision. Each lot had to contain eight different vegetable crops, sufficient to feed a family of five, as well as a ten-foot flower bed across the front of the lot. A frequently repeated anecdote was that the project produced so much food that local grocers complained of unfair competition. Because the vacant lots were often gardened without the owners' consent, the club discouraged any permanent structures and instituted a policy of evacuation within five days of an owner's request. It provided similar materials and support to home gardeners as well. The club also provided junior memberships through which children under sixteen could pay fifty cents for seeds, instructions, and two fruit trees. The Garden Club employed a garden superintendent and six assistants to advise the gardeners. The work cost the club $3,584 and produced a crop estimated to be worth $11,802.[39]

The Minneapolis Garden Club and other organizations described in this chapter are illustrations of a much broader movement that promoted gardening for the sake of civic improvement. Popular magazines and organizations' annual reports praised these efforts with anecdotal accounts. It is difficult to evaluate the results of these programs in terms of civic beautification or social change. More important, perhaps, is the evidence they offer that gardening was consistently viewed as a valuable activity, whether as a temporary gesture on a vacant lot, at an institution such as a school, or at home. The general goodwill toward gardening engendered in the late nineteenth and early twentieth centuries served as a critical foundation for the development of urban garden programs thereafter.

PART II

National Urban Garden Campaigns, 1917 to 1945

Introduction

SIMILAR TO THE URBAN GARDEN programs at the turn of the century, the programs that developed between 1917 and 1945 were promoted for multiple beneficial outcomes, including food security, nutrition, and recreation. Whereas earlier programs had targeted particular groups—the poor, immigrants, and children—the later garden programs, arising out of national crisis, sought popular support from all people. This was especially the case during the two world wars. The garden programs of the Great Depression of the 1930s, on the other hand, were intended primarily for the unemployed; however, since so much of the population was affected, they can for all intents and purposes be considered popular programs. This popular appeal shifted the weighing of the various benefits associated with urban gardening. Since the goal was to involve the household and family, propaganda and promotion focused on gardening's recreational, nutritional, and social benefits. Financial incentives were less important than the promotion of gardening as a healthful activity. The psychological and recreational benefits associated with gardening were particularly important to offset negative effects of war and depression.

Urban gardening continued to be considered a restorative activity to counteract urban conditions. But whereas earlier garden programs often had waxed nostalgic about agrarian living, the garden programs that emerged in the mid-twentieth century had far fewer connections to rural concerns. Clear distinctions were made between agriculture and gardening.

Agricultural food production could increase through smart business, mass production, and technological advances. Meanwhile, gardening provided household food supplements and recreation. The shift away from agrarian motifs might have been partly due to the gradual loss of personal or familial connections to agriculture among the urban population. By 1920, approximately 51 percent of the U.S. population was urban, and by 1940 this figure had reached 56.5 percent. But probably more important was the increased accessibility of the suburban lifestyle. Streetcars, trains, and eventually automobiles made it easier for more people to live in suburbs and commute to the city. While the earlier vacant-lot cultivation programs proposed to decongest cities by shifting immigrants into agricultural professions, programs during this period envisioned gardening as a hobby in the suburbs. As each crisis subsided, advocates hoped that gardening would continue, not as community gardens but under the auspices of home ownership and as backyard gardening for recreation and family life.

Probably the most marked distinction between the three phases of urban gardening described in part II and the phases that preceded them is the shift to an organizational approach that blended top-down guidance and bottom-up action. Although the urban garden campaigns of these decades still relied heavily on civic associations, women's clubs, garden clubs, and other groups, governmental agencies had an increased presence. Federal involvement had already been established with the U.S. Bureau of Education's Office of School and Home Gardening in 1914. With the beginning of the World War I campaign, this program was transformed into the U.S. School Garden Army. Gardening was included in the domestic conservation efforts conceived by the United States Food Administration and the Council of National Defense. Voluntary organizations, including the National War Garden Commission and several national women's federations, provided the pivotal connection between federal agencies' technical support and local implementation by volunteers. Promotion by these national groups inspired city-appointed war-garden committees, parks departments, and civic groups to participate. However, because much of the leadership and organizational capacity was borrowed from federal agencies and national organizations that had other priorities when the war ended, these programs dissipated at the end of the war crisis.

The Depression of the 1930s brought a new round of urban garden programs that also relied on governmental support. As during the 1897 depression, in the 1930s gardening was seen by many as a logical and immediately tangible means to counteract local manifestations of the economic downturn. Vast unemployment overwhelmed existing charitable resources, so communities had to improvise new programs to serve and occupy the idle. Often, garden projects started by local civic groups and municipalities were eventually subsumed under state and eventually national programs. Two types emerged: subsistence gardens in which households grew food for their own consumption, and work-relief gardens that employed men at an hourly rate to grow food that was given to institutions and the poor. Federal agencies coordinated these efforts and provided funding for a brief period until other New Deal programs began. While work-relief gardens largely disappeared, some local and state agencies continued to support subsistence gardens as part of the package of relief.

World War II victory gardens generally followed the patterns of earlier garden programs but had some distinct differences. While the World War I food campaign had stressed patriotic self-sacrifice, the World War II food campaign focused on maximizing agricultural efficiency and mandated rationing and price controls. Gardening was not considered a food source as much as a way to promote nutrition and recreation and to maintain households' quality of life. Gardening was validated as a recreational family activity with nutritional, psychological, and social returns for the individual and family. After the war, advocates hoped that people would continue to garden in their suburban backyards. Nevertheless, as the war crisis subsided and government promotion ended, the structure that had supported urban-garden organizing disappeared.

All three of these phases of urban gardening showed an increased reliance on directed organization. To orchestrate national campaigns for local garden programs, a hierarchical format was established. National agencies and institutions provided expertise and technical information, while citywide activities were frequently organized by local chapters of national clubs, such as women's clubs and garden clubs. The campaigns were organized for efficient dispersal of information and resources. Through existing channels that were temporarily involved in urban gardening, individuals could acquire the necessary land, seeds, and instruction.

Because of their hierarchical structure, these programs neither cultivated local leadership nor established gardening as a sustainable resource in the community. While each phase promoted community gardening, each did so only as a temporary gesture during the crisis. The borrowed leadership shifted back to its previous activities or progressed into new areas, and the land that had been donated for the war effort became more valuable for other forms of development.

FOUR

Patriotic Volunteerism

The War Garden Campaign

> Probably no other appeal to the patriotism of the American people ever met with more widespread and generous response than "war gardening." It set the great heart of America beating from coast to coast. Inspired by the excellent showing made last year and spurred on by the knowledge that "food will win the war" men, women, and children all over the United States took up war gardening this year. Both as individuals and as members of various organizations they have gone about this as true soldiers of the soil, in the same spirit with which their husbands, fathers, brothers, and friends went into the army and the navy.
>
> *Charles Lathrop Pack, "Making a Nation of Garden Cities,"* Garden Magazine *27, 4 (May 1918): 183*

ON APRIL 1, 1918, the *San Francisco Chronicle* announced the city's official opening of its war garden campaign with the proclamation that "the first food gun of the nation" had been fired. Festivities on that day, designated by the mayor as War Garden Day, began with a parade up Market Street to the Civic Center. Battalions of soldiers, sailors, and marines marched with civilians alongside floats laden with vegetables grown in the backyards of San Francisco. Twenty girls danced "the dance of war gardens and victory" in front of city hall. The ceremony continued with the dedication of a demonstration garden established by the local High School of Commerce on a vacant lot near the Civic Center, at Hayes Street and Van Ness Avenue. Eleven hundred students participated in the event, along with city and school officials, veterans, and civic groups.

San Francisco's war garden effort was not necessarily the "first food gun," since similar efforts were being started across the nation, but it does illustrate the patriotic, participatory nature of the war garden campaign during World War I. The campaign was a voluntary effort to promote domestic food production so more farm-raised food could be sent overseas. Guided by government agencies and volunteer organizations, American households in 1917 planted approximately 3.5 million gardens, produced $350 million worth of food, and canned 500 million quarts of fruits and vegetables. The program expanded, so that in 1918, 5,285,000 million gardens were planted, $525 million in food produced, and 1.45 billion quarts of fruits and vegetables canned.[1] Besides household gardens—located both in backyards and in community plots—the war garden effort included gardens grown by organizations, such as the Boy Scouts, the Girl Scouts, and churches. School garden programs, both existing and newly formed, supported the war garden campaign under the collective affiliation of the U.S. School Garden Army. Even the military met some of their own food needs through gardens, both in the United States and internationally.[2] While many companies provided land on factory and office grounds for their employees, railroads offered land along their rights-of-way for their employees. The war garden campaign promoted participation by people from all walks of life as an expression of patriotic duty during this period of international turmoil.

NATIONAL FOOD SECURITY AND THE WAR GARDEN CAMPAIGN

When the United States officially entered World War I on April 6, 1917, Europe was already facing a severe food crisis. Since 1914, Europe's agricultural production had decreased as farmers left their land to fight and fields were destroyed during battle. Poor weather conditions had reduced both European and American crops, and supplies sent overseas still faced the threat of German submarines. While the European countries on both sides of the war were attempting to compensate for food shortages through home gardens and changes in eating habits, the Allies looked to America to supplement their food supplies. To meet this demand, not only would American farmers need to grow more food, but part of the domestic food supply would also have to be diverted for export.

To increase food exports, the federal government placed the highest priority on orderly agricultural development. Having identified the world food shortages—primarily in wheat, livestock, and dairy products—experts from the U.S. Department of Agriculture and agricultural colleges proposed to increase American production through educational programs, expanded extension services, and government intervention in the purchasing, distribution, and sales of seeds and fertilizer. However, the necessary shift would take time, especially in the first year, since by early April many farms had already been planted. To meet immediate needs as well as to complement ongoing agricultural efforts, another tactic was to reduce the domestic demand for agricultural products. The experts hoped that local food production could meet some of the domestic food demand.

The European food crisis caused many Americans to reflect on domestic reliance on food grown far from the place of consumption, whether in other parts of the United States or elsewhere in the world. Dependency on imported food was criticized as wasteful and potentially dangerous. One war pamphlet, for example, began with a description of Massachusetts's tenuous food supply based on imports, warning that "[if] Massachusetts were suddenly cut off from the outside world and left to depend upon the foodstuffs produced within the boundaries of the state, four-fifths of her people would starve."[3] Furthermore, the burden of a dispersed food system on transportation infrastructure was an immediate drain on the war effort. In 1916, for example, an estimated 750,000 to 1,000,000 railroad cars were required to carry fruits and vegetables for domestic consumption. This capacity could be directed to the war effort if food were grown locally. Similarly, experts cited a 1915 survey that found only one-third of 584 cities with populations of 10,000 or more had farmers' markets, many of which were reported to be of minimal use because of poor management. Promoting and improving farmers' markets, local truck farms, and other local means of food production and distribution were part of the domestic war campaign.

While farmers geared up to export the necessary commodities, all Americans were asked to voluntarily reduce their consumption of exportable foods through conservation, substitution, buying from local growers, and gardening. So that more wheat, beef, pork, dairy products, and sugar could be sent overseas, Americans were encouraged to change their eating habits by consuming vegetables and fruit that they could grow themselves. As promotional materials frequently reminded their readers, every meal raised at home meant one more ration for the soldier. In light of the war effort, "[The] hoe is ranked almost with the gun and the fertilizer bag with the high explosive shell."[4] President Woodrow Wilson supported this appeal, stating that "everyone who creates or cultivates a garden helps, and helps greatly, to solve the problem of the feeding of the nations."[5] Through gardening, the individual could personally assist in America's war effort while also ensuring personal comfort. The war garden was "a form of self-taxation which the citizen can impose on himself as a patriotic act which will afford him pleasure and profit as well as a feeling of duty performed," said Henry Griscom Parsons, whose experience in school gardening prepared him to be a spokesperson for the war garden campaign.[6] The various benefits previously associated with urban garden programs—health, beautification, economic opportunity, and others—now found expression under the aegis of patriotism. Furthermore, gardening helped people feel that they were contributing and drew their attention away from worries about their loved ones overseas. In fact, the therapeutic nature of gardening benefited both civilians and recovering soldiers in military hospitals.[7]

THE NATIONAL CAMPAIGN

The war garden campaign was primarily a propaganda campaign to excite citizen participation. Instead of mandating programs or rationing food, the federal government chose to involve citizens through voluntary drives. The goal was to lead individuals, families, and institutions in a coordinated volunteer effort that produced food with minimal waste or redundancy. The campaign symbolically combined every individual's effort into one national cause. Philosopher and educator John Dewey declared, "Organized work will bring the greater moral advantages of developing the power of concentration along with the interest in national and community service."[8] To accomplish this feat at the national level, several governmental and nongovernmental organizations were established to lead local efforts; these included the Food Administration, the National War Garden Commission, the Woman's Committee of the National Defense Council, and the U.S. School Garden Army.

The Food Administration

In August 1917, President Woodrow Wilson established the Food Administration under the leadership of Herbert Hoover.[9] The administration's mission was to protect both producers and consumers from hoarding and monopolies and to support efficient conservation, production, and distribution. Its functions included exerting government control over the supply, distribution, and movement of food, fuel, fertilizer, farm equipment, and tools. Along with coordinating transportation and regulating prices, the program enlisted popular support through education programs that gave "every man, woman, and child an intimate understanding of the objects of the government and the duty of its citizens."[10] The Food Administration established divisions to target specific groups, including women, children, African Americans, immigrants, laborers, farmers, and religious organizations. It also provided lecture notes to local organizations, such as the series "Conservation and Regulation," which covered food conservation, production, and substitutions, as well as other topics. In addition, it published a guide for graphic exhibitions that could be displayed at fairs and expositions. While not directly involved in the war garden campaign, the Food Administration did encourage Americans to "eat all the vegetables you want" and to put vacant lots and land at the urban periphery into production for household nutrition and recreation. As a result of its efforts, nearly 10 billion pounds of meat, fats, dairy products, and vegetable oil, and 1.25 billion pounds of cereals and cereal products, with an estimated value of over $3 billion, were delivered to Europe.[11]

The Food Administration centralized its planning but decentralized implementation. The national administration in Washington, D.C., was charged with studying the country's food and conservation needs and planning appropriate programs. It distributed information to state and local groups and then collected resulting data in order to report on the national campaign's success. Each state had a food administrator and a home economics director, both of whom were volunteers appointed by the national office. State administrators communicated information to the necessary governmental agencies as well as to county and city administrators. They worked closely with the cooperative extension services of the U.S. Department of Agriculture and the state agricultural colleges to provide practical directions and to adjust national program directives to local conditions.

The National War Garden Commission

As a national promoter and information clearinghouse, the National War Garden Commission was a leader in the war garden campaign. Initially founded in 1917 as the National Emergency Food Garden Commission, the commission was organized, funded, and led by Charles Lathrop Pack, one of the five wealthiest men in the United States prior to World War I. The commission included influential government, intellectual, and business leaders, such as U.S. Commissioner of Education Philander P. Claxton, horticulturalist Luther Burbank, and Charles Eliot, president emeritus of Harvard University.[12] Its mission was to convince Americans of the need for war gardens, provide promotional and instructional materials, and encourage communities to organize war garden associations. Toward this end, it produced numerous pamphlets, newspaper and magazine articles, and promotional graphics that provided basic gardening information while conveying enthusiasm, patriotic duty, and urgency. The commission contracted with commercial artists to develop posters and graphic materials. For example, it called on James Montgomery Flagg, designer of the famous "I Want YOU for the U.S. Army" poster, to design a war garden poster that showed a flag-bedecked Lady Liberty sowing a field. Artist Frank V. Dumond evoked the sublime in his poster of the Goddess of Victory floating over a vacant-lot garden. Other artistic posters included Maginal Wright Enright's image of a boy marching with a hoe over his shoulder and J. Paul Verrees's image of the German Kaiser in a fruit jar. The commission's 1918 publication *A War Garden Guyed* provided patriotic stories, quotes, and cartoons to be used by local media to encourage gardening.

The Call to Women: National Women's Organizations and Clubs

Because women controlled an estimated 90 percent of food consumption in the United States, promotional materials about household food conservation and production were directed at women and women-dominated institutions and clubs.[13] Federal offices such as the Food Administration and the Council of Defense actively recruited women's cooperation, mainly through outreach to women's clubs. For instance, the Food Administration established a division solely responsible for providing articles and promotional materials to women's clubs and journals. Through a door-to-door campaign, the Food Administration asked women to display posters in their windows and sign pledge cards whereby they promised to follow the Food Administration's advice on food substitutions, such as preparing meals in compliance with wheatless, meatless, and porkless days. Although the pledge cards did not specifically mention gardening, they did ask women to expand their use of fruits and vegetables and to patronize local food sources.

Another organization that targeted women regarding conservation and food production was the Woman's Committee of the National Defense Council. Created just fifteen days after the war was announced, this organization's goal was to tap into women's organizations and provide a link between women's activities and various government agencies.[14] The national office coordinated the efforts of state divisions, which in turn oversaw local branches. Through this structure, the national office encouraged various

Figure 28. The Goddess of Victory rises above a vacant-lot war garden. Poster by artist Frank V. Dumond produced by the National War Garden Commission, circa 1917–19. Courtesy of the National Archives, photo no. NWDNS-4-P-147.

women's clubs to conduct door-to-door pledge drives for the Food Administration, register women for volunteer work, and start projects such as community gardens and children's war gardens. For instance, the New York City Women's Committee had a Children and Adult War Garden Committee, chaired by Fannie Griscom Parsons, founder of the DeWitt Clinton Farm School.[15] The organization secured appropriations from the board of education to pay public school teachers for their war garden work during the summer months, resulting in forty-three teachers' supervising one hundred gardens and a range of other school garden and adult projects in all of the New York boroughs.

Various preexisting national women's clubs took up the call as well. For instance, just three weeks after war was declared, the Women's National Farm and Garden Association volunteered its members to assist in civilian work suggested by the government so "that they were doing as patriotic work as sailors and soldiers at the front."[16] The national office provided educational lectures and materials, established the Land Service League to link farm women with urban women consumers, and helped to establish courses at Columbia University for women to study agriculture and gardening. Local branches assisted in the management of community and children's gardens, canning and preserving campaigns, home and canteen cooking, sewing, and collecting recyclable materials.

Another booster was the Garden Club of America, a predominantly upper-middle-class women's organization. While the organization had traditionally focused on ornamental gardening for home and civic improvement, the board realized that its members could provide useful skills in the food campaign. Its May 1917 *Bulletin* stated, "Ours is a peaceful craft but a useful one and when the war comes a task of real importance and a service of true patriotism confronts us."[17] Through its newsletter, the organization provided members with information on the status of food and gardening in Europe, expert advice, and suggestions for club activities, such as establishing children's gardens and community gardens and organizing units of women workers for agricultural work. Wealthy garden club members donated land and other resources to community efforts. One member of the North County Garden Club of Long Island dug up her polo field to plant potatoes, while another loaned land to her employees for gardens. The Cincinnati Garden Club loaned, plowed, and harrowed eleven acres of land that were subdivided into community gardens plus a four-acre garden for the Boy Scouts. The Club in Newport, Rhode Island, started community gardens, established a market where surplus from gardens could be sold, assisted with canning, and initiated a small unit of women workers for agriculture.

The various women's organizations coordinated their efforts through the National League for Woman's Service, organized in 1917 to aid the military and provide civilian services to help the war effort. The league included among its directors the presidents of the Garden Club of America and the National Women's Farm and Garden Association. One of their standing divisions was an agricultural division dedicated to promoting agricultural training and home gardening. The various women's groups worked together to create the Woman's Land Army, through which groups of women were hired to help farmers and oversee truck farms.[18]

THE UNITED STATES SCHOOL GARDEN ARMY

Over there! Over there!
Send the word, send the word over there,
That the lads are hoeing, the lads are hoeing,
The girls are sowing ev'rywhere,
Each a garden to prepare;
Do your bit, so that we can all share
With the boys, with the boys, the brave boys,
Who will not come back 'til it's over, over there!

Excerpt of "The New Garden Song" (to the tune of "Over There"), composed by Joe Lee of Davis Junior High School, Lexington, Kentucky

America's children were valuable contributors to the food campaign, and involving them in gardening also supported the pursuit of educational and health goals. John Dewey commented: "There will be better results from training drills with spade and hoe than from parading America's youngsters up and down the school yard."[19] Considering children an untapped labor resource, the National Child Labor Committee developed a plan—endorsed by the chief of the Children's Bureau, the Secretary of Labor, the Commissioner of Education, and various governors and state officials—to engage children in farm work and gardening.[20] Boys over fourteen years of age could be hired out to farmers, and groups of boys could be sent to work in farm districts. Both boys and girls under fourteen years could cultivate gardens under the supervision of schools and clubs.

Adopting the motto "A garden for every child, every child in a garden," the United States School Garden Army was organized as an extension of the Bureau of Education's Office of School and Home Gardening, which had been established in 1914.[21] The program had two purposes: to increase food production and to train children in thrift, industry, service, patriotism, and responsibility. By using the existing school structure and coordinating school activities with the Department of Agriculture, the Council of National Defense, and the National War Garden Commission, "the United States School Garden Army [gave] an opportunity for a more effective appeal to the patriotism of American youth than [was] possible through any purely local organization acting alone, even though it be statewide in scope."[22] Justifying the effort, a partial report made in July 1918 showed that 1.5 million boys and girls were enlisted and twenty thousand acres of unproductive home and vacant lots had been converted to gardens.[23]

National leaders promoted the children's garden effort. For example, President Woodrow Wilson wrote a letter of support that stated, "The movement to establish gardens, therefore, and to have the children work in them is just as real and patriotic an effort as the building of ships or the firing of cannon."[24] Herbert Hoover, head of the Food Administration, also provided a letter of support that praised youths' contribution to the war effort by allowing more food to be exported and minimizing the domestic demand on railroad transportation: "The example set by you in your undertaking has stimulated and inspired others to produce where they had not produced before."[25] Commissioner of Education Claxton emphasized that children's garden efforts helped in the fight for freedom. Warning that "if we lose, all the world will soon be in bondage to the autocratic German Government and the freedom for which our fathers fought will be gone," he added:

> Most important of all is food. Without it soldiers cannot fight, workmen can not produce ships, guns and shells; and men, women, and children will die. The people of the United States must this year produce more food than they have ever produced before. The President of the United States is therefore asking all boys and girls from 9 to 16 years of age in cities, towns, and villages to join the United States School Garden Army and grow vegetables, berries, fruits, and poultry. There are 7,000,000 such boys and girls. If 5,000,000 of these volunteer, it will be the largest army ever raised in the United States and larger than all other boys and girls' clubs combined. By hard work and with wise direction they can produce food enough to feed all the hungry children of Belgium. Will you join the United States School Garden Army? Your teachers will tell you about the plan.[26]

Figure 29. Poster for the United States School Garden Army. Reprinted from *Fall Courses in School-Supervised Gardening in the Northeastern States* (Washington, DC: GPO, 1919), cover.

The School Garden Army's chain of command moved from the national to regional to citywide level. Federal funds were provided for staff and technical materials. President Woodrow Wilson gave $50,000 from the national security and defense appropriations to promote school gardens in the first six months, later adding $200,000 more to carry out the work for another ten months. Under the guidance of a national director, four regional directors oversaw the northeast, central, southern, and western parts of the country. These offices distributed promotional materials, manuals, and course outlines to school systems and individual teachers. The Bureau of Education urged teachers to participate, reassuring those without gardening experience that the resources from the Department of Agriculture and the School Garden Army would provide adequate technical support. Teachers were encouraged to build upon their existing nature-study and gardening programs, and to correlate war gardening with regular studies. Some school districts designated a garden supervisor to coordinate the School Garden Army in their city or town.

The children in the School Garden Army were organized in military companies, each with a captain and two lieutenants and a maximum of 150 members. As part of the military motif, privates received bronze service bars with one star and lieutenants received ones with two stars. Participants pledged an oath: "I consecrate my head, heart, hands, and health, through food production and food conservation, to help win the world war and world peace."[27] Each child signed an enlistment sheet provided by the bureau agreeing to raise one or more food crops, keep records of his or her work and results, and report them to the teacher or garden supervisor. Promotional materials were infused with war metaphors. For example, the Northeastern Division's lessons described the garden toad as "the Garden Tank" and the mole as "the Tunnel Maker." There was even a program to provide children with hand-grenade banks as rewards for buying war savings stamps with their proceeds.

Cincinnati's school garden program illustrates how one war garden program was organized within the school district.[28] Its administration included a director, eight supervisors, five assistant supervisors, and approximately thirty-five regular teachers who received extra pay for their summer and weekend garden work. The city's board of education allocated over $15,000 toward school gardens in 1918. Reports from about eleven thousand of Cincinnati's Junior War Garden Volunteers indicated that 160 acres of school and home gardens raised nearly $40,000 worth of vegetables. There were six hundred children in Market Garden Clubs who cultivated at least one-twentieth-acre-size plots through the season (representing a total of 45 acres) and produced a harvest of $14,500 worth of vegetables, or $322 per acre. Because some garden sites were in suburban locations, the board and donors furnished car tickets to the value of $225 so downtown children could travel to suburban gardens, where 216 children cultivated nearly eleven acres and produced more than $1,600 worth of vegetables.

LOCAL ACTION AND ORGANIZATION

With the federal government's encouragement, and with instruction manuals available from federal agencies, civic groups, individuals, and municipal agencies were encouraged to start local campaigns for home, commu-

Figure 30. Children displaying a large head of cabbage raised in the war garden at Public School 88, Queens, New York, circa 1918. Photo by J. H. Rohrbach. Courtesy of the National Archives, War and Conflict no. 562, photo no. NWDNS-165-WW-172.

nity, and school gardens. Most local campaigns involved collaborations between city agencies, local clubs, parks departments, school boards, local newspapers, and other groups. Through their combined efforts, garden programs were organized, funds and in-kind contributions collected, and technical assistance provided. Funds and land for community and vacant-lot garden programs came from both public and private sources. In Dayton, Ohio, the city council funded a vacant-lot garden program, which cost approximately six hundred dollars. In other cities, such as Nashville, the entire project was supported through voluntary contributions from citizens and in-kind donations from local businesses.[29] Once the land and resources were gathered, various local organizations took the responsibility to allocate plots, supervise, and provide assistance.

To encourage participation and spark healthy competition, local groups organized contests, exhibitions, and demonstration gardens. Often located in prominent public locations in downtowns, demonstration gardens provided exposure and served as instructional tools. For instance, a demonstration garden in New York City's Bryant Park was established in 1918 by the War Garden Committee of Manhattan under the direction of the parks commissioner. The garden was manned by a supervisor who provided daily public demonstrations of gardening and harvesting. Bulletin boards provided information on insect control, weeds, spraying formulas, seed varieties, plant diseases, and other garden data. Manhattan authorities reported a 70 percent increase in war garden activities in the garden's first year compared to the previous year.[30] Similar demonstration gardens were also installed at New York's Union Square and the Boston Common.

WAR GARDENS IN CHICAGO AND ILLINOIS

Collaboration was essential to the success of the garden campaign in Chicago, and in Illinois generally. Chicago's program was directed by the Chicago War Garden Committee, which included members of the business community, truck farmers, seeds men, social service providers, a school commissioner, a parks commissioner, a representative of the International Harvester Company, and staff members from public welfare departments. The committee's first project was to map all the vacant-lot garden and school garden locations in order to organize efforts and delegate the responsibilities to various agencies and volunteer groups. Approximately 2,200 acres of vacant or idle land were donated, and an estimated 90 percent of the vacant land in Chicago's suburbs was cultivated. Nine hundred individuals were assigned land by the committee, and two thousand others were referred to garden organizations overseeing the territory in which they lived. The committee also provided seeds, fertilizers, tools, and materials for spraying at wholesale prices. It distributed over one million seedling vegetable plants grown by the Chicago Florists' Club, the Parks Department, and the House of Correction. In some cases, it provided free or at-cost plowing and harrowing to citizens.

The committee publicized gardening efforts through newspapers, garden guidebooks, posters, lectures, and contests. It also furnished the press with daily official bulletins on gardening, which were sent to all Chicago news-

Figure 31. The demonstration war garden in New York City's Bryant Park. Courtesy of the Prints and Photographs Division, Library of Congress, photo no. LC-USZ62-105593, lot 11310 G.

papers except the *Tribune* and *Daily News,* both of which had their own garden writers doing publicity work. It distributed 150,000 copies of educational primers, one on how to grow and maintain a garden and one on controlling insects and fungus diseases. In 1918 the committee distributed one hundred thousand war garden posters, designed by the *Chicago Evening Post,* which read:

> Registered War Garden, under Protection of State Council of Defense. All the Ammunition doesn't come from the Powder Factories. Feed Yourself. A War Garden Will Do it. Be A Soldier of the Soil. Exempt No Land. To destroy the food supply is to give comfort to the enemy. Food will win the war. War Garden Committee, 120 W. Adams Street.[31]

In addition, the committee established seventeen demonstration gardens in the Chicago area, with one in each of the city's large parks. Park staff gave

Figure 32. Poster from the War Garden Committee of the Illinois State Council of Defense. Artist: J. N. Dingo, engraved by Barnes-Crosby Company, 1918. Courtesy of the Library of Congress, photo no. LC-USZC4-7869.

weekly demonstrations and instructions to classes of children and adults. The committee also offered a short gardening course that was attended by 340 participants, of whom eighty volunteered to pass along the instruction by delivering talks before clubs and factory groups. Lunchtime speakers informed audiences in factories and at public schools, improvement associations, and neighborhood organizations. By July 1918, 682 lectures had been given to audiences that totaled 366,635 people. Over three hundred improvement associations, garden clubs, YMCA clubs, Boy Scout and Girl Scout troops, and others requested information, of which sixty affiliated themselves with the efforts and carried on gardening campaigns. Approximately 8,422 gardeners were registered in this way.

School gardens were an important part of the effort. Ninety thousand Chicago children participated as members of the U.S. School Garden Army. Advocating at least one garden for each school, the committee worked with the Women's Directing Committee for Children's Gardens to organize women volunteers for each school garden. In acknowledgment of their participation, U.S. School Garden Army insignia bars as well as bronze medals provided by the *Chicago Tribune* were given to every youth gardener who received high marks on records kept by the teachers, with silver medals given to the best boy and girl gardeners in each school district, parochial school, or juvenile institution.

Parks departments were also involved in the campaign. For example, in West Chicago, the Parks Commission provided approximately one hundred children's vegetable gardens, five feet by six feet, at each of its three recreation centers—Eckhart, Dvorak, and Harrison Park. The commission prepared three demonstration gardens at three large parks—Garfield, Humbolt, and Douglas Parks. The Parks Commission's head florist and his gardeners maintained the gardens and were available to visitors for garden advice. The produce grown was distributed to charitable institutions, such as the Home for Destitute and Crippled Children, the Industrial Home for the Blind, and orphanages. The Parks Commission also provided a community garden that served seventy men and women—mostly Polish immigrants who had had agricultural experience in their homeland. Thirty acres were secured outside of the city, a little over an hour's ride by streetcar. Each family received a 25-foot-by-125-foot plot and discounted seeds and plants. According to T. J. Smergalski, Superintendent of Recreation Centers in West Chicago, the garden became a social outlet and led to group activities and picnics. "Suffice it to say that they considered their garden work in the nature of wholesome recreation and also the source of healthy exercise and amusement, not mentioning the success they met with in securing a satisfactory yield of garden products."[32] Table 3 summarizes the results of the Chicago war garden effort in 1918.

COMMERCE JOINS THE EFFORT: COMPANIES AND RAILROADS

Various companies aided the war garden campaign by providing land and materials to workers so they could garden for their households.[33] Following procedures similar to those of the earlier vacant-lot cultivation associations, many company garden programs developed formalized registration

TABLE 3

Chicago's War Garden Campaign Results for 1918

	Acres	*Participants*	*Value of Crop*
Home yard gardens	3,850	140,000	$2,800,000
Vacant-lot community gardens	774	8,422	673,760
Children's gardens	206	90,000	55,620
TOTALS	4,830	238,422	$3,529,380

SOURCE: Table based on information in J. H. Prost, "The War Garden Activities in Chicago," *Parks and Recreation* 1, 4 (July 1918): 33–41.

systems, assigned plots, and provided supervision and instruction. For example, the Brown and Sharpe Manufacturing Company in Providence, Rhode Island, supported five hundred gardens on thirty acres of leased or donated land.[34] The Whitcomb and Blaisdell Machine Tool Company in Worcester, Massachusetts, started a cooperative venture called the Paxton Potato Syndicate, whereby workers bought five-dollar shares in a twenty-acre farm and received a portion of the harvest.[35] Some companies also inspired war gardening through exhibits and contests, such as the garden contest at DuPont plants that offered prizes totaling $10,000 in thrift stamps for the best canned vegetables from a war garden.[36]

In the process of supporting the growing food for the war effort, companies discovered other, more self-serving benefits. The superintendent of Foster, Merriam and Company stated that his plant's war gardens not only produced food but also "promoted a fine spirit of democracy and fellowship among the men. Everybody, from the president to the humblest employee, had a garden plot. And officers and employees, working together as they did, found mutual interests and fellowship there."[37] Although not specifically about war gardens, a timely report produced by the Department of Agriculture identified the benefits to the employer from industrial workers' home gardening in terms of health, positive attitude, sobriety, and income subsidy.[38]

Railroad companies also provided land for war gardens.[39] Previously, several railroads had allowed employees to garden on their rights-of-way, but with the outbreak of the war, they opened their land to the public. For example, the Buffalo, Rochester, and Pittsburgh Railroad provided 1,543 applicants with land, seeds, and instruction to produce $51,431 worth of crops.[40] According to an account in *Garden Magazine,* "right of way gardens" proliferated in the South and could be seen from almost any passenger train.[41]

HOUSEHOLD PARTICIPATION

The war garden campaign was directed at all income levels and social groups and often focused on the household as the point of intervention. An *American City* article titled "A Special Message from the Food Administrator to City and Town Officials" encouraged family participation, defining "family" as "every social group subsisting together, whether patrons of restaurants, boarding-houses, hotels or public institutions. Grudging support will not be successful. The support must spring to the breach."[42] To illustrate the popularity of the drive, L. L. Sutton, secretary of the Commercial Club of Excelsior, Minnesota, wrote in a 1917 *American City* letter to the editor, "Businessmen now take their garden tools instead of their golf clubs to the office and hurry to the municipal gardens after office hours."[43] However, with many men either at war or busy in war-related industry, much of the promotional material was directed to children and women. Suggesting that a working man could devote only two hours a day to his garden, Edward Farrington, garden writer for the *Boston Globe,* emphasized the use of Saturdays as gardening days and the contributions of a man's wife and children in the management of larger plots.[44]

Whereas there was debate regarding women's participation in other aspects of the war effort, such as working in defense plants or filling traditionally male occupations, food conservation was regarded as a traditional responsibility for women. Pamphlets and articles by the National War Garden Commission assumed that gardening, canning, and food drying would fall to the woman who oversaw household matters. The hurdle that women did face was gaining support from family members who would grumble about food substitutions and garden chores. To encourage women to "practice economy without parsimony," as one National War Garden Commission pamphlet phrased it, advice and recipes were offered to help women satisfy family food preferences while at the same time substituting beans and peas for meat and potatoes. However, other advocates framed women's role in wartime household conservation in more complex terms. For example, author and feminist theorist Charlotte Perkins Gilman suggested that given the wartime shortage of labor and the efficiency of mass food production, food preparation should not be addressed at the level of the

Figure 33. "Mother Makes Her First Appearance in Her New Overalls" says the original caption. Cartoon from *The War Garden Guyed,* a booklet of poems, pictures, and cartoons produced by the National War Garden Commission to help with local promotion of the war garden campaign, 1918.

household but should be done collectively by trained specialists.[45] This proposal was radical for its time and was generally ignored, so most food production remained the responsibility of the household and the homemaker.

SLACKER LAND MADE INTO WAR GARDENS

The war garden campaign targeted any "slacker" or idle land that could be gardened, including backyards, playgrounds, vacant lots, railroad rights-of-way, and other public lands. Newspaper articles with headlines such as "Patriotism Commences at Home," "Make Use of Your Waste Land and Be Patriotic," and "Grass and Weeds Won't Feed an Army" appealed to landowners to garden on unused properties. To identify likely parcels, cities and organizations conducted surveys. In Des Moines, for instance, city commissioners first surveyed vacant lots and then drafted an ordinance that made it possible to seize vacant lots to be allotted to gardeners rent-free. As

Figure 34. Any unused land was "slacker land" that could be gardened. Cartoon reprinted from Charles Lathrop Pack, *War Garden Victorious* (Philadelphia: J. B. Lippincott Company, 1919), 174.

a result of these aggressive efforts, nearly five thousand new gardens were added within the city limits.[46] According to Charles Lathrop Pack, 1.5 million acres of city and suburban land nationwide were cultivated in 1917, much of which had not previously been under cultivation.[47]

The backyard war garden was particularly convenient for the household. A Columbia University War Paper titled *Mobilize the Country-Home Garden* urged intensification of suburban home gardens and summer home gardens. Noting that such gardens already had well-conditioned soil, often under the care of professional gardeners, the author admonished that "the most inexcusable of Idle Acres is the fertile and tended acre that fails to contribute its share to the nation's staple food supply at a time of national need."[48] Various sources encouraged aesthetic considerations along with productivity, citing the ornamental qualities of practical plants or the interspersing of flowers amid vegetables.[49]

While the backyard garden was convenient, vacant-lot and community gardens brought larger pieces of land into production. For those without backyards or local vacant lots, there were plots of land on the outskirts of town that could be shared. Families living in apartments, shop workers, and employees of large manufacturing plants might find that these "community gardens," as Department of Agriculture horticulturalist W. R. Beattie described them, reduced costs in time, labor, and money through the sharing of responsibilities and tools.[50] They also provided a central location for instruction by municipal authorities or the local garden organization. Only a few basic rules were considered necessary for efficient communal efforts. The rules generally discouraged people from taking products from other gardens or disturbing other gardens by walking through them or flooding them, while encouraging weeding and tidiness and not letting vegetables go to waste.

INSTRUCTION

Assuming that most of the war gardeners would be amateurs, a wide variety of sources provided general gardening instruction. Lessons from past urban garden programs were made available through a bibliography produced by the Russell Sage Foundation, which included articles about vacant-lot cultivation associations, civic garden campaigns, and school gardens.[51] Government agencies such as the Food Administration, the Bureau of Education, and the Department of Agriculture, as well as colleges, individuals, and companies, produced instruction manuals on topics such as soil, location, planning, fertilizers, composting, seedbeds, hotbeds, cold frames, plant varieties, cultivation requirements, and tools. The Department of Agriculture also produced technical resources, such as two Farmers' Bulletins appropriate for the war garden campaign: *The Small Vegetable Garden* and *The City and Suburban Vegetable Garden.*[52] Magazines also promoted the national effort. In 1917, *American City* offered itself as a clearinghouse of "what to do and how to do it." *Garden Magazine* had a regular section titled "The Patriotic Garden," with monthly advice on what to grow and how best to cultivate crops, and another column, "Uncle Sam's Gardening," which provided updates on local activities around the nation. It also sponsored an experimental garden to ascertain how modern scientific methods could increase yields in a patriotic garden.[53] Other magazines providing technical and promotional articles included *House Beautiful, New Country Life, Scientific American, Industrial Management, National Geographic,* and *House and Garden.*

Promotional literature conveyed urgency and suggested dire consequences for underutilized land, misguided efforts, or lack of participation. Any "slackers" who did not garden were equated with the enemy for contributing to the starvation of allies in Europe and American soldiers. Military themes and metaphors heightened the association between gardening and patriotism. Gardens were described as munitions plants and gardeners as soldiers of the soil. Posters and newspaper articles promoted the effort with slogans such as "Can the Kaiser," and "Plant a War Garden for Life and Liberty." *Garden Magazine* articles and school garden leaflets described garden pests as "Huns" and enemies, and sequential planting as "new recruits." The consequences of ignorant gardening were portrayed in severe terms as comparable to aiding the enemy. In a *Garden Magazine* article on pest management, in which pests were described as "Kaiser's allies," F. F. Rockwell wrote:

> The man who permits insect pests and destructive diseases to become established in his garden may be looked upon by his neighbors as a garden pest himself. Anyone who uses valuable seed and fertilizer and garden space, only to feed the Kaiser's allies in the garden, is not only a traitor to the Free for All Fraternity of Gardeners, but also an undesirable citizen giving aid, comfort and good grub to the enemy! Ignorance and inexperience are no excuses for negligence. The garden "rookie" must assume his share of responsibility in the food trenches along with the veteran.[54]

On the other hand, these materials also strongly discouraged overzealous gardening. Many sources advised new gardeners not to try too much or to destroy ornamental gardens unnecessarily in their anxiety to contribute to the war effort. The goal was to intensify production but not by wasting energy or resources. For war gardening to be a "pleasure, not a burden," Henry Griscom Parsons encouraged people to plan their gardens based on the amount of time they had available to cultivate. "Make the garden a real help to your family. If you want to make it a success, never let the labor become a burden."[55] Edward Farrington was even more direct, suggesting that the inefficient use of seed, tools, labor, and enthusiasm was an unnecessary waste of the nation's resources.[56] Advising school teachers on how to increase productivity, Susan Sipe Albertis discouraged any plans that would destroy existing ornamental plantings or educational elements: "However, with all the vacant lots at schools' disposal it is both extreme and unnecessary to change a lawn into a potato patch or to destroy the perennial border—the garden of old friends, that has taken so long to become established—in order to raise a few additional cabbages."[57]

For the sake of efficiency and to minimize waste, the literature generally suggested starting small with simple rectangular plots. Many manuals and pamphlets oriented their information to gardens that would fit in a standard suburban backyard or on a shared vacant lot. Recommendations varied, from the ten feet by thirty feet proposed by the U.S. School Garden Army to a quarter-acre garden suggested by the Bristol County Agricultural School in Massachusetts. However, no garden was too small. Several promotional pamphlets encouraged porch and window gardens in populated sections of larger cities.

Many guides included planting plans and tables that took into account nutrition, taste, and preservation. A war garden pamphlet produced by University of Illinois professor J. W. Lloyd recommended planting vegetables high in food value, vegetables that could be stored fresh (potatoes, beets, carrots, onions), vegetables suitable for canning, and vegetables for summer use.[58] Guidelines differentiated between planting for summer use, in which the goal was to maximize variety and sequential planting, and planting for canning, which required everything to ripen at the same time in larger quantities. Although the priority was food production, some advocates suggested a limited amount of flower cultivation as well. In an article in *Garden Magazine* Iowa landscape architect Leonidas Ramsey suggested ways to make the vegetable garden more attractive through parterre geometry, mixing flowers and vegetables, and architectural features such as bird baths, garden seats, and sun dials.[59] In her regular *Garden Magazine* series, "Uncle Sam's Gardening," horticulturalist and landscape writer Frances Duncan justified growing flowers that could be sent to the wounded soldiers, citing such a program initiated by the American Florists and Ornamental Horticulturists.[60]

THE CRISIS ENDS . . . WILL GARDENING?

As the end of the war came into sight, the call to garden remained strong. Europe would continue to need imported food supplies through a reconstruction period that was estimated to last five to ten years. The war garden was renamed the "victory" garden. In terms of sheer enthusiasm as well as financial results, the war garden campaign had been very successful. Families had been fed, children had been able to contribute to the war effort, and

TABLE 4
Estimated Value of War Garden Crops in Certain Cities, 1918

Denver, Colorado	$2,500,000
Minneapolis, Minnesota	1,750,000
Indianapolis, Indiana	1,473,165
Washington, D.C.	1,396,500
Los Angeles, California	1,000,000
Grand Rapids, Michigan	900,000
Louisville, Kentucky	750,000
Salt Lake City, Utah	750,000
Worcester, Massachusetts	750,000
Oklahoma City, Oklahoma	500,000
Scranton, Pennsylvania	450,000
Rochester, New York	350,000
Dallas, Texas	300,000
Burlington, Iowa	250,000
Pittsburgh, Pennsylvania	250,000
Newark, New Jersey	160,000
New Orleans, Louisiana	125,000
Atlanta, Georgia	100,000

SOURCE: These numbers were reported by Charles Lathrop Pack in *The War Garden Victorious* (Philadelphia: J. B. Lippincott Co., 1919), p. 98.

American agriculture had thrived. To illustrate how much the garden campaign contributed to the war effort, the National War Garden Commission listed the financial value of some cities' efforts, ranging from $100,000 in Atlanta, Georgia, to $2.5 million in Denver, Colorado.[61] Table 4 lists cities that provided the commission with estimates of the monetary value of food produced in war gardens.

Looking beyond the war crisis to consider some of the other benefits associated with gardening, many garden promoters echoed the same logic of earlier civic-improvement gardening campaigns in their hope that the experience of gardening would inspire people to take it up as an avocation or hobby. Writing at the beginning of the war, G. W. Hood, from the State University of Nebraska, described gardening as an enjoyable and "profitable recreation" and commented that *"the products of the home garden represent the labor income in otherwise lost time."*[62] After visiting many backyard and vacant-lot gardens, *Garden Magazine* writer F. F. Rockwell remarked that the well-planted and cared-for garden was proof that gardening was not like other "waves of popular enthusiasm." He continued, "The vacant lot campaign and similar enterprises have merely opened the door of opportunity to those who were waiting to push through!"[63] Edward Farrington believed that war gardening awakened people to the pleasure of gardening and the joy of fresh homegrown produce, stating, "Certain it is that thousands of people who have learned the advantages of a home garden will never be content again to eat the stale and withered produce which comes from the stores."[64] As her monthly *Garden Magazine* column, "Uncle Sam's Gardening," came to a close, Frances Duncan envisioned a continuing interest in gardening, especially among suburban women who would enjoy healthy competition by raising flowers and specialty crops such as strawberries or asparagus.[65] However, as time passed, less attention and support were given to urban garden programs. Although some communal gardens continued in specific locations or institutions, most sites were reclaimed for their previous purpose or development. With the end of the crisis, federal agencies and voluntary leadership shifted to other activities. Advice and stories about community gardens and vacant-lot gardens eventually disappeared from the pages of civic and garden magazines.

Some efforts, such as children's gardens, did continue, but on a smaller scale. With the School Garden Army literature on hand, some teachers and schools continued school gardening with minimal assistance. For instance, a 1918 U.S. School Garden Army bulletin addressed to teachers observed, "It has become the privilege of every teacher to serve humanity by leading her pupils to sympathetic consideration of garden problems to the end that they may become intelligent producers of food and beauty, not only in these years of their youth but also throughout their lives."[66] However, by the early 1920s the Bureau of Education was no longer involved in promoting gardening as part of school curricula. Some hoped that children would retain gardening as a hobby.

Nevertheless, the war garden campaign did evolve into a postwar promotion of backyard gardening, particularly in the context of suburban home ownership. A *Garden Magazine* article titled "America a Nation of Gardeners" mentioned the national "own a home" campaign in which the home garden was perceived as an asset by builder and real estate agent alike.

"The home is the cornerstone of the nation, and the garden is the cement which helps to hold this cornerstone in place."[67] The war campaign had contributed knowledge and a familiarity with gardening that many people could apply in their suburban homes. Few voiced concern about how urban dwellers who had little access might continue to garden, for the assumption was that they would soon move to the suburbs.

FIVE

An Antidote for Idleness

Garden Programs of the 1930s Depression

THE GODMAN GUILD COMMUNITY GARDEN in Columbus, Ohio, started as a World War I war garden and, based on local interest, continued through the 1920s. It was still active when families struggling with unemployment at the onset of the Great Depression of the 1930s decided to grow their own food to offset household expenses. In response to local need, the Godman Guild settlement house, which managed the garden, expanded the existing program to serve 270 families of limited income. Both experienced gardeners and novices, both White and African American, made use of this thirty-five acres of Olentangy River bottomland. The settlement house took applications, oversaw site maintenance, employed a site supervisor, and kept records. In return, the gardeners paid a three-dollar fee to help with expenses, signed a contract, and were given an identification card with the assigned plot number. Although the garden was intended to be self-supporting, hardship during the Depression meant that many gardeners received their plots gratis and the settlement house covered the deficit. In 1931, the garden cost $984 to maintain and produced an estimated $13,000 worth of produce that helped feed the participating families. As an example of local initiatives to compensate for the Depression, the Godman Guild Community Garden was highlighted in governmental materials, including the Community Plans and Action series of the President's Emergency Committee for Employment, and *Subsistence Gardens,* a report prepared for the President's Organization on Unemployment Relief.[1]

Elsewhere, garden projects started by local groups slowly gained municipal and state support as relief programs. On October 27, 1932, the *East Side Journal,* a newspaper serving communities along the eastern shores of Seattle's Lake Washington, included as front-page news "Unemployed Garden Yields Big Supply of Welcome Food." The article described a volunteer workday when over 175 sacks of potatoes were harvested, as well as beets, carrots, and other crops. The garden, located on donated land and cultivated with volunteer labor, produced sixty-five tons of vegetables to serve the poor during winter. The program expanded in 1933, with the King County Welfare Board starting a countywide seed distribution program. To facilitate this and other such local efforts, a state program was established.

Garden programs were most popular at the beginning of the Great Depression, from approximately 1931 to 1935. They were called by an assortment of names—thrift gardens, self-help gardens, subsistence gardens, employment gardens, industrial gardens, and community gardens—but they followed essentially the same logic as earlier programs: providing land, materials, instruction, and supervision for food production. As during depressions of earlier decades, garden programs arose during this depression with the simultaneous goals of constructively occupying the unemployed and increasing their access to healthy food while sustaining their self-respect. Financial relief was also on the agenda, but whereas past programs had encouraged the sale of food for income, the Depression-era garden programs discouraged individual entrepreneurship in favor of gardening for family food needs and collective gardening as a works project, with the produce generally used by relief organizations.

Many cities had garden programs already in place when state agencies stepped in with state and federal grants to support relief efforts. As early as 1931, federal agencies highlighted gardening projects as appropriate local relief venues and provided technical assistance. Federal funding was made available for garden projects from 1934 to 1935. The resulting garden programs balanced federal and state requirements with local organization. In 1934, the Federal Emergency Relief Administration reported 1,820,372 gardens planted on 328,456 acres through organized programs in forty-three states, costing $5.5 million and producing food worth over $47 million.[2] These gardens supplied over 36 percent of all the fresh fruit, canned fruit, and vegetable requirements used for relief throughout the country. Approximately 2,387,240 families received food from garden programs. In twenty-eight states, besides individual plots for households, there were collective garden projects that employed relief clients to supply fresh produce for local food relief.

However, the national exposure given to gardens as relief programs was short-lived. As more New Deal programs came into effect to address unemployment and provide farm subsidies (such as the Works Progress Administration in 1935 and the Food Stamp Program in 1937), the role of gardens diminished. While providing seeds, advice, and land where families could grow food for household consumption remained a valid form of relief, changes in federal funding policies curtailed work-relief garden programs. Some programs did continue locally under the auspices of civic groups or local governments. In the cycle of popular support for urban garden programs, attention once again waned.

SITUATING URBAN GARDENS WITHIN RELIEF EFFORTS

To understand why gardens were an initial but not sustained response to the national depression requires a look at changing economic conditions and strategies for charity or relief.[3] Between the depression of 1893 discussed in chapter 1 and the Great Depression of the 1930s, there had been several economic depressions of national scope during which the majority of relief came from churches, charitable organizations, trade unions, and local governmental agencies. Even in times of economic prosperity, these organizations and institutions provided aid to the chronically poor, such as the elderly, infirm, and otherwise unemployable. During times of elevated need, these organizations found it necessary to coordinate with each other to avoid duplicating services and to assure efficiency. Relief efforts became professionalized, with trained social workers employed by public and private agencies to conduct orderly inquiries into requests for aid and produce individualized assistance packages to suit each case. Experts collected neighborhood data on unemployment and relief needs so that local planning agencies could organize the activities of philanthropic groups, charities, and social services for maximum efficiency. Nevertheless, although centralization was useful for efficient planning, the actual work of relief remained staunchly decentralized so it could be responsive to neighborhood conditions and individual needs.

The high unemployment and related social and economic problems that began in the late 1920s, however, overwhelmed the capacity of local relief agencies. In 1931 there were six million persons out of work, a number that grew to fifteen million by 1933. Given that approximately one-third of the potential working population was unemployed or underemployed, previous conceptions about poverty as a character flaw had to be reconsidered.

Rather than a matter of personal weakness, unemployment was now viewed as reflecting the instability of industrial economic systems that had been developing for fifty years. Less concern went into separating "the worthy from the unworthy poor," which had preoccupied earlier philanthropic debate, and discussion turned instead to how to provide relief on a large scale without discouraging self-help and personal initiative. Relief agencies and philanthropic groups were alert to the negative effects of idleness, which was considered to be demoralizing and to raise the risk for social disturbances. The challenge faced by relief agencies was to simultaneously resolve financial needs while also providing healthy activities to occupy the unemployed.

With the dual interests of providing relief and keeping people busy, the preferred form of aid was work relief, in which clients were temporarily employed in public-service activities and received wages or goods for their labor. This was not a new idea, having been promoted since the late nineteenth century as a way to aid the poor without destroying their self-esteem.[4] Generally, the wages earned through work-relief programs were not sufficient to meet a family's entire financial needs but were part of a larger relief package determined by a welfare caseworker.[5] A family might receive a work assignment along with direct monetary relief and in-kind contributions of food, fuel, and supplies. In addition, the family was expected to make its own sacrifices, such as moving in with other family members, seeking employment for wives and children, and gardening for food.

A survey of cities conducted by the Association of Community Chests and Councils found that almost two hundred cities had some form of work relief in 1930–31.[6] To appease concerns that work-relief programs might threaten regular jobs, most work projects were seasonal, temporary, and unskilled, such as street cleaning, snow removal, vacant-lot clearance, road grading, traffic improvements, tree planting, and parks development. Other programs employed people to produce goods, such as sewing clothes and mattresses or producing and processing food. Although such programs generally cost more than giving money directly to the poor, advocates rationalized that it strengthened the morale of the unemployed while also providing a public service to the taxpayers.

To further complicate matters, malnutrition and hunger were serious concerns for the increasing numbers of unemployed. While the unemployed needed access to food, contemporary charity-organization literature discredited food-relief programs such as soup kitchens or bread lines as degrading and reactionary. Instead, food production—through gardening, cooperatives, and work-relief farms—was considered an appropriate approach because it inspired self-help and initiative.

WORK-RELIEF AND SUBSISTENCE GARDENS

During the heyday of urban garden programs at the beginning of the Depression, two types of programs arose: the work-relief garden and the subsistence garden. The work-relief garden, much like a truck farm, employed relief clients to garden collectively for wages. The harvested produce belonged to the overseeing agency and was typically distributed as food relief to the needy or to institutions. In 1934, there were 17,196 acres of work-relief gardens reported in twenty-two states. By 1935, this had grown to 26,531 acres in twenty-eight states.[7]

Far more common were subsistence garden programs that provided garden plots, seeds, technical assistance, and occasionally tools and fertilizer so households could grow their own food. FERA maintained data on subsistence gardens based on four categories—home or vacant lot, community gardens either divided into individual plots or maintained collectively, municipal gardens organized by city agencies for individuals, and industrial gardens sponsored by a factory or business for the use of former employees.[8] Table 5 provides a summary of subsistence gardens in 1934 according to these types. Again, as in the past, the strengths and limitations of each type were acknowledged. While the home garden was convenient and made use of otherwise unproductive land, the large tract of land with multiple household plots provided a central location where aid in the form of demonstrations and supervision could help produce larger yields. In his 1932 report, New York State agricultural advisor W. E. Georgia concluded that better results came from projects where individuals had their own plots in a shared tract of land, whether run by a municipality, industry, or other group. Individual plots "not only provided a stimulus in competitive gardening but gave the individual something that he actually called his own."[9]

Work-relief and subsistence gardens served different intentions and yielded different results. The advantage of the work-relief garden program was that food could be grown efficiently and thus more could be distributed. The benefits of subsistence gardens included a sense of personal responsibility, positive competition with fellow gardeners, and healthy family recreation. In a Russell Sage Foundation report on garden programs, the authors raised the concern that work-relief gardens might jeopardize the

TABLE 5

Types and Number of Subsistence Gardens in a 1934 National Survey

	Number of Sites	*Acres*
Home and vacant-lot gardens	1,673,173	343,643
Community gardens	107,205	44,707
Municipal gardens	31,396	8,415
Industrial gardens	8,889	1,828
TOTAL	1,820,663	398,593

SOURCE: Ernest J. Wolfe, *Industrial and Agricultural Work Relief Projects in the United States* (New York: Governor's Commission on Unemployment Relief, 1935), 9.

NOTE: Although based on FERA data, Wolfe's report included some totals different from those found in FERA, *The Emergency Work Relief Program of the FERA, April 1, 1934–July 1, 1935* (Washington, DC: GPO, 1935), 6.

sense of independence associated with individual gardens. "When men are mobilized to work in gangs, it makes little difference whether they are set to dig ditches or weed crops, as far as their own participation or interest in the project is concerned. . . . When, however, the actual return to the individual gardener depends on his own management of his own plot, we might expect to find more of the men taking a vital interest in the task to be performed."[10] Citing the personal satisfaction inherent in gardening, the report concluded with a preference for subsistence garden projects. Indeed, as the Depression continued, state and federal reports focused increasingly on the subsistence garden format, and work relief shifted to other types of projects, although in some cases work-relief crews prepared sites for subsistence gardens as one of their assignments.

AT THE GRASS ROOTS: LOCAL GARDEN PROJECTS

In the early stages of the Depression, voices in cities, towns, and villages suggested garden programs as an immediate means to address the crisis locally. City agencies, charitable organizations, corporations, and civic clubs initiated projects, either individually or in collaboration. The Godman Guild Community Garden described at the beginning of this chapter, for instance, was run by a settlement house. In other cases, a civic club might start the program, such as in Louisville where one was established by the women's club. Municipal agencies also initiated projects, such as those in Detroit and Cleveland. To facilitate collaboration, many cities formed garden committees that included experts, influential business leaders, and representatives of educational and philanthropic groups. For example, Akron's Citizen Garden Committee consisted of the president of the city council, the secretary of the Association for Colored Community Work, the parks director, the county extension agent, and representatives of leading industries in the city.[11]

Funding for such programs initially came entirely from local sources through fundraising, community chests, local club sponsorship, and in-kind donations. Quite often, local projects made use of donated goods, volunteers, and public equipment and resources to keep costs down. Responsibilities were shared among different agencies and groups. For example, the garden project in Minneapolis was organized so that the university farm school analyzed soils and provided plowing and harvesting assistance; the family welfare association provided supervision; the council of social agencies took applications; and local landowners, businessmen, and city agencies provided land, seeds, tools, and other equipment. Thanks to in-kind contributions and donations, Cleveland's 1932 garden program, estimated at a cost of $25,000, was able to cover 90 percent of the expenses for its home garden program and 42 percent for the community garden program. Table 6 provides a breakdown of expenses for 1932, showing how much was covered by donations or cash.

Unlike the children's school garden movement or the World War I war garden campaign, no national advocacy group came forward to spearhead a national campaign for gardening during the Great Depression. Given the widespread promotion of war gardens only ten years earlier, it is probable that most communities had some organization or individual who felt competent to start a local program. A few individuals from past garden movements resurfaced to play leading roles within this framework. For example, Charles Lathrop Pack, previously president of the National War Garden Commission, encouraged a community garden for unemployed residents of the resort community of Lakeside, New Jersey.[12] One national organization that did promote gardening in several communities was the Red Cross, which initiated and funded garden projects in drought-hit states, particularly in the South.[13]

The idea of starting local urban garden programs spread quickly, aided by publications that praised the approach as both economically and socially

TABLE 6

Expense Estimates for Cleveland's Garden Programs, 1932

	Home Garden Program Expenditures			*Community Garden Program Expenditures*		
	Donations	Cash	Total	Donations	Cash	Total
Supervisors and foreman	—	—	—	—	$2,160	$2,160
Office space	$305	—	$305	$150	459	609
Printing, postage, and telephone	—	—	—	25	326	351
Land conditioning	—	—	—	6,797	—	6,797
Seeds	—	$264	264	18	726	744
Plants	119	33	152	110	123	233
Fertilizers	930	—	930	800	1,710	2,510
Tools	1,250	—	1,250	370	908	1,278
Transportation	175	—	175	1,177	6,751	7,928
Miscellaneous	—	—	—	211	81	292
TOTAL	$2,779	$297	$3,076	$9,658	$13,244	$22,902

SOURCE: Table based on Joanna Colcord and Mary Johnston, *Community Programs for Subsistence Gardens* (New York: Russell Sage Foundation, 1933), 51.

beneficial. In 1932, the President's Organization on Unemployment Relief published *Subsistence Gardens,* which reported on industrial, municipal, and philanthropic garden programs in five states. The next year, the Russell Sage Foundation produced a pamphlet, *Community Programs for Subsistence Gardens,* that described how to organize a garden committee, assign gardeners, find land, secure materials, distribute information, and encourage participation. These sources repeatedly stressed that gardening was not considered the sole means of relief but rather a supplement to other efforts. In the Russell Sage Foundation report, authors Joanna Colcord and Mary Johnston weighed the multiple benefits associated with gardens against their capacity as a relief strategy, commenting that "few other forms of community enterprise can be made to appeal to the general public as much as a garden project." They identified as advantages the participatory nature of gardening, the minimal start-up costs, and the quickness and tangible results and concluded that the "success of such a project should be measured in terms of general community participation as well as in families provided for and garden produce raised, for it is an opportunity to interest the general public in its unemployment relief program which no municipality can afford to lose."[14]

COMPANY AND RAILROAD GARDEN PROJECTS

In addition to municipal and philanthropic garden programs, many companies initiated gardens to compensate for worker layoffs, reduced hours, and wage cuts.[15] Companies provided community gardens on their property, leased land for gardens, or provided resources and incentives for employees who cultivated backyard gardens. Such projects were promoted not only by urban enterprises but also by rural industries such as lumber and mining. Some companies built their programs based on previous projects, such as the Firestone Tire and Rubber Company in Akron, Ohio, which had maintained gardens since World War I. The National Cash Register Company in Dayton, Ohio, which had garden experience from its Boys' Garden, established in 1893, as well as war gardens, developed a thirty-five-acre site that provided 260 gardens for employees.[16] Corporations with multiple factory locations, such as the United States Steel Corporation, occasionally mandated garden programs at all their business locations. In this case, the corporation furnished a plot for every employee who expressed the desire for one, totaling 73,511 gardens, of which one-third were small individual gardens and two-thirds were community gardens. Similarly, the Consolidated Coal Company sponsored subsistence gardens in five of its divisions in West Virginia, Kentucky, and Maryland, totaling 3,674 gardens on 2,125 acres. The Ford Company started a "balanced work and food production plan" at its plants that furnished several thousands of acres of land to an estimated fifty thousand gardeners, with plots estimated to be big enough to yield a year's supply of vegetables for a family.[17]

Since World War I, gardening along railroad rights-of-way had become a more or less permanent policy of railroad companies. A 1932 survey by the President's Organization on Unemployment Relief reported that more than forty railroad companies encouraged their employees to garden on land owned by the railroad, or that they intended to do so during the economic depression.[18] The majority of railroads provided land only to their employees, although some also leased land to other residents or cooperated with other civic or charity groups that were initiating garden programs. For example, in 1931 the Chicago and Illinois Midland Railroad collaborated with the United Charities of Springfield, Illinois, to develop 126 gardens on twenty-three acres. The project cost the railroad company $1,250 for seeds, tools, insecticides, supervision, and a cabin on site for the supervisor, and produced food that was valued at over ten thousand dollars.[19]

THE B. F. GOODRICH COOPERATIVE FARM

The B. F. Goodrich Company in Akron, Ohio, initiated its garden program as a way to sustain the over five hundred workers who had been dismissed, as well as many who had been reduced to part-time employees. Although it also provided some seed packets for employee home gardens, the company decided to develop a cooperative farm, in part because most of the workers did not have previous garden experience but also to illustrate the company's commitment to mass production principles. The Akron Community Garden was founded on two hundred acres of leased land. The company consulted with county agricultural agents, the agriculture department of Ohio State University, and several farmers. A former Goodrich engineer, B. E. Seaver, who had previous farming experience, served as supervisor.

Over nine hundred workers participated in the program from May through October 1932. Because the site was five miles from the factory, workers were brought to the site in company trucks. Their hours were carefully documented as the basis for the proportion of the harvest each would receive. A labor rotation scheme was devised so that each man could work one eight-hour day per week at the garden. The average participant worked twelve eight-hour days over the season, the minimum being four hours and the maximum 290 hours. Workers were divided into squads, each with a leader. After the first year, the company found that between 125 and 150 men per day was the practical limit for effective supervision. Only a quarter of the men had any farming background, but the company reported that everyone willingly learned new skills, just as they would if assigned an unfamiliar job in their regular industrial employment. As reported in a 1933 pamphlet produced by the company: "It was noted throughout the season that the men who had the best shop records, in almost all instances, made the best garden workers. This project provided a striking demonstration of the flexibility and adaptability of the industrial worker."[20]

The company paid for supervision, plowing, seeds, lease, fertilizer, and

Figure 35. The B. F. Goodrich Cooperative Farm, also known as the Akron Community Garden, as envisioned on the cover of the company's brochure *Industrial Cooperative Gardening: The Story of a Cooperative Garden Plan* (Akron: B. F. Goodrich Company, 1933), frontispiece.

other materials, at a total cost of between nine and ten thousand dollars. It engaged nine hundred workers and produced over one million pounds of vegetables. At a distribution house at the center of the garden, bags and baskets of produce were labeled with the names of the workers and placed alphabetically on marked shelves for weekly pick-up. Families took home their percentage of the harvest and were responsible for their own storage and canning. The company calculated that families on average received food worth $1.60 per day. By comparing this figure to the $1.60 to $1.76 that was the standard wage for farm workers in northwestern Ohio, company officials were able to claim their venture in cooperative farming as a financial success. The company produced a report, *Industrial Cooperative Gardening: The Story of a Cooperative Farm,* in 1933 that included the previous year's results as well as plans for 1933. Reviewing the benefits of the project, the report stated: "Cooperative gardening is not championed by its industrial sponsor as the whole answer to present conditions, but [it] is contended that such projects, properly operated, can be of fundamental value in communities facing substantial requirements of charity with purses flattened by [the] unprecedented demands of the past three years."[21]

The report included a letter of support from one of the gardeners, who wrote, after receiving his final share of garden produce: "If given opportunity to work for Goodrich in the future I think we will be better employees and take much interest in our work because the company has surely proved that it has our interest and welfare at heart."[22] The report confirmed that gardening would continue in 1933 as a coordinated effort with three components: the Red Cross would facilitate backyard gardens; community gardens for the unemployed would be provided through the Akron City Gardens, a city relief organization; and industrial gardens like the Goodrich Cooperative Farm would expand to engage other Akron industries.

STATE RELIEF AGENCIES GET INVOLVED IN GARDENING

In 1932, local programs received a boost when federal policies made more relief funds available to support garden programs. Quite often, statewide procedures were established for fund requests, seed distribution, and technical support from the state agricultural college. Generally, while the state provided some structure, the local municipality or garden committee retained control. By 1934, forty-five states plus Washington, D.C., reported subsistence garden programs, with results as shown in table 7.

In New York, a statewide subsistence garden program was established in

TABLE 7

Garden Program Acreage by State, 1934

State	*Home or Vacant-Lot Gardens*	*Community Gardens*	*Municipal Gardens*	*Industrial Gardens*	*Total Acreage*
Alabama	9,000	294	—	—	9,294
Alaska (not yet a state)	—	—	—	—	—
Arizona	—	270	—	—	270
Arkansas	37,411	931	—	—	38,342
California	—	—	—	—	—
Colorado	1,500	100	—	—	1,600
Connecticut	197	378	—	—	575
Delaware	—	—	—	—	—
District of Columbia	—	—	77	—	77
Florida	2,213	—	—	—	2,213
Georgia	7,000	75	—	—	7,075
Hawaii (not yet a state)	—	—	—	—	—
Idaho	—	76	—	—	76
Illinois	4,215	3,180	—	—	7,395
Indiana	11,610	3,278	1,731	581	17,200
Iowa	6,054	387	—	—	6,441
Kansas	1,825	—	—	—	1,825
Kentucky	19,340	—	—	—	19,340
Louisiana	7,230	100	—	—	7,330
Maine	520	—	—	—	520
Maryland	—	—	—	—	—
Massachusetts	—	—	119	—	119
Michigan	6,521	—	2,593	560	9,674
Minnesota	9,221	—	—	—	9,221
Mississippi	4,844	2,458	—	—	7,302
Missouri	6,134	2,642	33	—	8,809
Montana	1,200	941	—	—	2,141
Nebraska	3,227	—	—	—	3,227
Nevada	—	—	—	—	—
New Hampshire	400	246	—	—	646
New Jersey	4,484	498	—	—	4,982
New Mexico	3,480	142	—	—	3,622
New York	2,000	180	1,483	171	3,834
North Carolina	53,632	—	—	—	53,632

TABLE 7 *(continued)*

State	*Home or Vacant-Lot Gardens*	*Community Gardens*	*Municipal Gardens*	*Industrial Gardens*	*Total Acreage*
North Dakota	1,046	396	—	—	1,442
Ohio	10,990	2,170	1,554	—[a]	14,714
Oklahoma	6,868	4,732	—	—	11,600
Oregon	487	—	—	—	487
Pennsylvania	19,683	3,960	—	—	23,643
Rhode Island	255	—	—	—	255
South Carolina	11,720	3,041	—	—	14,761
South Dakota	7,853	2,043	—	—	9,896
Tennessee	8,000	1,370	400	—	9,770
Texas	3,401	1,554	—	—	4,955
Utah	568	82	—	—	650
Vermont	1,163	23	—	—	1,186
Virginia	6,250	16	—	—	6,266
Washington	703	7,873	—	—	8,576
West Virginia	55,852	733	—	—	56,585
Wisconsin	4,778	433	—	—	5,211
Wyoming	768	105	425	516	1,814
TOTAL	343,643	44,707	8,415	1,828	398,593

SOURCE: Table based on Wolfe, *Industrial and Agricultural Work Relief*, 9–10.
[a] Ohio had several industrial gardens, but the precise number is not provided in Wolfe.

1932, based on the previous efforts of municipalities and charitable organizations.[23] Harry Hopkins, then chairman of the New York Temporary Emergency Relief Administration (TERA), sent a support letter to mayors, commissioners of public welfare agencies, chairmen of emergency work bureaus, and others, advising communities that if they included seed and garden tools in Home Relief orders, they could receive a 40 percent reimbursement from the state. By 1933, New York state had 41,149 families with subsistence gardens as well as seven garden work-relief programs employing 850 men. TERA coordinated the program using established procedures and centralized resources, including statewide contracts for seeds and fertilizers that municipalities had to use in order to receive reimbursement.

TABLE 8

Results from the New York State Garden Program, 1932

	Number of Sites	*Acres*	*Gross Returns*	*Cost*	*Net Return*
Work-relief gardens	—	115.0	$14,202.48	$8,271.93	$5,930.
Municipal gardens	9,005	1,225.5	217,723.10	39,534.31	178,188.
Industrial gardens	4,367	530.5	101,153.44	16,861.43	84,292.
TOTAL	13,372	1,871.0	$333,079.02	$64,667.67	$268,411.

SOURCE: Table based on information in W. E. Georgia, *Subsistence Gardens in New York State in 1932,* rep prepared for the New York Temporary Emergency Relief Administration (Ithaca, NY: Department of Vegeta Crops, New York State College of Agriculture, Cornell University, 1933), 21.

Data collected by TERA revealed the cumulative results for work-relief, municipal, and industrial gardens for New York state in 1932, as shown in table 8.

FEDERAL ENCOURAGEMENT AND LIMITED FINANCIAL SUPPORT

In the early stages of this unprecedented national unemployment, the federal government was reluctant to get directly involved with relief at the state and local levels. President Herbert Hoover's initial approach was to encourage volunteer-led activities to address local manifestations of the depression. Having successfully garnered volunteer support in his work as Food Administrator during World War I, Hoover once again appealed to citizens to voluntarily aid in the national crisis. For instance, in 1931, the President's Emergency Committee for Employment encouraged local civic groups to hire the unemployed in public and home improvement projects. The committee promoted various civic campaigns, such as the "Spruce Up Your Garden" campaign, which suggested that civic clubs initiate labor-intensive beautification projects such as street-tree planting, clean-up campaigns, new sewer construction, gardens, and park improvements.[24] In other bulletins produced by the President's Emergency Committee for Employment, home gardens were promoted as "a natural reserve which may be drafted in time of emergency," and it equated local garden-employment campaigns with the war garden campaign.[25]

However, as unemployment persisted, the government acknowledged the need for more federal organizational support and funding. Initially, this was in the form of loans to states, then grants to both public and private agencies, and finally through direct federal-city coordination of programs. With the election of Franklin Delano Roosevelt as U.S. president in 1932 and the passage of the Federal Emergency Relief Act in May 1933, the federal government took a more active role.[26] From April 1, 1934 to July 1, 1935, the Federal Emergency Relief Administration orchestrated the federal campaign for emergency relief and employment. The purpose of FERA was "to aid the states in meeting the cost of furnishing relief and work relief, and in relieving the hardship and suffering caused by unemployment."[27] Structured as a decentralized system, the agency had a national administrator, appointed by the president, who oversaw the release of grants to states based on federal policies. Each state set up its own Emergency Relief Administration to oversee the distribution of combined state and federal funds through both direct and work-based relief. The state administrations, in turn, oversaw county and local relief administrators who carried out the actual relief.

With federal support available, state and local administrations initiated and expanded existing work-relief projects. Federal policy stipulated that projects had to be of a public character and of economic and social benefit to the general public or public institutions—such as construction and renovation of public buildings, road improvements, development of recreational facilities, education programs, and the production of clothing and food. The federal government also established a production program to promote activities that produced household goods, such as clothes, fuel, construction materials, and food. Gardening fit within the requirements of such funds because it could employ people and produce food. In 1934, garden projects represented 7.5 percent of an estimated 7,650 relief projects designed to produce goods.[28] Harry Hopkins, Federal Emergency Relief Administrator, reiterated the various attributes associated with gardening that supported its use as relief: "When you get a man out of the house into the open, with a spade and rake and hoe, you lift him out of a bad mental state into which enforced idleness inevitably plunges him. His improved mental attitude reacts favorably upon his health, and so does the increase of fruits and fresh vegetables in his diet."[29]

Under FERA-supported garden programs, two million families were helped with gardening and food preservation. However, federal financial support for gardens was short-lived. On July 1, 1935, the Federal Emergency

Relief Administration was replaced by the Works Progress Administration, which provided direct federal employment, mainly in public works and conservation, including substantial development of the nation's recreational systems. The federal government ended all financial support for garden programs in 1937 when it established the food stamp program for distributing farm surplus. Garden programs continued as local efforts on the periphery of much larger federal relief efforts.

No longer part of the public strategy for economic relief, community gardens did continue at the local level, based on resident interest and available land, but with less public attention or documentation. For example, the Workaneat Group continued to support community subsistence gardens in and around Seattle in 1937 after the state and federal programs had ended. The group considered the provision of subsistence gardens as "food insurance" and anticipated that five thousand households would continue to seek garden plots.[30] In Monroe County, New York, garden programs that had been established in 1930 continued to furnish Works Progress Administration (WPA) workers with subsistence plots in lieu of supplementary relief.[31] A later report on the project written up in the Department of Labor's *Monthly Labor Review* noted positive results in terms of physical and mental health as well as the opportunity to teach more people about gardening and encourage self-help.[32]

Even though the work-relief and subsistence garden programs were not sustained by the federal government after 1937, gardening was present in other New Deal programs to a lesser extent. The government's housing experiments repeatedly provided some level of self-sufficiency through household production, and gardening was one of the key manifestations of this. Plans for the three Greenbelt towns made gestures to incorporate gardening as a complement to industrial employment. In Greenbelt, Maryland, the site plan included not only community gardens but also small truck farms to sustain the community. The Farm Security Administration often included allotments or gardens in their housing schemes. And perhaps most directly, the federal Subsistence Homestead Program completed thirty-three homestead projects and the Federal Emergency Relief Administration another thirty-four projects that provided single-family homes with gardens, orchards, and/or wood lots in both suburban and rural contexts.[33]

RELIEF GARDENS IN WASHINGTON STATE

Garden projects in Washington state—both work-relief gardens and subsistence gardens—nicely illustrate the evolution of garden programs. At the

Figure 36. The federal Subsistence Homestead Program encouraged backyard gardening for income generation as well as household consumption. From the father's wages as a streetcar conductor—$100 per month—this family paid $16.20 monthly toward the purchase of this four-bedroom house plus garden in El Monte, California, in 1936. Photograph by Dorothea Lange. Courtesy of the Library of Congress, photo no. LC-USF-34 RA 1714-C USDA FSA.

onset of the Depression various municipal and philanthropic groups in Washington developed garden programs to feed the hungry. Most were on donated land and used volunteer labor. By 1933, there were 26,927 relief garden plots in the state.[34] Over 770 acres were cultivated, and the average household plot was 25 feet by 50 feet, or 1,250 square feet. With the establishment of the Washington Emergency Relief Administration in 1934, the state developed the Garden and Food Preservation Program to provide seeds, gardens, canning equipment, and advice to people receiving public assistance. To justify the investment, one of the organizers extrapolated the returns from one large community garden in Seattle to all the land under subsistence garden cultivation throughout the state and concluded that there would be an estimated $100,000 net return to the state for an initial investment of $30,424 for seed purchases. Seed packets were produced as part of a women's work-relief project and distributed through county welfare offices. Counties also oversaw land allotment and supervision. Canning centers were developed, primarily in rural areas where people had larger gardens. In 1935, the program produced over three-quarters of a million dollars' worth of fresh vegetables, an estimated average of forty-four dollars' worth of food for each family.

There were 815 acres of work-relief garden projects in 1934, employing a minimum of 500 workers seasonally.[35] In Seattle, for instance, two work-relief garden projects provided employment. The Airport Farm was a 55-acre site that had previously been used as a community garden.[36] In 1934, it became a work-relief site where an average of fifty men were employed eight hours per day, six days per week. The total cost for the year was $67,840. The food produced was distributed to workers, as well as to agencies such as the King County Welfare Board, the Salvation Army, the YMCA, hospitals, and children's homes.

In 1936, when the Washington Emergency Relief Administration became the State Department of Public Welfare, no specific funds were designated for the Garden and Food Preservation Program. However, counties continued to support gardens by utilizing networks of community support. Newspapers and radios kept advertising the program, while leftover seeds from the previous year were used along with donations from seed companies. The Home Economics Department of Washington State College and Agricultural Extension provided technical assistance with canning and food preservation. In her 1936 report on the status of the Garden and Food Preservation Program, the program's supervisor, Genevieve Saxe, noted that through community cooperation the program could continue without special state funds. However, whether it did or not is unclear because no further documentation has been found.

PARTICIPATION AND THE DIFFICULTY OF COMPULSORY GARDENING

Most publicly supported garden programs were intended for people who had been found in need by relief agencies. However, the subsistence gardens were also considered a means to keep families off relief lists. As the Depression continued, several programs expanded to include borderline cases and part-time workers to keep them off the lists too. Participants generally did not lose their plots if they found employment later. Unlike past pro-

Figure 37. FERA Airport Farm, Seattle, September 25, 1934. Courtesy of Manuscripts, Special Collections, University Archives, University of Washington Libraries, UW 18910.

grams, the undertone of the Depression-era garden programs was that they were mandatory rather than voluntary. Slogans such as "no garden, no relief" and the policy to eliminate produce from grocery orders in the summer months exposed limited choices available to relief applicants.[37] In some places, grocery orders and work assignment cards were delayed until the applicant had started a garden. In Indiana, state law allowed trustees to withhold relief if clients refused to work, which included making a garden.

The garden work-relief program in Muncie, Indiana, exemplifies a garden program in which labor was associated with relief allotments. According to a 1932 account, the township trustee and the local social service agency required that people applying for food relief had to work for the amount of their relief orders at the rate of fifty cents per hour.[38] The "Muncie Plan" involved a thirty-five-acre community garden and over one thousand home gardens. Technical assistance was provided by the Purdue University Horticulture Department, and a full-time garden supervisor was on site to assign workers and assure that clients were working satisfactorily. Food grown at the community garden was canned for winter relief. In 1931, 1,836 bushels of potatoes, 300 bushels of carrots, 159 bushels of beets, 54 bushels of navy beans, 30 tons of cabbage, 34 bushels of turnips, 125 bushels of parsnips, and the canned produce from ten acres in tomatoes were supplied to the winter relief program. Muncie officials developed similar programs for fuel relief and medical services, whereby recipients paid for services through street cleaning, park maintenance, and other civic maintenance duties. The Muncie Plan was frequently cited as a model program and was duplicated in other Indiana cities, including Wabash and Kokomo.

Reports generally stated that those who were interested and/or experienced in gardening made use of the resources provided, but compulsory participation policies often met with resistance. In the 1933 Russell Sage Foundation report on subsistence gardens, Colcord and Johnston described a survey conducted by a state agricultural extension service that reported only 15 of 700 respondents were indeed interested in gardening. Rather than accept this resistance as a reason to discourage gardens, the authors acknowledged that garden programs require nurturing and care. "A garden is not a mechanical thing; a subtle co-operation between the soil and those who till it seems to be necessary to success; and unless interest and inclination are present in the gardeners, or can be taught and developed by the supervisors, the project will not prosper."[39] A report by the Federal Emergency Relief Administration noted that while inexperienced gardeners were unwilling, clients accustomed to gardening were usually glad to have the opportunity to raise their own food. "Compulsion, of course, could not in the long run produce efficient subsistence gardening, and in fact the unwilling gardeners neglected their gardens when they did not become fond of the work or were not sufficiently impressed with the results of their efforts to continue."[40] It was hoped that skepticism could be overcome through supportive education. Participation was encouraged through advertising, leaflets, lectures, contests, competitions, and social activities. In some cases, for instance, supervisors highlighted the best-kept plot in a community garden plot with flags or notices to draw attention to it and praise the effort.

LAND TO GARDEN

One of the justifications for gardening during the Depression was that it made use of otherwise idle land. Similar to the World War I war garden campaign literature, Depression-era manuals suggested that organizations conduct surveys of vacant land and seek cooperation from real estate boards, industry, railroads, and public agencies. Although the ideal was to find sites close to residential communities, availability and the need for large sites to serve many people often meant that gardens were located at the city's edge. Transportation to the garden was a frequently mentioned problem, especially at harvest time when gardeners had large loads of produce to bring home. Solutions to the transportation problem included borrowing trucks to transport workers and produce, reducing fares on railroads, providing two free tokens per week for carfare, and providing shelters so families could temporarily stay at the garden.

A plot designated for a family was generally quite large in order to provide adequate harvests—forty feet by one hundred feet, or fifty feet by one hundred feet, was often recommended. Minneapolis provided gardens at either forty-five feet by ninety feet, or ninety feet by ninety feet, while Cleveland's gardens were forty feet by fifty feet. Work-relief gardens that employed workers in farmlike production also ranged in size.

GARDEN RULES AND PROCEDURES

Gardeners generally filled out contracts in which they agreed to abide by rules of garden maintenance, security, and respect for other people's plots. While the issue of site security had been occasionally mentioned in earlier phases of urban garden programs, it was a frequently raised concern during this period. Most programs required gardeners to wear badges or carry identification cards. During harvests, some programs hired watchmen, scheduled gardeners for day-long occupation of the site, and allowed people to temporarily camp at the gardens.

Most programs hired full-time garden supervisors who were knowledgeable in gardening, good at working with people, enthusiastic, and diplomatic. The number of supervisors varied, from one person per tract in some places to one supervisor per one hundred to two hundred gardens, as for instance in Rochester, New York. Supervisors provided instructional materials, conducted demonstrations, and provided model gardens. Gardeners could also gather information from written materials produced by the Department of Agriculture, state agricultural colleges, and other sources. Some cities, such as Chicago, encouraged volunteers from local garden clubs and women's clubs to visit plots and provide instruction and encouragement.

To increase yields, some well-informed supervisors as well as advisors from state colleges and county extension offices developed planting plans. While some programs required following an exact planting plan, others allowed more latitude. For instance, Indiana had an exact garden plan for a fifty-by-one-hundred-foot plot that was developed by H. E. Young, community and industrial garden leader at Purdue University. This garden was designed to accommodate seed packets produced by W. Atlee Burpee Company that could be purchased for 22¾ cents and included beans, beets, carrots, lettuce, parsnips, radishes, spinach, corn, chard, and turnips.[41] In Akron, Ohio, foreign-born gardeners were encouraged to plant according to the methods of their homeland as a way of developing personal pride and interest in the garden.[42] Some programs provided packets of seeds selected by experts. Tools were provided by either the gardener or the committee. Some programs also provided instruction and equipment for canning and food preservation.

The Detroit Thrift Gardens program, started in 1931, is a good example of an organized program. The real estate board arranged for donations of twenty-seven garden parcels, totaling three hundred acres, to be leased for the entire growing season in order to insure gardeners against losing crops.[43] There were 2,765 gardens of forty by one hundred feet. Under the supervision of an experienced gardener, each participant received a packet of assorted seeds and was expected to follow a garden diagram. The plan determined the entire garden except for a small area left to the gardener's own choice; most often it was planted in potatoes. The program provided seeds to community gardens and to 1,604 home gardens as well. Canning demonstrations assured safe food storage for winter months. Each gardener signed a pledge that stated:

> I agree to plant a garden according to the prescribed garden diagram and to keep it in good condition, free from weeds, by cultivating it at least once a week during the season.
>
> I agree to keep a record of the amount of produce harvested and will make a summary report of my garden at the close of the season and send it to the Detroit thrift gardens.
>
> I agree to consider the rights of others and to do all in my power to protect my neighbor's garden from harm as well as my own, and further agree to avoid damage to sidewalks, trees, or any other improvement.
>
> I agree not to sell or transfer my garden privilege.
>
> I agree that I will not offer for sale on the general market the products of my garden.

DETROIT THRIFT GARDENS, 176 E. JEFFERSON AVE.

		GARDEN NO.
1. PRINT NAME		
2. MARRIED SINGLE WIDOWER	ADDRESS	FIELD NO. APP. NO.
3. HOW MANY IN FAMILY?		
4. HOW MANY WORKING?	NEW ADDRESS	CASE NO.
5. WHERE?	DATE OF APP. BY	DATE ASSIGNED BY
6. HOURS PER WEEK?		
7. HOW MANY SCHOOL AGE?	HOME GARDEN, LOCATION	
8. HOW MANY CAN GARDEN?		
9. HOW LONG IN DETROIT?	LOCATION OF GARDEN ASSIGNED	
10. HAVE YOU USE OF CAR?	WHAT EXPERIENCE { FARM TRUCK HOME	HOW LONG
11. NATIONALITY? AGE?		
12. HOW LONG IN U.S.A.?		
13. ARE YOU A CITIZEN?	DID YOU HAVE A THRIFT GARDEN IN 1931?	
14. HAS WELFARE EVER HELPED YOU?	I HAVE READ THE RULES AND REGULATIONS OF THE DETROIT THRIFT GARDENS AND WILL ABIDE BY THEM TO THE BEST OF MY ABILITY.	
15. ARE YOU NOW RECEIVING HELP?		
16. EDUCATION?	SIGN HERE ______ FULL NAME	

Figure 38. Detroit Thrift Gardens application form, reprinted from Joanna Colcord and Mary Johnston, *Community Programs for Subsistence Gardens* (New York: Russell Sage Foundation, 1933), 65.

I agree to wear my badge, which shall be provided by the Detroit thrift gardens, in a conspicuous place, at all times that I am working on my garden.

I agree to forfeit all rights and privileges of my garden if I fail to comply with the above rules and regulations.[44]

Unlike the relief gardens of the 1893–97 depression, which promoted food production for sales as well as consumption, the 1930s relief gardens focused entirely on consumption. In fact, attentive to the plight of farmers and the nation's food surplus, most garden programs prohibited sales of surplus produce. To counter any criticism that the garden was competing with truck farmers, administrators insisted that gardening provided food to the poor who could not afford to buy produce. The Russell Sage Foundation report commented: "The utmost that can be demanded of such a project is that the goods so produced be rigidly kept off the market; and it is the duty of any group sponsoring a subsistence garden plan to take every precaution that legitimate trade, for which a market exists, be not interfered with by the sale for cash of any of the foodstuffs produced."[45]

JUSTIFYING GARDENS FROM A FINANCIAL PERSPECTIVE

As a form of relief, gardening in the Depression era was once again championed as a multi-benefit activity that was also low-cost and easily implemented. To promoters at this time, gardening—particularly subsistence garden programs—provided an economical form of relief. A nationwide analysis in 1934 reported the total cost for garden and canning programs to be approximately $8 million, and the total return was valued at $37 million, representing a net return of $29 million.[46] That is, on every one dollar invested there was a $3.45 return. Of course, the return was in the form of food, not dollars, since sales were usually prohibited. A report by the Federal Emergency Relief Administration acknowledged that the return estimates did not include the gardener's labor, and that the produce raised in these gardens cost much more than similar produce raised in large-scale truck farming. The report questioned the economic return but lauded the beneficial health returns:

> The economics of [gardening] are of course still in dispute between those who believe that such family food-raising will continue to have a significant place in our industrial civilization, and those who find efficiency only in large-scale production. However, the social and educational effects of the program are not involved in that argument. By raising a wide variety of vegetable foodstuffs with high nutritive value, many people have been enabled to change and improve their habitual dietary range to their permanent advantage.[47]

In the minds of many supporters, the greatest success of the gardens was that they kept up morale and stabilized families. The strain of "enforced idleness" due to unemployment or reduced employment was lessened by purposeful visits to the garden. Gardening produced healthy food even as it encouraged exercise and recreation.

EFFORTS DWINDLE AS FEDERAL SUPPORT SHIFTS

When federal New Deal programs revised the relief structure, many urban gardens lost their federal funding and with it their organizational structure. Local efforts may have continued, but the national drive subsided, as did promotional materials and reports. The image of a squatter's garden in 1939 (figure 39), along with occasional notes in local papers, confirms that some gardens continued even without public funding. Like the urban garden

Figure 39. This April 1939 image of gardens near Newark, New Jersey, captioned "Gardens for the Unemployed. Squatters' Houses in Jersey Meadows, on the City Dump," documents the existence of gardens—whether programmed or not—after the removal of federal support. Photo by Arthur Rothstein. Courtesy of the United States Farm Security Administration Collection, Prints and Photographs Division, Library of Congress, photo no. LC-USF-34-27166-D USDA FSA.

programs of earlier decades, both work-relief and subsistence gardens relied on borrowed land, usually at the city's edge. These sites were never discussed as permanent resources for the urban poor. The land was too valuable for development once the economy picked up, and its distance from the city was inconvenient for urban dwellers who would not always have idle time to commute. Thus the 1930s gardens, while gaining public recognition for nutritional, recreational, and social benefits, did very little to establish gardening as a sustained community resource.

SIX

Victory Gardens of World War II

Vegetables for Victory
Give vitamins for health—
Add riches to the nation—
Augment our country's wealth.
Now go to work good neighbors
With busy rakes and hoes
So that garden diligence
Will help defeat our foes.
The vegetables have registered—
Each root and bulb and leaf—
All stand awaiting your command—
Their General-in-Chief.

L. Young Correthers, Vegetables for Victory
(San Diego: E. P. Wilson and Company, 1942)

ON DECEMBER 19, 1941, just twelve days after the attack on Pearl Harbor, citizen groups and garden advocates initiated an American victory garden program.[1] While victory garden promoters hoped to show their patriotic support through gardening, as had happened during World War I, the federal experts initially hesitated to include urban gardening in their national policy for wartime food export and domestic security. Citing the inefficiencies of small-scale urban gardens and the likelihood of fertilizers and seed being wasted by inexperienced gardeners, the experts proposed programs to increase food production through improved agriculture and promotion of rural gardens. However, public desire to assist in war preparedness—plus justifications that victory gardening was not just about food but was also for health, recreation, and morale—succeeded in convincing federal officials to support urban garden programs as well. Citizens fulfilled expectations, with an estimated fifteen million victory gardeners in 1942 who produced 7.5 billion pounds of food. By 1944, M. L. Wilson, director of extension programs for the U.S. Department of Agriculture, could report that between eighteen and twenty million families had victory gardens that collectively provided 40 percent of the total American vegetable supply.[2]

DIFFERENCES BETWEEN WORLD WAR I AND WORLD WAR II FOOD STRATEGIES

Even before the United States officially declared war, there had been some domestic preparations to protect food supplies. As had been the case in World War I, victory in the Second World War depended on a secure domestic economy and a healthy flow of exports as well as military might and strategy. American food was needed overseas for allies, for American soldiers and sailors, and for the liberated areas, as well as for American civilians at home. As Marvin Jones, War Food Administrator, stated, "[Food] is just as necessary as guns and tanks and planes."[3] However, social and technological changes during the twenty-four years between the world wars led to a different food-security strategy during the second war. The Department of Agriculture had expanded its programs in education, technical assistance, and soil conservation services. Farmers were producing more food than they had previously. In 1942, for instance, farms produced 26 percent more food than the average annual production of the previous five years.[4] Thus, experts were less concerned about food scarcity than about increasing efficiency through centralized planning, technological innovation, and improved agricultural practices. With the goal of growing "the right amount of the right things in the right places," the federal government established policies in production selection, price controls, and rationing of materials.

Efficiency through planning was applied to distribution as well as production. It was the responsibility of the War Food Administrator to allot the food supply to the various claimants, including the Allies, the military, and civilians. As a result, approximately 75 percent of food was targeted to civilians and 25 percent to the war campaign. The Lend-Lease program, established in March 1941, coordinated aid and services, including food exports and imports between Allied countries. Shipments of American wheat, flour, sugar, canned and cured meats, dairy products, peas, beans, canned vegetables, and vegetable oil were sent to the United Kingdom and the Soviet Union. In 1943, for example, the allocation of food for the war campaign distributed approximately 13 percent of the total food produced in the United States to military forces, 10 percent to Lend-Lease, and 2 percent to other special needs associated with the war.

A major difference between World Wars I and II lay in the government's approach to civilian supply and volunteerism. Whereas during the First World War the aim was to reduce the domestic drain on the food supply through a campaign of voluntary food substitution and home production, during World War II the federal government implemented price controls and food rationing. Between the wars, purchasing power had increased such that wealthier citizens could afford higher prices for food while the poor could not. And as a result of the war, there had been a 13 percent rise in food prices between September 1942 and May 1943. To address these issues, the Office of Price Administration established a program of domestic rationing and a point system for food purchases as well as price controls. The foodstuffs included in the program varied over the course of the war, depending on availability and demand. However, with many specialty goods available on the black market, civilian cooperation still needed to be encouraged.

THE FOOD FIGHTS FOR FREEDOM CAMPAIGN

The Food Fights for Freedom campaign engaged several federal agencies collectively to inform and encourage citizens to comply with national policies and to promote community activism. The federal campaign provided national publicity while also linking individual programs to each other so that everyone involved could see the larger impact of their work on the war effort. Campaign organizers relied on state and local institutions to implement policies, acknowledging that each community had its own food concerns, leadership, and ways of doing things, and that "better than anyone else the people in the community itself will know what can and should be done."[5]

The theme "Food Fights for Freedom" and its emblem were used in pamphlets and promotional materials developed by federal, state, and local groups. In addition, the campaign spearheaded several national drives, such as a pledge drive to comply with rationing and price controls and the declaration of November 1943 as "Food Fights for Freedom Month." The campaign was conceptually structured along four general guides to action: produce, conserve, play square, and share. To meet production needs, the campaign urged citizens to volunteer as farm laborers and grow food for the household. To conserve, community groups were encouraged to promote nutritious eating habits, economical food substitutions, food preservation

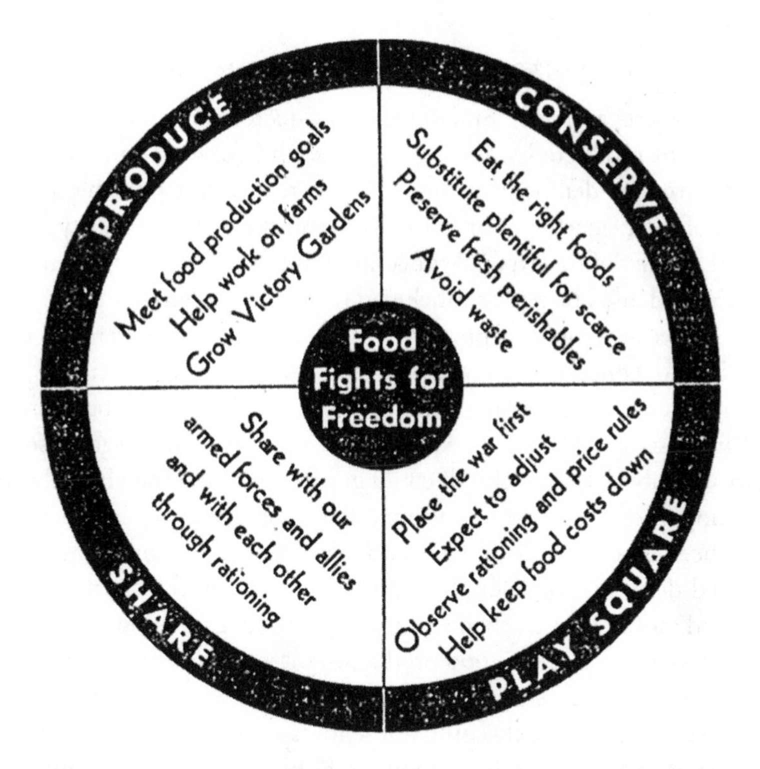

Figure 40. The Food Allocation Chart of the Food Fights for Freedom campaign displayed the four guidelines for action: produce, conserve, play square, and share. Reprinted from Office of Program Coordination, Office of War Information, and Office of Price Administration, in cooperation with the War Food Administration, *Food Fights for Freedom* (Washington, DC: GPO, 1943), 34.

through canning, and to avoid wasting food and resources. "Playing square" referred to cooperation with rationing and pricing rules. Lastly, citizens were reminded that they were rationing their food in order to share with the armed forces and Allied countries. The campaign also organized nine individual informational programs on the following topics: farm production goals, victory gardens, home food preservation, nutrition, food conservation, farm labor programs, rationing, food price controls, and home front pledges.

THE VICTORY GARDEN CAMPAIGN BEGINS

Initially, experts and government officials directed war food preparedness toward farm productivity, rural gardening, and conservation and downplayed citizen gardening campaigns. A 1941 report developed by subcommittees of the Department of Agriculture and the Federal Security Agency expressed many of the experts' concerns about the inefficiency of small individual gardens. The report concluded that farm gardens—and large suburban gardens—should be promoted while urban gardens were less desirable due to concerns about conserving seed supplies, fertilizers, and insecticides. However, representatives of garden clubs and other horticultural interests quickly urged expansion of the domestic program so that urban and suburban households could participate in the war effort through gardening. Promoters of a new war garden campaign cited similar benefits as had been used to validate the World War I war garden program, such as increased domestic food supply, reduced pressure on transportation, and improved morale. Rather than considering civilian gardening as part of the direct program to increase food production, it was justified as a way to satisfy household tastes while also putting less demand on resources needed for the war effort. Not only could households enjoy more food diversity on their plates, but growing fresh foods at home also meant that more tin, labor, power, and machinery currently being used by the food industries could be directed to the war effort.

To discuss the benefits of a war garden program and to strategize the best means of orchestrating a national campaign, a National Defense Gardening Conference was held on December 19, 1941.[6] Over three hundred horticultural experts, business leaders, educators, newspaper editors, and representatives of garden clubs, youth organizations, and federal and state agencies attended the conference. With an eye to conserving seeds, fertilizers, tools, and other materials as well as to avoid disappointing results, conference attendees developed a victory garden program that was focused mainly on vegetable gardens on farms, in small- and medium-sized towns, and in suburban areas. While city backyards were not considered practical for intensive vegetable gardening, enough interest was voiced at the conference that eventually urban community gardens for vegetable production were also encouraged. Ornamental gardens in backyards and vacant lots were also encouraged for their morale-building and recreational benefits. Conference speakers Claude R. Wickard, secretary of agriculture, and Paul

V. McNutt, director of the Office of Defense, Health, and Welfare Services, described the goals of the victory garden campaign as follows:

> Increase the production and consumption of fresh vegetables and fruits by more and better home, school, and community gardens, to the end that we become a stronger and healthier nation.
>
> Encourage the proper storage and preservation of the surplus from such gardens for distribution and use by families producing it, local school lunches, welfare agencies, and for local emergency food needs.
>
> Enable families and institutions to save on the cost of vegetables and to apply this saving to other necessary foods which must be purchased.
>
> Provide, through the medium of community gardens, an opportunity for gardening by urban dwellers and others who lack suitable home garden facilities.
>
> Maintain and improve the morale and spiritual well-being of the individual, family and nation.[7]

A National Advisory Garden Committee was formed to coordinate the participating agencies and organizations. The federal Office of Defense, Health, and Welfare Services was given responsibility for the defense activities of all agencies dealing with health, nutrition, recreation, and welfare services. This included nutrition programs, school garden programs, and a few remaining relief and subsistence garden programs that were managed by the Works Progress Administration. In coordination with this office, three federal agencies contributed substantial support for the victory garden campaign: the Department of Agriculture, Office of Civilian Defense, and Office of Education. In addition, a National Victory Garden Conference was held each year as a forum where information could be shared and national guidelines developed.

Each year of the campaign, the secretary of agriculture established a goal for the number of victory gardens that the various agencies could publicize as a means to spur participation. For example, the 1942 target was fifteen million gardens—five million on farms and ten million in towns, which was accomplished during that growing season. In 1943, the goal was increased to six million farm gardens and twelve million nonfarm gardens, which was again met. This national goal served as a catalyst to expand federal, state, and local efforts while also reminding everyone that all the individual efforts combined to contribute to a vast national front. More concretely, the Department of Agriculture also provided substantial technical

Figure 41. Poster from U.S. Information Services, Division of Public Inquiry, Bureau of Special Services, 1941–45. Artist: Robert Gwathney. Courtesy of the National Archives, photo no. NWNDS-44-PA-368.

support and direct assistance, particularly through its cooperative extension services. With more than three thousand agents in forty-eight states, extension services had a ready network of local representatives to promote gardening in rural and farming regions.[8] Extension agents distributed bulletins and circulars, set up demonstrations and classes, and organized adult clubs and youth 4-H clubs. They also worked with school gardens and lunch programs and provided technical and financial support for canning centers in conjunction with the War Food Administration. However, while they were effective at rural organizing and at producing regionally appropriate instructional materials, extension services had less presence in urban areas. The department's published materials, on the other hand, served gardeners everywhere. Their *Victory Garden Kit,* for instance, though intended primarily for rural gardeners, included facts about the national food program, encouraging quotes from national leaders, and suggestions for kick-off campaigns, as well as information on staging harvest shows, publishing flyers, conducting radio interviews, and acquiring promotional films that were equally useful in urban contexts.

Much of the urban victory garden campaign was facilitated by the Office of Civilian Defense. Through the office's local defense councils and volunteer headquarters, local citizens could be directed to a range of patriotic volunteer positions, from air-raid warden to garden coordinator. Local councils often worked with garden programs organized by agricultural extension offices, schools, municipalities, or other institutions—or organized a garden program themselves if none existed.

Although no School Garden Army was organized during this war, the Office of Education did encourage victory gardening as a school or extracurricular activity. Articles on school-directed gardens were regularly included in the Office of Education's biweekly publication *Education for Victory,* which was sent to approximately sixty thousand universities, colleges, libraries, and other educational institutions. Topics included garden and canning instruction, lesson plans, instructor training in agriculture, summertime garden supervision, parental involvement, and farm volunteering. The Office of Education also encouraged coordination of school gardening, food conservation, and school lunch programs. School-lunch garden programs often focused on fresh vegetables and fruit that could be harvested for use during the school year, although they also preserved food by canning, drying, salting, pickling, and freezing, often done in home economics courses. The Office of Education also collaborated with the Department of Agriculture on a Victory Farm Volunteer program to recruit urban youth to help on farms during summer holidays.[9]

STATE AND LOCAL CAMPAIGNS

Federal agencies provided guidelines and information but relied on state and local organizing for the actual development of victory gardens. To encourage state planning, the USDA extension service office in Washington, D.C., instructed state directors to call state victory garden conferences, to be attended by representatives of state, federal, and local agencies as well as other interested parties. Based on federal policies but attentive to state conditions, each state then developed its own victory garden program and established goals for the desired number of gardens statewide. In similar fashion, the state victory garden councils then directed county committees to coordinate local efforts, hold public meetings, organize lectures, and distribute information from extension services and other sources.

Local victory garden committees usually included influential community members from real estate boards, chambers of commerce, and service clubs, as well as municipal agency directors, educators, and other advocates. These local committees often relied on subcommittees to coordinate more specific activities, such as land procurement, registration and assignment of plots, insect and disease control, and conservation and canning. Many groups organized neighborhoods through block captains who supervised community and vacant-lot gardens and provided individual encouragement to residents in a neighborhood. Through registration campaigns, the local committees kept records of civilian contributions to the national food supply. They supported individual gardeners and operated community gardens, demonstration gardens, and canning centers. They also worked with company-sponsored and school garden programs.

Each state organized its victory garden program differently. In Massachusetts, for example, the structure was largely as described above.[10] The state committee that administered the Home Garden and Food Preservation Program made broad policies to increase the acreage under cultivation, provide technical information, and encourage community and allotment gardens. These policies were then translated at the local level into outreach through bulletins, meetings, home visits, and work with schools and clubs, and into support for new gardens through surveying available land, equipment, and technological assistance. A different system was developed in North Dakota, where garden programs were categorized into three distinct groups, each with a different leadership. Farm gardens were the responsibility of the federal Department of Agriculture and county extension agents who worked with local leaders.[11] Town and city garden programs were organized by the state and local defense councils. Community gardens were the responsibility of municipalities working with local businessmen, service clubs, and others. The goal in all states was efficiency as information traveled from guidance at the federal level to local implementation.

Because much of the victory garden campaign utilized existing avenues of outreach—federal agencies, schools, and local organizations—there was little dedicated funding per se. Local implementation was often led by volunteers, and in cases where a supervisor was needed, either the position was funded through local fundraising efforts or the duties were taken on by parks staff or schoolteachers. Some projects compensated for expenses by charging rental fees for plots. Whereas past garden programs had often quantified results to prove their effectiveness, the victory garden campaign literature did not focus on the monetary value of the food produced. It was assumed that gardening helped the national defense campaign, and that financial returns were experienced as subsidies to household budgets that allowed more purchasing power in other areas.

GARDEN CLUBS AND WOMEN'S ORGANIZATIONS

As they had in the past, women's and garden clubs played particularly important roles in the World War II garden campaign. The American Women's Voluntary Services, which was established to facilitate women volunteers in a variety of war-related activities, included a Food for Victory committee that promoted gardening, food preservation, and nutrition education. The organization was structured in a top-down fashion, with a national leadership directing the efforts of state and local groups. In addition, garden clubs and women's clubs came to the fore once again to promote local gardening. As an article in *Horticulture* stated: "It is the privilege and duty of garden clubs to see that food is grown in suburban sections and small towns throughout the country under conditions which will avoid waste and produce the vegetables that are most important in supplying vitamins and minerals, as well as calories."[12]

Garden clubs disseminated advice, established demonstration gardens, taught gardening and canning, surveyed available land for community gardens, sponsored garden shows, paid supervisor salaries, and donated plants. For instance, as soon as war was declared, the National Council of State

Garden Clubs officially pledged support for the victory garden campaign, with members offering technical assistance, pamphlets, and other aid to local efforts. The council also held victory garden schools to instruct members and civic leaders in how to organize local garden programs. The Garden Club of America supported the war effort by planting gardens at army installations, sponsoring horticultural therapy programs in army hospitals, teaching classes on gardening and preservation, buying mobile kitchen and canning units, and sending seeds to European countries. The Woman's National Farm and Garden Association, under the presidency of Mrs. Henry Ford, resolved at its 1943 annual meeting to support all aspects of the production and preservation of foodstuffs as part of the war and postwar efforts. Their local club projects included gardening along with other activities such as volunteering at canteens, selling war bonds, working on ration boards, and making presents for servicemen.

SAN FRANCISCO VICTORY GARDENS

San Franciscans, accustomed to abundant fresh produce and specialty crops grown in the nearby Salinas Valley, were prepared to sacrifice for the war effort when USDA policies asked growers to switch from "unessential" crops such as lettuce and artichokes to less perishable crops, such as peas, carrots, and potatoes. Agricultural labor shortages during the war impacted food availability as well as prices. Food prices rose; lettuce, for example, jumped from four cents a head before the attack on Pearl Harbor to twenty cents by January 1943. To offset both the rising prices and the limited availability of fresh produce, the San Francisco Garden Advisory Committee was formed in November 1942.[13] Under the chairmanship of John Brucato, agricultural economist and vintner, the committee included representatives from local colleges and universities, the American Women's Voluntary Services, the school board, park commission, chamber of commerce, newspapers, and garden clubs. This organization served as a clearinghouse for information on gardening and kept tabs on the various garden activities being developed by neighborhood groups, institutions, schools, and civic organizations. They also persuaded the Board of Supervisors to lower the water rate for households that had gardens at least 100 square feet in size, to ease regulations to permit raising chickens and rabbits in the city, and to authorize use of unoccupied city-owned and county-owned land for community and vacant-lot gardens. They also organized and publicized various educational opportunities provided by San Francisco Junior College staff, University of California extension services, garden clubs, and others.

San Franciscans established gardens in backyards, vacant lots, playgrounds, parks, and schools. The settings ranged from a steep hillside vacant lot to open land adjacent to public housing on the outskirts of the city. In January 1943, a victory garden was established in Golden Gate Park to serve nearby residents. The site—over 120,000 square feet of land—was located at Main Drive and Park Presidio Driveway. Very quickly, over 500 applications were received, in response to which approximately 250 plots were assigned. The gardeners were given twenty-by-twenty-foot plots plus free water, seeds, access to tools, and instructions. Over 10,000 children participated in victory gardens on school property or in nearby vacant lots.

To spark public interest, demonstration gardens were established in Union Square, the Civic Center, Balboa Park, and Golden Gate Park. Citizens kept abreast of opportunities and technical information through local newspapers such as the *San Francisco Chronicle,* which ran a weekly "Homes and Gardens" section to support victory gardening. Leslie Layton, the newspaper's garden editor, also hosted a weekly radio show. To create a festive and social atmosphere and encourage friendly competition, neighborhoods around the city held garden fairs to display their produce. On June 27, 1943, a citywide Victory Garden Fair was held at Golden Gate Park. Thousands of San Franciscans gathered to see exhibits, take in the entertainment, and enjoy picnic lunches. Exhibits by over fifty organizations, seed companies, and individuals showed the potential of victory gardens and provided advice. Entertainment, which included Scottish bagpipes, a Russian balalaika orchestra, traditional Mexican folk songs, and Irish pipers, reflected the city's multicultural population. The fair was heavily promoted by radio broadcasts at the event and by subsequent newspaper articles and photographs. The fair continued as an annual event. In 1945, two days after that year's fair, the official opening of the victory garden season was celebrated at City Hall as the Victory Garden Queen, sixteen-year-old Colleen McInerney, was presented with a crown of parsley and radishes.

GARDENS AT WORK AND THE NATIONAL VICTORY GARDEN INSTITUTE

As they had before, for-profit businesses and industries contributed to the wartime garden campaign. Industries related to gardening and farming

Figures 42, 43, and 44. *Above:* The slope of this vacant lot in San Francisco did not stop the victory gardeners who terraced it for their garden plots (1945). *Opposite, top:* Victory gardens at the Sunnydale Housing Project, San Francisco, 1943. *Opposite, bottom:* Victory gardens were laid out in public spaces, such as Golden Gate Park in San Francisco (photo taken in 1948). All photographs courtesy of the San Francisco History Center, San Francisco Public Library.

contributed to the campaign through in-kind and technical support. Florists', nurserymen's, and seedsmen's associations encouraged their members to provide land, greenhouses, seed, plants, and technical assistance to local campaigns. And, as had been the case in the First World War, some companies encouraged their workers to garden by providing land, seeds, technical assistance, and incentives. Industrial companies, utilities, and railroads all participated in the campaign. Some, such as the Whiting Corporation, Goodyear Tire and Rubber Company, and Firestone Tire and Rubber Company, had supported gardening during the Depression and World War I. Others, such as the Sears Roebuck Chicago plant, Texaco, General Electric, and Edison, were new to the campaign. A few companies established farms to supply their cafeterias and employees, such as the 75-acre farm run by Continental Machines in Minneapolis, Minnesota, and

the 650-acre farm belonging to Denison Engineering Company, in Columbus, Ohio. Railroad companies once again provided land along their rights-of-way for gardens.[14] Usually the land was free to employees, while other participants paid a minimal fee.

The Firestone Tire Company's program is a good illustration of how one company supported the campaign.[15] The company encouraged workers to garden at home by providing soil tests and recommending soil amendments. Employees without home gardens were supplied with 3,500-square-foot community-garden plots, a packet of seeds sufficient to plant half the plot, insecticides, and fertilizer. Approximately 2,500 employees gardened on 150 acres either owned or leased by Firestone and produced $310,000 worth of produce. The company also offered three hundred dollars in cash prizes to be awarded for soil quality, garden layout, disease and insect control, variety of produce, tillage, and general garden practices. Other methods of promotion included an annual harvest show and a demonstration garden. To encourage food preservation, the company loaned one hundred pressure cookers to employees. The company was still fostering garden efforts in 1948, when it received a Green Thumb medal, the highest award of the National Garden Institute.[16]

Companies also supported the victory garden program by donating their advertising budget and/or space in newspapers and magazines to victory garden promotion. Along with providing tractors to prepare community gardens in Chicago parks, the International Harvester Company donated fifty thousand dollars of direct advertising in national magazines, newspapers, and posters, and produced an eighty-four-page booklet titled *Have a Victory Garden.*[17] Other companies, including Standard Oil and General Mills, also published victory garden manuals. The Beech-Nut Packing Company published a pictorial workbook that also served as company advertising. The cover stated: "Growing superior quality vegetables has always been of primary interest to Beech-Nut Packing Company because only the finest foods go into our products. We realize the enjoyment that comes from eating good food; we know the importance of a healthy, well-fed nation; we wish you every success with your garden."[18]

The efforts of industries and companies received substantial technical support and publicity from the National Victory Garden Institute, which was formed in December 1942 as a nonprofit educational enterprise financed entirely by industry to promote victory gardening. Its advisory committee included leaders in the victory garden campaign as well as important figures in horticulture, business, and industry.[19] With regional offices in New York

City, Chicago, Detroit, and San Francisco, the institute helped to coordinate local, governmental, and industrial efforts. The institute served as a clearinghouse for information on procedures to set up victory garden organizations and to publicize them through conferences, radio, newspapers, newsletters, posters, contests, and awards. Its strategies included recognition of exemplary companies through an actively publicized awards program.[20] It also published *Gardengram,* a free monthly newsletter written specifically for company-employee garden programs. In addition, the institute lobbied in Washington to change ration policies in order to release metal for home pressure cookers and garden tools, chemicals for fertilizers and insecticides, and sugar for home canning, and it requested gasoline-ration exemptions for gardeners who had to travel to their community-garden plots. While the institute had minimal influence over these government policies, its campaign to increase participation was more effective.[21]

The National Victory Garden Institute encouraged companies to initiate garden programs not only as an act of patriotism but also as a way to improve employee morale and performance. Gardens improved company-employee relations and were credited with reducing absenteeism and increasing employee satisfaction. In an address at a National Victory Garden Conference, E. J. Condon, assistant to the president of Sears, Roebuck and Company, praised victory gardens for producing better, more reliable workers: "Company gardens make for excellent employee relations. Nerves in a war plant get pretty threadbare. The noise, the strain and the constant pressure make for short tempers. Take a couple of riveters off the assembly line. Put a hoe in their hands and put them into a garden patch. It's just plain good for a man's soul."[22]

VICTORY GARDENS TO ASSURE TOTAL HEALTH AND ENJOYMENT

Whereas much of the World War I promotion of gardening appealed to American selflessness in the face of the European food crisis, the World War II campaign often highlighted the tangible benefits to individuals and households who joined the victory garden campaign. For instance, M. L. Wilson, of the Federal Security Agency, equated war-preparedness with personal health for every man, woman, and child: "[One] cannot expect to be physically fit, mentally alert, and ready to 'take it' unless a well-balanced diet, including plenty of fruits and vegetables, has provided that energy and fuel which is necessary to keep in topnotch condition all of the time."[23]

With newspapers and bulletins popularizing recent scientific reports that found Americans had poor eating habits, and with the National Draft Board reporting that 40 percent of youth did not pass their physical examinations because of undernourishment, Americans were encouraged to increase their intake of minerals and vitamins by consuming fresh fruits and vegetables.[24] This campaign was intended to not only help the war effort but also change people's eating habits after the war.

Besides good food, civilians needed activities that would reinvigorate body and soul so they could keep up with the demands of the domestic war effort. Because gas and rubber restrictions limited the ability to travel to recreational outlets, Americans needed local leisure activities—and what could provide a more pleasurable and healthy outdoor experience than a victory garden? Various writers supported the idea of gardening as recreation by joking about the growing preference among office workers for the rake and hoe over the golf club and tennis racket.

In addition to nutritional health and physical exercise, gardening also promoted psychological health. Statements regarding the therapeutic nature of gardening by Frederick P. Moersch of the Mayo Clinic in Rochester, Minnesota, were frequently cited in promotional literature. He asserted that gardening eased the emotional unrest caused by war and uncertainty and that physical and mental health went together: "For the person who is on edge, anxious and sleepless, and has a heavy heart, there is no more hope-inspiring, restful, healthful recreation than gardening. One might speak properly of gardening as a 'work cure.' "[25] As a restorative hobby, gardening occupied the mind and body and thus relieved stress felt by families who had loved ones in the war. Illinois Food Director Lester Norris noted that "many parents with sons somewhere in foxholes in the South Pacific or North Africa have found solace by working in their gardens—close to nature—feeling that they too were contributing something personally to the effort to win the war."[26] Gardening also provided an outlet for patriotism as well as rehabilitation for those injured in the war. For example, concerning her work at Camp Kilmer in New Jersey, Mrs. Stephen Van Hoesen reported that men with mental and physical disabilities were frequently sent to the garden where they could casually talk about their problems while occupied in gardening activities.[27]

In general, victory gardening was promoted as a pleasant preoccupation rather than a burden. Not all memories from the World War I war garden campaign were positive, and many sources tried to avoid the overzealousness of the earlier era that had resulted in wasted seed, sore backs, and poor

Figure 45. Victory garden for patients at the Mather Field Station Hospital, 1943. Courtesy of the San Francisco History Center, San Francisco Public Library.

harvests and had left a bad taste for gardening. While growing food was important, so was the enjoyment of gardening as recreation and respite. Marvin Jones, War Food Administrator, characterized the symbiotic relationship of gardening to the war economy and personal needs in his speech before the National Victory Garden Conference of 1944:

> Working in a garden for an hour or two at the end of a busy day spent in an office or factory has provided a wonderful balance wheel to millions who have worked day after day at war jobs with little or no vacation or

recreation. Contact with the earth, and with growing things, is good for all of us, especially in times like these when we are all working so hard in the jobs assigned to us.[28]

ENGAGING AND INVOLVING EVERYONE

The victory garden campaign used many of the same outreach methods as past garden movements, as well as some new ones. The public learned about victory gardens through written materials, participatory events, and films. Books and pamphlets were augmented by handouts and bulletins, which proliferated due to the availability of mimeograph machines. Garden and home magazines along with local newspapers published advice columns and special interest stories. For example, *House and Garden* produced two victory garden supplements in 1942 and monthly calendars of garden activities and frequently published gardening stories, such as a description of the kitchen garden at Mount Vernon, pictures of people's victory gardens, and stories about children's gardens. Newspapers, chambers of commerce, and industries provided regionally specific garden advice in articles, books, bulletins, and advertising.[29] Posters with mottoes such as "Vegetables for Victory" and "Food for Freedom" were displayed in store windows, libraries, gardens, and homes to encourage support.

Potential gardeners were also encouraged through demonstration gardens, classes, and events. Boston Common and other popular parks and plazas were plowed for demonstration gardens that provided visual information and classes. Harvest shows were a popular forum for publicity and fundraising.[30] Contests and exhibitions were organized to spark local, regional, and national competition. Friendly competitions were proposed, such as the one between the governors of seven Southern states. An innovative approach, described in a 1946 issue of *Gardengram,* was a Victory Garden Special Train, sponsored by the National Garden Institute and the Chicago and North Western Railway, that toured Illinois, Iowa, Nebraska, South Dakota, Minnesota, and Wisconsin with exhibition cars on gardening and food preservation. Approximately twenty thousand people visited the exhibit.

The use of radio and film was new in garden campaigning. Short film clips on gardening, often produced by government agencies, businesses, universities, and colleges, were both educational and promotional. A notable one included a humorous lesson in proper gardening that featured comic Jimmy Durante and his straight-man son.[31] Several pamphlets pro-

Figure 46. This obviously posed photograph was included in an April 29, 1942, newspaper with the caption "Mrs. Francis Schwab of Culver City, wife of an aircraft worker, is one of the many who have joined the 'victory garden' drive. Like other gardeners, she has her share of problems. Her biggest, she says, is to keep down the bugs. She's shown with her first crop of radishes." Courtesy of the Herald Examiner Collection/Los Angeles Public Library.

duced by the Department of Agriculture and other institutions included lists of movies and how to acquire them for local screenings.

Observing the potential of community gardens to break down class or occupational distinctions and build friendships, garden advocates often lauded victory gardens as a democratizing experience that brought together people from all walks of life. A good example is the following description of a Pittsburgh garden in a guide to victory gardening:

> In one plot, for instance, there was to be found an unusually interesting cross section of American life, including seven doctors (one a woman), two dentists, several lawyers, a college professor, a research man from the Mellon Institute, the president of one of Pittsburgh's largest utility companies, a member of the Stock Exchange, a councilman, a man in the City Water Bureau, a street car conductor, a newspaper man, two Negro families, one Czech family, several mill workers, white collar workers, housewives, and garden club members.[32]

A different angle was taken in Los Angeles, where promotional literature often highlighted the opportunity to possibly rub elbows with movie stars and celebrities in a victory garden. The public was also encouraged by stories of victory gardens tended by the military stationed in the Pacific Rim, by Japanese-Americans in relocation camps, and by Native Americans on reservations. Pamphlets on victory gardening provided for Native Americans by the Bureau of Indian Affairs voiced the standard appeal to patriotism, health, self-reliance, and reduced food costs: "Indian families are hoping and praying that peace will come soon. They are hoping that their boys will soon be home, and when they come home, there must be feasting and gladness and good times. When our boys come home, there must be more good food than they ever had before. By this they will know that we at home have carried on while they were away at war."[33]

CHILDREN AS VICTORY GARDENERS

Once again, children were identified as an untapped resource for domestic food production. Children were encouraged to garden at school, home, and elsewhere through their churches, 4-H clubs, scouting programs, parks and recreation programs, and other local civic organizations.[34] For example, the Boy Scouts of America had a Food for Victory program to promote home and camp gardens, and to encourage youth to help farmers in harvesting crops, gathering fruit, and raising poultry and rabbits and other livestock. In some cities, such as Chicago and Detroit, park and recreation departments provided children's gardens in parks and playgrounds.

Although the national school garden movement had dwindled after World War I, some school districts and communities still included gardening as a curricular or extracurricular program. For example, the preexisting school garden program in Los Angeles developed its own Food Fights for Freedom program. According to an *Education for Victory* article, prior

Figure 47. This April 19, 1943, image from the *Herald Examiner* was captioned "Beverly Hills has turned to victory gardens in a big way, and many cinema stars are joining in the campaign. Marion Talley, right, opera star and a resident of Beverly Hills, is shown as she supervised a group of neighbors in a victory garden lot." Courtesy of the Herald Examiner Collection/Los Angeles Public Library.

to the war approximately two-thirds of the 190 elementary schools in Los Angeles had gardens.[35] As of 1942, 258 schools had gardens and 15,603 students were cultivating home gardens. Besides growing vegetables, there were animal-raising projects in thirty-six schools, and school programs supported over seven thousand animal-raising efforts in homes. One high school, probably Gardena Agricultural High School, was highlighted in the article for its many garden programs, including a model garden and animal-raising demonstration area at the school, over five thousand home gardens planted, and a school community garden. In addition, the school's public speaking department provided garden lectures, the art department made posters and exhibits, and the printing department produced informational sheets, while the agricultural classes and Future Farmers of America club provided technical assistance to other schools and to the community.

Figure 48. These young victory gardeners from Bunker Hill, Los Angeles, tend their garden at Sunue and Boyleston Streets. The garden was sponsored by the city's recreation department. *Herald Express* photo, May 29, 1945. Courtesy of the Herald Examiner Collection/Los Angeles Public Library.

Children were also encouraged to garden at home. *House and Garden* included articles to inspire children to plant home gardens and provided parents with suggestions for age-appropriate activities, garden design, and plant selection.[36] The Department of Agriculture's 1945 *Victory Garden Kit* included several items to promote children's and family gardens; one piece designed for radio advertising described the garden as a family training ground where a child "learned fascinating lessons about Nature and developed healthful ways of occupying his time that is [*sic*] denied kids who never had an opportunity to work and play with their elders in a family garden."[37]

The National Victory Garden Institute established a national MacArthur medal program to acknowledge children's achievements in their victory gardens with ribbons, certificates, and prizes. The national winner received a $500 war bond. To participate, children were required to keep records of their garden and have it visited by a local garden supervisor. The following message from General Douglas MacArthur was read at a luncheon during which youth groups were awarded General MacArthur medals for excellence in victory gardening:

> Please convey to the youth of America . . . my admiration for the magnificent spirit and tireless energy which characterized their efforts to increase our agricultural production to meet the pressing demands of war. Tell them that they have met the challenge to patriotism in the tradition that has carried our country forward in the world to its present exalted position. I am fully confident that as with the passing years the mantle of responsibility for guiding its future destiny falls upon their capable shoulders they will meet such tests with more determination and vigor.[38]

A PATCHWORK OF VICTORY GARDENS

The desire to encourage efficient gardens during wartime is encapsulated in the title of a 1942 article in the USDA's *Land Policy Review:* "Gardens, Yes, But with Discretion." Most promotional literature emphasized larger suburban home gardens and community gardens. For example, a 1943 pamphlet by the New York State College of Agriculture stated that backyard gardens in built-up areas like Manhattan were unduly expensive and inefficient while gardens in suburban homes or community gardens were more feasible. For urban residents who wanted a victory garden, experts urged that people participate in community gardens, vacant-lot gardens, school gardens, and company-run industrial gardens.

Like previous urban garden programs, the community garden was desirable because of its efficiency in layout, the availability of expert instruction, and the convenience of tool lending arrangements, as well as its social aspect. Community gardens were often located on larger pieces of property, often at the city's edge or on underutilized public land. The gardens, managed by municipalities, victory garden committees, colleges, or voluntary associations, were organized so that urban and suburban families could acquire a plot of land. Similar to programs in preceding decades, interested citizens applied for a plot and signed an agreement that they would maintain it. In some cases, there were fees for the use of land or water. Some gardens provided tools and supervision, while others did not. Along with the benefits, community gardens also posed some risks. Whereas earlier phases of community garden promotion had occasionally included descriptions of security measures, literature from the World War II era often acknowledges incidents of theft and vandalism as well as measures to assure security. For instance, in an effort to deter theft, Chicago officials set up fines of fifty dollars for trespassing and up to two hundred dollars for willful damage or theft in a community garden.

Figure 49. Victory gardening where lawn used to be in an apartment courtyard, Los Angeles, April 1942. Courtesy of the Herald Examiner Collection/Los Angeles Public Library.

Figure 50. Victory gardening at the Charles Schwab estate, New York City, June 1944. Photographed by Howard Hollem for the U.S. Office of War Information. Courtesy of the Library of Congress, photo no. LC-USW 3-42632-D.

Vacant-lot gardens were arranged on a case-by-case basis in a more informal manner than community gardens. Typically, a vacant-lot garden was the project of one or more households who claimed a lot near their homes for gardening. As had been the case in past garden movements, advocates validated both community gardens and vacant-lot gardens as means to beautify otherwise neglected land. At the 1943 National Victory Garden Conference, held in Chicago, Fred Heuchling, assistant director to Chicago's program, was eager to show conference participants the transformation of "ugly weed-infested, rock and brick-strewn" vacant lots through volunteer efforts into "neat, orderly, and productive gardens."[39] Not only were these gardens good for food production, but they also facilitated beautification. As Heuchling noted, "In practically every case the victory gardener adds flowers, shrubs or some other decorative plant material."

As in the urban garden programs of the past, the assumption was that any available land would be willingly donated. Between the limit on travel imposed by gas and rubber rationing and the hectic work schedules of war workers, community gardens needed to be close and preferably accessible by mass transit. To thwart possible misunderstandings, several sources suggested making legal contracts between the landowner and the gardeners or garden organization. A bulletin from the District of Columbia Victory Garden Office reminded gardeners that the land they were cultivating had been donated for the war effort.[40] In return for its use, gardeners were advised to maintain the property, haul away refuse, and not build any permanent structures on the land.

The Chicago victory garden campaign is a good illustration of the patchwork of victory garden options. As of November 1943, Chicago had 44,578 gardens, covering 848 acres, registered within its city limits.[41] In the suburban areas, registration totaled 32,609 gardens on about 1,750 acres. The figures did not distinguish between community gardens and backyard gardens, but the assistant director estimated that twenty thousand families gardened on community plots, including one community garden site that served over eight hundred families. The Chicago Parks Department and International Harvester Company collaborated in plowing 1,500 vacant lots for community gardens. Similar to the World War I effort, the parks district provided children's gardens in parks and play areas, amounting to over thirty thousand children's garden plots, each five feet by twelve feet. Demonstration gardens were installed at many of Chicago's larger parks and other public locations, including a demonstration hotbed and cold frame installed on Michigan Avenue next to the Chicago Art Institute. The Chicago Parks Department received a plaque from the National Victory Garden Institute in 1943 in recognition of its support for Chicago's exemplary victory garden program.

THE GARDEN PLOT: DESIGNED FOR NUTRITION, BEAUTY, AND RECREATION

To keep the garden manageable and to discourage waste, most sources advised gardeners to plan their gardens based on nutritional value, anticipated shortages of certain foods, family tastes, and how much time could be devoted to gardening. USDA materials advised gardeners not to destroy lawns or flowerbeds, but instead to start vegetable gardens in the corners or backs of their backyards. Community-garden plots varied in size according to site and demand, but here too gardeners were encouraged not to "bite off more than they can chew." With the aim of minimizing waste and maximizing nutritional output, the Department of Agriculture's 1943 *Victory Garden Leader's Handbook* provided guidelines for calculating a family's nutritional needs and planning a garden accordingly. Articles in *Horticulture* and *House and Garden* provided information on planting and suggested plant varieties for nutrition, taste, variety, and utility. Many manuals suggested planting strategies to get multiple crops, ways to extend the fresh vegetable season, and schemes to maximize yields for canning and preservation.[42] In order to encourage new eating habits, victory garden literature often included cooking advice, recipes, and storage tips. Promotional materials suggested a range of garden plans to fit a variety of household types, locations, and levels of gardening experience. Along these lines, *House and Garden* produced sample garden plans for a family of five with a maid, two families sharing a garden, and two single career women. Instead of being laborious, victory gardening was intended to fit into the busy lives of war workers and to be enjoyable. H. W. Hochbaum, chairman of the Department of Agriculture's Committee on Victory Gardens, considered an hour a day adequate for maintaining a backyard garden.[43] Similarly, advocates of community gardens encouraged gardeners to use plots convenient to their homes so they could tend them two or three times a week, either after work or on weekends. Figure 51, which shows children in a playpen while their mothers garden nearby, was meant to illustrate how gardening could be a pleasant and efficient addition to the daily routine.

In the 1943 edition of *The New Garden Encyclopedia,* E. L. D. Seymour listed the priority victory garden crops: beans, cabbage, swiss chard, lettuce, onions, peppers, rhubarb, root crops (beets, carrots, kohlrabi, parsnips, and turnips), spinach, and tomatoes. Crops to omit due to limited space, difficulty of cultivation, or ready commercial availability included cauliflower, celery, corn, peas, potatoes, and vine crops such as cucumbers, melons, squash, and pumpkins. While efficiency and maximizing yield were important considerations, the victory garden campaign did not exclude flowers and ornamental plants but instead encouraged a balance of vegetables and ornamental plants. Flowers were "spiritual food" for building morale, and growing them could be a form of recreation.[44] *House and Garden*'s two special editions devoted to victory gardens also included information on lawns and flowerbeds. Flowers not only brightened the home but could also be sent to local hospitals or to decorate United Service Organization (USO) buildings. Creative aesthetic mixtures of edible and ornamental plantings were also frequently mentioned in weekly victory garden sections

Figure 51. This 1943 photograph in a San Francisco newspaper included the caption "There's no absenteeism, because of worries over children, among women who are developing a community Victory Garden. . . ." Creative solutions to supervision were coupled with camaraderie among the mothers. Courtesy of the San Francisco History Center, San Francisco Public Library.

of newspapers. One *San Francisco Chronicle* issue, for instance, suggested growing chives as a border along a driveway and taking advantage of the foliage effect of rhubarb and artichokes.

RATIONING GARDENING SUPPLIES

At the 1943 National Victory Garden Conference, David Burpee of the W. Atlee Burpee Seed Company assured the public that while the demand for American seeds was at an all-time high, there would be sufficient seed for victory gardens in 1944. However, there were federally imposed restrictions on other supplies, such as fertilizers, especially synthetic nitrogen. To rectify this, the Department of Agriculture promoted the use of an official

Figure 52. This artful victory garden of cabbage and potatoes was at the rear of the Metropolitan Life Company, at Stockton and California Streets in San Francisco (1943). Courtesy of the San Francisco History Center, San Francisco Public Library.

"Victory Garden Fertilizer—for Food Production Only" that contained 3 percent nitrogen, 8 percent phosphoric acid, and 7 percent potash, to be sold at or below a ceiling price.[45] Besides synthetic fertilizers, gardeners were encouraged to use readily available organic amendments, such as manure, bone meal, and compost. Similarly, because insecticides and fungicides were available in limited amounts, manuals advised gardeners to respond to the first signs of damaging insects or disease. Restricted access to metals for new gardening tools meant that tools required good care. When it came to preserving garden produce, gardeners faced sugar rations and a limited availability of rubber seals for canning jars. Communities and industries opened canning kitchens or arranged for equipment to be available on loan to compensate for the limited production of pressure cookers.

POSTWAR APPEALS TO GARDEN FOR GARDENING'S SAKE

As the war progressed and victory seemed just around the corner, government and civilian groups urged the public to continue gardening as part of the postwar reconstruction. Home food production was still necessary, since it would take time to repair the damages of the war to European agricultural fields and manufacturing. At the 1945 National Victory Gardening Conference, sponsored by the War Food Administration, experts acknowledged the continued need for twenty million victory gardens. There was a note of urgency in the postwar appeals that had been largely absent from the victory garden campaign during the war. President Harry Truman was quoted as stressing the need for victory gardens after the war:

> The United States and other countries have moved food into war-torn countries in record amounts, but there has been a constantly widening gap between essential minimum needs and available supplies. The threat of starvation in many parts of the world and the urgent need for food from this country emphasize the importance of continued effort to add to our total food supply this year. A continuing program of gardening will be of a great benefit to our people.[46]

Even before the war ended, some advocates were looking to sustain interest in gardening as a hobby and as part of a balanced home life. In a 1942 *House and Garden* article, editor Richardson Wright pressed this vision of the long-term benefits of gardening for the American people: "Let us not, come peace, drop this effort to produce bodily and spiritual food, considering it merely an emergency measure. We can never go back to the old ways."[47] H. W. Hochbaum proposed that the victory garden campaign might change Americans' expectations for their homes and communities:

> We need to create a deep desire for beautiful home grounds, for less congested housing areas, for more sunshine and greenery, for less grimy, cheerless, sordid city areas. We need to build a love for gardening and horticulture in the fibre of our people so that our home grounds will be made more beautiful and livable, our communities improved, our beautiful countryside preserved, parks, playgrounds, and better housing developments provided for many of our city people now living in undesirable and unhealthy situations.[48]

At the 1945 National Victory Garden Conference, M. L. Wilson, director of the USDA cooperative extension service, justified the continuation of gardening for gardening's sake. First and foremost, he noted the health objectives met through gardening, including nutrition, exercise, and emotional restoration. Commenting that garden campaigning had stimulated a solid interest in gardening, he concluded, "Therefore, let us encourage the enthusiasm unloosed by war and direct it toward a greater appreciation of gardening for gardening's sake."[49] His suggestion was made into a recommendation of the conference. He argued that gardening should be considered in real estate development and general beautification around homes, businesses, highways, railroads, and parks in both urban and rural contexts. He recommended continuing the work of the Department of Agriculture in promoting rural gardening. Furthermore, he recommended the appointment of urban extension agents to serve urban gardeners even after the war. The conference concluded that victory gardens should be promoted in 1945, even if the war ended, and they set a goal of twenty million victory gardens for 1945. The conference's other recommendations included expanding support for children's school gardens, for home gardens, and for civic beautification efforts. To achieve these goals, committees recommended continued technical support from the Department of Agriculture and from the Office of Civilian Defense, among others. No specific mention was made of community gardens. However, as I note in chapter 7, the proposal to provide guidance for urban gardening was realized in 1976 with the development of the USDA Cooperative Extension Service Urban Garden Program.

As the war neared its end, some garden supporters scrambled to assure the continuation of a national garden campaign, yet only a few advocates encouraged the continuation of community gardens. Canadian H. Tuescher, curator of the Montreal Botanical Garden and Montreal's city parks, argued for the value of community gardens as part of the parks system in an article in *Parks and Recreation,* the journal of the American Association of Park Superintendents, serving both the United States and Canada. He promoted the idea of permanent community gardens at the outskirts of the city where urbanites could rent space to pursue their newfound interest in gardening. Providing public support for gardening as recreation expanded the role of parks officials. Furthermore, by "aiding in making a large part of the public more garden conscious, we reap the additional reward that our public parks will be better appreciated and, with public cooperation, will be made safer against thoughtlessness or willful vandalism." Tuescher proposed self-governed community gardens as a way to foster "tolerance, good will, cooperation and neighborliness and . . . therewith help to create better citizenship."[50] However, while gardening continued to be desirable, it was not so clear how people without their own land might continue. For example, as soon as the war seemed about to end, the District of Columbia Victory Garden chairman Franklin Mortin began to warn gardeners of the potential conflict between more relaxed postwar lifestyles that would allow more time for gardening and the removal of construction restrictions that would threaten the security of community garden sites. Although the demand for urban garden space was still high, the district's commission reportedly stated that "[victory] gardening has no place as a 'proper peacetime municipal function.' "[51] While some victory gardens did evolve into postwar community gardens, this was not the norm. On the other hand, increasing numbers of middle-class Americans were acquiring private backyards as large-scale suburban housing developments were on the rise.

The victory garden campaign is warmly remembered by many people today as a national effort that blended patriotism with personal motivation. Not only did gardening provide an outlet for citizens to aid in the war effort, but it also provided a local solution to recreation, health, and morale needs. The structure of the national program formed around the efficient dispersion of information. Instead of starting anew, the campaign made use of existing governmental agencies, institutions, and clubs to orchestrate local gardening efforts. Designed to address the immediate war crisis, the organization that supported victory gardening was not meant to be permanent. By getting millions of households to garden, however, the victory garden campaign did reinforce an interest in gardening. While most of the land that had been borrowed for community gardens was returned to its previous use or developed, the ideal of the home garden continued to influence the planning of housing developments in the postwar years.

PART III

Gardening for Community, 1945 to the Present

Introduction

AFTER THE NATIONAL EXPOSURE OF the victory garden campaign during the Second World War, gardening transitioned from a citizen's duty to a personal hobby and recreational outlet. The expansion of postwar suburban housing made it possible for more families to enjoy backyard gardening. For those without access to yards who still wanted to garden, a few remaining victory gardens and school gardens were available, although this was not the norm. Overall, attention to communal gardening alternatives had waned. Special-interest stories describing an urban garden project would occasionally appear in newspapers and magazines during the 1950s and 1960s, but in general, public attention was minimal.

This changed in the mid-1970s when renewed interest in urban gardening grew as an expression of self-reliance amid the energy crisis, rising food prices, and an emerging environmental ethic. In the early 1970s, stories were beginning to be heard of residents and garden activists transforming a New York vacant lot or a Seattle truck farm into a community garden. Momentum grew as these stories inspired people in other neighborhoods to start their own gardens. To accommodate the growing interest, citywide organizations formed to assist in the establishment of new gardens in other neighborhoods, schools, and institutions. Organizations such as New York City Green Guerillas, Seattle P-Patch, Boston Urban Gardeners, and Philadelphia Green provided volunteer and staff assistance in land acquisition, garden construction, education programs, and other activities. The United States Department of Agriculture also promoted urban gardening through the Urban Gardening Program, which ran from 1976 until 1993. With projects across the nation, the community garden movement has continued to gain strength to the present.

The term *community garden,* which had been used in earlier phases of urban garden promotion, took on a broader meaning in the 1970s. Previously, the term had referred to a type of garden site—a large property divided into individual gardens. As previous chapters have described, the community garden was frequently credited with important educational and social outcomes because it allowed for public demonstrations and promoted good-natured competition among gardeners. With the resurgence of urban gardens in the 1970s, the community garden was not only a type of garden, but also stood as an expression of grassroots activism. Faced with racial tension, a declining urban population, abandoned properties, and urban renewal projects that were tearing neighborhoods apart, local residents and activists sought to reclaim and rebuild communities and expand the open-space resources in their neighborhoods through gardening. Individuals could take personal steps to address inflation, the environment, and social anomie while also contributing to a neighborhood renaissance of sorts. While many of the same impulses—gardening for beautification, nutrition and health, income generation, and education—that had spurred efforts in the past still impelled people to garden on vacant lots and other underutilized land, these outcomes were now acknowledged as part of the social process inherent in negotiating communal garden space and procedures. The focus was on community—the community of gardeners who designed and maintained the garden, as well as the impact of the garden on the neighborhood, city, and larger society. Gardens added resiliency to deteriorated social and physical infrastructure through new social networks that worked toward physical and social reclamation.

In comparison to earlier phases of urban garden promotion, the community garden movement of the 1970s and 1980s revealed a shift toward more user involvement in planning and development. Once again, much of the support for local projects came from "experts" and was shaped by top-down policies. However, while in the past an outside organization—a philanthropic group, women's club, or civic group—had developed the garden, so that local gardeners simply had to apply for a plot when the garden was ready, community gardens of the 1970s and later tended to rely on local control and maintenance of the garden. Community leadership developed around gardens, not only through the interactions necessitated by daily maintenance, but also through the processes of community outreach, negotiating with city agencies for resources, and, potentially, fighting to protect a garden site from destruction. This shift is made evident by looking at a few of the transitional programs that existed between World War II and the advent of the community garden movement of the 1970s. These programs were still sustained by the motivations of earlier urban garden campaigns—such as gardening for food, civic beautification, and education. However, they also began to transition toward the self-management and local responsiveness that characterize contemporary garden programs.

THE FENWAY COMMUNITY GARDEN

In cities such as Washington, D.C., New York, and Boston, some victory gardens evolved into recreational programs after the war. One of these was the Fenway Victory Garden, established in 1943 along one edge of the Fens (part of Boston's "emerald necklace" park system), one of approximately fifty victory gardens in the Boston area during the war, many of which had been developed on park properties. Initially, the Parks and Recreation Department and the school board oversaw the victory garden, but in 1944 management of the garden shifted to a volunteer committee, the Fenway Garden Society. This group proposed to manage the site so they could continue to garden after the war. In the words of long-time gardener Richard Parker, "When peace was declared, and the need abated, the urge to garden had become so firmly embedded, and the demand for continuance so obvious, that the Boston Park Department agreed to it."[1] Over time, the garden continued to evolve and expand. Though development—including proposals to establish a hospital, a school, and, on three separate occasions, an automobile parking lot—has threatened the garden from time to time, newspaper and radio publicity have combined with political support at the city and state levels to save the garden.

The Fenway Community Garden still exists today. Walking through its rows of gardens is like visiting a garden exhibition. During the war and for a year or two afterward, the Parks Department plowed the complete area in spring, so that all materials had to be removed and the plots restaked annually. Today, each plot is maintained independently, so many of them include not only vegetables and flowers but also perennial plantings, small fruit trees, furniture, and statues that express the individuality of the gardeners and give the garden a feeling of permanence. Asphalt paths running through the garden allow park strollers and the nongardening public to look into the gardens. Extending over nine acres, the garden currently has a six-month waiting list of people who are eager to acquire one of the 450-square-foot plots for their own use.

THE NEIGHBORHOOD GARDEN ASSOCIATION OF PHILADELPHIA

Another example of a transitional garden program is the Neighborhood Garden Association of Philadelphia, which ran from 1953 to 1977. The impetus for the project came from Louise Bush-Brown, who had retired from her position as director of the School of Horticulture for Women at Ambler and was interested in addressing the visual conditions of Philadelphia's blighted neighborhoods.[2] Through meetings that included representatives from suburban garden clubs, settlement houses, community centers, public housing, and education, the Neighborhood Garden Association was formed with the mission to encourage and foster an interest in flowers and gardening in the congested areas of the city and awaken a sense of civic pride and fellowship through the organization of constructive garden projects for adult and youth groups.

The initial role of the Neighborhood Garden Association was to connect suburban garden clubs with local institutions in order to start beautification projects. In the first year, over 12,000 plants were brought into the city, 427 window boxes were installed and planted on seven blocks, and over 800 dooryard gardens were established at a housing project. These efforts evolved into the concept of Garden Blocks, in which partnerships were established between particular suburban garden clubs and neighborhood groups. To insure the partnership's longevity, the sponsoring garden club made a two-year commitment to provide planting materials and instruction, after which time the block was expected to take over responsibility. In addition, the Neighborhood Garden Association helped to establish 4-H clubs, a grants program, a demonstration garden, an annual recognition dinner, contests, and other resources. As of 1964, there were 385 Garden Blocks, 16 community gardens on vacant lots, 12 vacant-lot play areas, 1,200 dooryard gardens in two housing projects, and 297 individual 4-H club project gardens. The organization received national recognition through articles in newspapers and magazines, awards, and emulation in other cities. For instance, Lady Bird Johnson visited the demonstration garden as part of her beautification campaign. Summarizing the organization's work, its ten-year anniversary brochure stated:

> Where nothing but apathy had existed before neighbors have learned to work together to improve their community. Leaders have emerged in areas where no opportunities for leadership had existed before, and teen-agers have learned to work enthusiastically together for neighborhood betterment. And out of this upsurge of civic pride has grown a sense of individual dignity and worth. The touches of bloom and beauty brought into these blighted areas have fallen like dew upon ground long parched by drought and have stirred something in the hearts of people which might have remained forever dormant.

When Louise Bush-Brown died in 1971, the program lost its leadership and sense of purpose, and it eventually closed in 1977. However, the Neighborhood Garden Association had proved to be a pivotal program that acknowledged a growing concern about local participation and ownership of gardening projects. Its work influenced the development of the Pennsylvania Horticultural Society's Philadelphia Green program, which works with community groups. In 1986, the Neighborhood Garden Association, a community garden land trust, took the name in homage to the earlier program.

THE NEW YORK CITY HOUSING AUTHORITY TENANT GARDEN PROGRAM

Another project to receive public attention during the 1960s and early 1970s was the New York City Housing Authority Tenant Garden Program. The housing authority established the program in 1962 as a way to encourage residents of public housing projects to help keep up the appearance of their communities. The project started with 105 individual and group entrants who were each allocated a plot and given a twenty-five-dollar allowance for plants and seeds. The housing authority also offered educational programs and a manual on plants and design. The program grew so that by 1971, there were 283 entries, representing approximately three thousand of the six hundred thousand tenants. The focus was initially on flower gardens, but the program expanded to include vegetable gardens in 1974.

The program's success raised awareness of gardening's community-building capacity. When Charles Lewis, horticulturist for the Morton Arboretum in Illinois, was asked to be a juror in 1962 for a Tenant Garden Program contest, he came away from the experience impressed by the results. It was not so much the horticulture or garden design that interested him as the personal pride manifested in the gardens. Writing about the gardens in the early 1970s, he noted that while they certainly improved the aesthetic appearance of public housing, more important benefits were the development of social networks, the opportunity for self-expression, and other community improvement projects that the gardens inspired.[3] New York's program has inspired others. For example, the Chicago Housing Authority started a garden program in 1973, as well as a youth-garden program in 1977. In 1996, New York City Housing Authority staff estimated that approximately five thousand participants were involved in its various garden programs, tending around seven hundred gardens.[4]

SCHOOL GARDENS

Although there was no national appeal for school gardening in the period after World War II, some schools and districts continued to include gardens in their school curricula. Probably the two most acknowledged programs of this period, which served as inspiration and models for renewed interest in school gardens in the 1970s, were the Cleveland school garden program and the Brooklyn Botanical Garden's program. The Cleveland program had been established in 1904 and continued until the mid-1970s. The program included a centralized office of horticulture for the district, a greenhouse, twelve school gardens, and a summer program. While inspiring to garden activists, the program ended in the 1970s due to budget cuts. According to one report, the program's demise coincided with renewed interest in community gardens, so that school gardens at ten schools became community gardens.[5]

Advocates also looked to the Brooklyn Botanical Garden's teaching garden for inspiration. From 1914 to 1945, this program, under the direction of Ellen Eddy Shaw, was very structured, with time-tested procedures, standardized activities, and a system of certificates and awards.[6] During the winter, children attended Saturday classes. The garden included approximately two hundred plots. Younger children followed planting plans for plots, while older children were given more latitude in what they could plant and where. There was also a Boys' and Girls' Club, intended to inspire civic-mindedness through public speaking, practicing Robert's Rules of Order, community service work, and recognition of good work through the awarding of medals and trophies. In the late 1970s, the program was revised to reflect changing ideas about child development, which called for more independent exploration and experimentation. The Brooklyn Botanical Garden has continued to support children's gardens until the present. For instance, in 1993, it collaborated with the City Parks Foundation to develop a new program, Greenbridge, which works with local schools and helps to train teachers in school gardening.

These garden programs between World War II and the 1970s laid a foundation for the resurgence of interest in community gardening. They reveal continuing trends, such as interest in giving children exposure to gardening and using urban gardening to beautify low-income neighborhoods. As the Neighborhood Garden Association and Fenway Community Garden illustrate, they also show a shift toward more input by participants in garden development.

SEVEN

The Community Garden Movement of the 1970s and 1980s

ON JUNE 11, 1976, residents of Boston's Lower Roxbury and South End were probably surprised to see National Guard trucks delivering soil to vacant lots in their neighborhoods. That year, a state bill had made it possible to claim unused land for community gardens, and a group of activists and residents had gotten together to do just that.[1] Littered with trash, junked cars and remnant foundations from demolished buildings, the sites had to be cleared and new soil introduced. State representative Mel King and his aide, Judith Wagner, acquired three thousand cubic yards of topsoil from the Metropolitan District Commission, but then had to figure a way to transport it from twenty-five miles away. Local activist Charlotte Kahn negotiated with the National Guard to deliver the soil. That summer, six gardens were developed that, despite the late start, produced vegetables for gardeners' families to enjoy. The enthusiasm continued, and in 1977 garden activists and others formed a coalition, the Boston Urban Gardeners (BUG), that continues to promote community gardens and environmental education today. In a recent interview, Kahn, one of the founders of Boston Urban Gardeners, expressed her belief about the empowerment resulting from urban gardening in the 1970s: "For me, the gardens were a symbol of the opposite of what was going on—the possibility for a better city and a real centered community, an expression of people getting along together. In opposition to what was going on in Boston at the time, as in

Figure 53. Some people who pass Boston's Rutland/Washington Community Garden on the elevated rail line call it the Gazebo Garden. It was first established in the early 1970s, after the land had been cleared as part of urban renewal. The site is now owned and managed by the South End Lower Roxbury Open Space Land Trust. Photograph by author, 1992.

racial violence and divisions among people, and [a] top-down approach to urban renewal, the garden was an expression of what is best in people."[2]

The 1970s witnessed renewed interest in urban gardening. One of the first of this new crop of gardens was started in 1972 in Burlington, Vermont, when retiree B. H. "Tommy" Thompson campaigned to establish a forty-plot community garden. Within a year the garden had expanded to 540 plots, and eventually it became the basis for a membership organization called Gardens for All.[3] Over the next few years, neighbors and garden activists started community gardens in cities across the nation. In New York, the Green Guerillas were organized in 1973 to start gardens on vacant land. A year later, they received their first lease for a community garden from the city. On the other side of the country, a group of Seattle students and residents started a community garden at the old Picardo truck farm, which evolved into the P-Patch program that currently promotes gardens citywide. In between New York and Seattle, other projects got under way as well. In 1976, the federal government put its muscle behind urban gardens through the USDA Cooperative Extension Urban Garden Program, which provided gardening education and support in twenty-three major cities. Soon, garden organizations, which included nonprofit organizations, municipally run agencies, and cooperative extension programs, were actively promoting and facilitating community gardens nationwide, with names such as Philadelphia Green, Milwaukee Shoots 'N Roots, and San Francisco League of Urban Gardeners (SLUG). By 1978, with community gardens located in most major cities, gardeners and activists established the American Community Gardening Association (ACGA) as a national network and support organization.

For many readers, the evolution of the 1970s community garden movement described in this chapter will seem to have taken place just yesterday. Nevertheless, it is worth tracing this development as a history in its own right, one that forms the basis of the interest in urban garden programs and the organizations that support them today.

INSPIRATION TO GARDEN

In 1973, an estimated eighty million Americans were gardening as a hobby, and the Department of Agriculture sought to acknowledge this fact by choosing "Landscape for Living" as the theme for its 1973 *Yearbook of Agriculture.* Several articles credited World War II victory gardening as the impetus behind the increased interest in recreational gardening while also acknowledging that growing food at home was economical, aesthetically pleasing, and a satisfying form of family recreation. The *Yearbook* included "how to" advice for the home gardener as well as articles on "plants in action" that described civic garden projects, community gardens, school gardens, and other public, socially compelling gardening efforts. Author after author cited the health, recreational, and therapeutic benefits of gardening and proposed the community garden as a way to make these benefits available to people without backyards.

The impulse to garden, whether at home or in a community garden, grew out of multiple personal and social factors that coalesced in the 1970s. Gardening appeared to be a way to counteract rising food prices due to the oil embargo and other factors. Various publications by the Department of Agriculture and others estimated that a family could save between $200 and $300 by growing its own fruits and vegetables.[4] However, given that vegetables were a relatively small part of most families' household expenditure, this alone could not explain the popular interest in community gardening.

Increasing concerns about environmental conditions and dependencies on wasteful systems were also prompting people to take control of their immediate living environment through gardening.[5] Through exposés, such as Rachel Carson's 1962 book, *Silent Spring,* the public was becoming increasingly aware of the negative impact of agricultural technologies, pesticides, and herbicides on the environment. Equally influential was the growing public concern about the health consequences of pesticide residues on commercially produced foods. Rather than passively accepting these threats to their well-being or waiting for someone else to propose a solution, individuals began to take action at the local level, where many environmental problems were ultimately felt. To counteract a perceived loss of control over their daily lives, people looked for ways to break out of the consumer culture and be more self-reliant. The teach-ins and demonstrations of the first Earth Day, on April 22, 1970, were a national acknowledgment of the reality of environmental problems and the need for action at local, national, and global levels. Another means was gardening. A national poll published in 1976 found that 51 percent of all American families had vegetable gardens, and of these, 10 percent were in community gardens.[6]

Individuals and households seeking to reduce their impact on the environment turned to recently published books and to organizations that provided advice and technical information on home gardening as part of a more ecological, self-sufficient lifestyle. Based on their personal experience of transforming their Berkeley backyard into a food resource, biologists Helga and William Olkowski wrote *The City People's Book of Raising Food,* a "how to" book for both home gardeners and community gardeners. John Jeavons and his colleagues at Ecology Action in Palo Alto, California, conducted research on intensive food production methods that maximized nutritional and caloric value on minimal amounts of land. Experiments in ecological living, such as the Integral Urban House in Berkeley, California, and similar experiments in New York, Washington, D.C., and Minneapolis, explored comprehensive strategies that transformed a typical house into an ecological home through renewable energy sources, recycling, composting, and gardening and home food production.[7]

Living more benignly and taking some control over one's health added to the personal satisfactions of gardening long claimed by garden aficionados. To many people, the act of gardening felt good physically and emotionally. As a diversion or hobby, gardening relaxed people and helped to soothe the tensions inherent in busy lifestyles. While most gardeners could anecdotally claim that gardening was therapeutic and restorative, this connection between gardening and mental health received a boost from scientific study in the 1970s. Environmental psychologist Rachel Kaplan's 1973 study showed that gardening and viewing green spaces produced a restorative response. Including both community gardeners and home gardeners as subjects, her research found that people enjoyed gardening for the experiences it provided, such as working the soil, seeing things grow, learning new things about gardening, and being outside. In addition, gardens provided sustained diversion, aesthetic pleasure, an opportunity to relax, and a sense of accomplishment. The tangible results—the tilled earth, the food produced, beautiful flowers—also contributed to the satisfaction and restoration that people found in gardening. Time in the garden allowed a person to be immersed in nature and natural processes. As Kaplan stated, "The garden is a miniature, a slice of nature compressed in space and a pattern of information compressed in time. Rarely is so broad a spectrum of nature and natural processes found in so little area. Rarely are so much nature-based action and so full a view of the life cycle so vividly visible and so rapidly completed."[8]

Gardening's ability to heal the individual could be extrapolated to community benefits as well. Charles Lewis, horticulturalist for the Morton Arboretum outside of Chicago, reflected on gardening's capacity to soothe social tensions. As a regular advisor and juror for the New York City Housing Authority Tenant Garden Contest, he had witnessed the positive response of residents who participated in the garden:

> Plants are alive and are dependent on the gardener for care, if they are to survive. In a world of constant judgment, plants are non-threatening and non-discriminating. They respond to the care that is given them, not to the race or the intellectual or physical capacity of the gardener. It doesn't matter if one is black or white, has been to kindergarten or college, is poor or wealthy, healthy or handicapped, plants will grow if one gives them proper care. They provide a benevolent setting in which a person can take first steps toward confidence.[9]

Many viewed gardening in a similar way—not as a chore but as an satisfying recreational activity. A 1979 poll reported that gardening was the eighth favorite leisure activity, while a 1982 poll found vegetable gardening was the top outdoor leisure activity.[10] The same poll noted that eighteen million household respondents did not garden because they lacked suitable space. For these people, joining a community garden allowed access to the multiple personal and social benefits associated with gardening. Thus personal incentives coalesced with social agendas in the community garden.

GARDENING AS URBAN ACTIVISM

In the 1970s, some activists and gardeners used their energy and their faith in gardening as a healing activity to counteract the failing conditions of the city. While a seemingly small gesture, gardening facilitated community unity and transformed a detriment—the vacant lot—into a resource. Social conditions were ripe for this kind of grassroots action.

The urban decline that was so noticeable in the 1970s had been long in the making. The boom in suburban development since World War II had shifted much of the population to new communities that encircled central cities. Jobs had also shifted, in both their training requirements and their location. While there was an increase in skilled jobs, such positions were available primarily in suburban locations. Meanwhile, cities suffered from a loss of manufacturing and other blue-collar employment opportunities. The city was becoming poorer as those with the means to do so left the city for the suburbs and peripheral towns. Segregation policies, discrimination, and the central location of public housing and other services effectively trapped low-income and minority groups in the inner city. For instance, the African American population in cities had been growing since World War II. Unfavorable social conditions in the South and changes in agricultural technology that made much farm labor obsolete had made rural agricultural livelihoods difficult, and while the pull of economic opportunities had brought many African Americans to cities in the North, segregation policies effectively limited where they could live. By the late 1950s, nearly half of poor Americans lived in cities, and most were African American. With economic and physical development occurring in the suburbs and the urban population growing poorer, cities began to show the physical manifestations of neglect: tax-defaulted properties, deserted buildings, and arson. Frustrated residents clamored for some sort of municipal and federal response.

To many planners, politicians, and others, the solution was urban renewal through clearing blighted areas and developing them anew.[11] In the quest for comprehensively planned services, housing, schools, and transportation, large swaths of urban land were cleared of the houses, apartments, stores, churches, and businesses that had made up the neighborhood fabric. Housing was lost and people were displaced. And while models and drawings of newly designed and planned neighborhoods showed promising future communities, the development process often took longer than five years. In the meantime, land lay fallow, and the remaining urban residents were left with even fewer resources at their disposal.

Financially strapped municipalities struggled to maintain the abandoned buildings and acres of cleared land in the inner city. Desperate for someone to take over maintenance and hopefully return the land to the tax rolls, state and municipal officials developed rent-a-lot programs, squatter programs, and other incentives. For example, to address St. Louis's six thousand abandoned buildings and lots, the state of Missouri authorized a Land Reutilization Authority to oversee their redevelopment. The authority developed an adopt-a-lot program and leasing programs to maintain the properties and promote private development of housing, jobs, and open space. Similar programs were set up in New York City, Boston, and elsewhere, usually with the hope that the sites would eventually be reclaimed for development.

Confronted with poor environmental and social resources, some urban dwellers began to rethink their living environment in light of greater self-reliance at both the individual and neighborhood levels. While some retreated to a simpler, quieter life in suburbs, rural areas, or small towns, others could not or would not forsake the city and instead sought ways to make the city better.[12] In the spirit of self-help, some urbanites sought a less wasteful and more socially responsible lifestyle through such activities as gardening, recycling, and composting, and sought to take control of local conditions. Gardening was an immediate community volunteer activity that could claim deserted land and transform it into something useful. Local residents could start gardens without making additional demands on public agencies that were busily trying to address larger issues such as better park maintenance and infrastructure improvements. Community gardens not only provided new opportunities for social interaction, recreation, and access to food; they also showed resistance to the deterioration of the city and produced an activated citizenry.

While taking over a vacant lot and planting a garden might have seemed a small gesture in a neighborhood dealing with crime, drugs, and large-scale disinvestment, many considered it a pivotal first step towards community revitalization. In both literature from the period and later interviews, garden activists expressed the belief that people who participated in gardening later became involved in other neighborhood improvement projects, such as cleaning streets, repairing houses, and neighborhood-watch programs.[13] Their success in changing a vacant lot into a garden empowered them to address larger social, environmental, and economic inequalities. Summarizing this optimistic outlook, in its 1973 book on community gardening, Gardens for All promoted a "garden way of living" as a plan "for a greener, happier world, based on the proposition that the more people who have large fruit and flower gardens the better they, their community, their nation, and the world will be able to solve environmental, economic, and social problems of our time."[14]

FROM VACANT LOT TO GARDEN, AND THEN SOME

While community gardens surfaced in all contexts in the 1970s—cities, suburbs, and rural areas—community garden activism in its initial stage was best known for revitalizing derelict urban land into usable open space. Stories of garden projects in industrial cities in the early '70s describe an overwhelming number of available vacant lots that required extensive cleanup and restoration. Sometimes an energized group would just start gardening on a vacant lot; however, in most cases they sought formal permission from the city or landowner before beginning. The most common arrangement was a yearly lease for a nominal sum, such as one dollar per year, plus proof of insurance.

The typical account of an urban community garden's birth begins by describing neighborhood volunteers and garden activists clearing out a site full of garbage, abandoned cars, and debris. Blacktop and foundations might have to be cleared and topsoil reintroduced to the degraded and often sterile earth. Then they would start to cultivate the soil. Gardeners made use of any opportunity that presented itself. A little help meant a lot, whether it was having city garbage trucks pick up the collected debris or obtaining free manure from a local stable. City agencies and local institutions could help the process through in-kind assistance, such as the use of machinery or donation of excess supplies. In this spirit of opportunism, Jamie Jobb wrote in his 1979 guide to community gardening, "just about everything a neighborhood garden needs may be donated to the project or found nearby."[15]

As a result, most community gardens did not look like a typical park or backyard. Generally gardens were planned to provide both individual garden plots and communal areas for picnic tables, children's play areas, and communal herb beds and orchards. "Making do" often resulted in creative applications of reused materials. A certain aesthetic emerged from the practice of using recycled materials—bathtubs and barrels as planting containers, forklift pallets for fencing and compost bins, bed springs as garden gates, wire spools as tables, and tires for retaining walls. When confronted with a blank wall adjacent to a garden, local artists and children often painted murals.

Gardeners had to adjust cultivation practices to suit urban conditions. The lack of vegetation—even weeds—on some vacant lots led to concerns about the possibility of toxins in the soil. While many people were growing food as a way to access healthy, fresh produce, others were raising the concern that growing in proximity to roads might add pollutants, especially lead, to the food. A 1975 article entitled "Poisoned Gardens," in the alternative newsletter *The Elements,* alarmed its readers by exploring the health risks associated with urban garden produce. It reported that one air pollution specialist said she would not eat produce raised in an urban garden.[16] In response to the rising concern, organizations such as the Boston Urban Gardeners and the Institute for Local Self-Reliance in Washington, D.C., began to test soils, conduct research, and disseminate information on appropriate garden practices. Studies showed that while some contamination was airborne, the main source of lead in the soil was lead-based paint, present in chips and dust from old buildings. Gardeners were advised to keep the soil pH between 6.5 and 7.0 and to increase organic matter in the soil in order to reduce lead intake, to cultivate vegetables grown for their fruiting parts (eggplants, tomatoes, peppers, squashes, etc.) rather than their leaves, and to wash vegetables prior to consumption.

Less potentially harmful but still of great concern to many gardeners was the threat of vandalism and theft. Reports of gardeners' efforts being destroyed by local youth or picked over by mysterious night marauders led to defensive measures such as fences and locked gates. Some groups set up a schedule to maximize the presence of gardeners at the site. To thwart temptation, gardeners were advised to plant tantalizing crops such as pumpkins, strawberries, and tomatoes out of view of the street. Another common approach was to encourage wide participation, particularly among youth and children in the neighborhood. Charles Lewis, curious as to why gardens in the New York City Housing Authority Tenant Garden Contest were spared the vandalism evident in the community at large, was told by residents that they knew who the troublemakers were and made sure they got involved in the gardening program, often as guards.[17] The local police reported anecdotally that the presence of gardens reduced vandalism both outside and inside the housing complex.

Building the garden together was only the first of many cooperative ac-

Figure 54. The Southwark/Queen Village Community Garden in Philadelphia was established in the mid-1970s on a site cleared for urban renewal. After several crises in which the garden was almost lost, it was assured more security as a demonstration garden of the Penn State Urban Garden Program, and the site was ultimately purchased by the city. Photograph by author, March 1999.

tivities inherent in community gardening. Gardens were usually self-managed, so the participants needed to work together to address ongoing matters such as plot assignment, paying for water and insurance, and making sure the site and the tools were in good condition. In some cases, everyone wanted to be involved, which meant regular meetings to negotiate and assign tasks. In other cases, gardeners just wanted a place to garden and were happy to defer to a coordinator or a committee. No matter how the garden was managed, communication was needed between the gardeners. Given that gardeners might visit their plots at different times of the day and week, common solutions to communication needs included mailboxes at each plot, bulletin boards, and newsletters.

Policies on fees, maintenance standards for individual plots, care of communal spaces, acceptable gardening practices, and codes of behavior had to be established. Most programs collected fees for plots—usually between ten and twenty dollars per year—to cover insurance, water, and shared materials. Most neighborhood community gardens were structured

Figure 55. The Southwark/Queen Village Community Garden transforms with the seasons. In winter, the site looks a little disorganized, with ad hoc plot edging and various containers used for water catchment. In spring, however, foliage and flowers take over. Adjacent to the garden is a colorful mural done with broken pottery, mirrors, and paint. Photograph by author, June 1999.

so that the gardener had his or her own plot and full control over what was grown there and how it was used. Some groups established policies that required organic gardening methods in all plots, while others took a more laissez-faire approach to garden practices. Most groups established rules on the maintenance of shared paths, watering, and mandatory plot cleanup after the final fall harvest. Because many gardens had waiting lists of prospective gardeners, most groups also developed procedures to reassign neglected plots after a grace period had passed.

A handful of books based on the experiences of successful garden projects were published in the 1970s and 1980s to pass on practical advice to people interested in starting a new community garden. Written for the layperson, these books provided information on plants and cultivation as well as organizational suggestions for acquiring land, coordinating volunteers, and promoting community involvement. The first such book, *Gardens for All,* published by the group of the same name in 1973, highlighted the successes of a number of programs. Mary Coe's 1978 book, *Growing with Community Gardening,* included advice on site selection, the duties of the garden coordinator, site planning, maintenance, and communication. Writing as someone who first tried rural communal living and then returned to the city, Coe reinforced gardening's value as a means for greater self-sufficiency and environmental, mental, and social health. The book also included very positive profiles of various garden programs and was intended to encourage others to initiate similar projects. *A Handbook of Community Gardening,* published by Boston Urban Gardeners in 1984, filled an important need by providing the layperson with a step-by-step guide for organizing a garden, from planning the first meeting to registering as a nonprofit organization.[18]

ORGANIZING CITYWIDE

While many community gardens started as very local volunteer efforts, they often had a wider influence by serving as an inspiration for other neighborhoods. Occasionally, groups organized into coalitions and nonprofit organizations to advocate for gardens across the city—outstanding examples would be the Boston Urban Gardeners and New York City's Green Guerillas. In other cases, community gardening was added as a program to park and recreation departments or other municipal agencies. For example, the Parks and Recreation Department of San Jose, California, established a community garden program in 1975. The first garden, Mi Tierra, consisted of two hundred plots on five acres.[19] Schools, universities, churches, garden clubs, service organizations, and other institutions also developed or promoted community gardens. Where a community had more than one organization involved in garden promotion, either collaborative or competitive relationships ensued.

Gardening organizations provided technical support and helped to streamline neighborhood organizing, site and materials acquisitions, and ongoing maintenance policies. Through good working relationships with various city agencies, these organizations could help cut through the red tape for the individual garden groups. Instead of each garden having to scramble for in-kind contributions and volunteers, the citywide organizations could coordinate distribution of surplus equipment, manure deliveries, garbage pick-ups, volunteer networks, and so on. They could also orchestrate citywide garden events and celebrations, garden tours, and other forms of outreach.

Over time, many community garden organizations broadened their scope to include environmental education, community art programs, neighborhood planning, and more. They promoted gardens in settings such as schools, senior centers, and hospitals. They targeted underserved neighborhoods and specific populations such as at-risk youth, seniors, and the disabled. To encourage participation by people of different ages, socioeconomic classes, and ethnicities, organizations sponsored educational and social events and sought collaborations with other organizations that worked with special groups. Some programs also evolved to support not just community gardens but also pocket parks, streetscape improvements, playgrounds, and other usages of open space.

To facilitate such a range of activities, garden organizations needed reliable funding and paid staff. Programs run through municipal agencies or larger institutions often had an easier time securing funds and materials. Smaller nonprofit garden organizations developed creative, garden-appropriate approaches to fundraising, such as raffles, bake sales, sales of cookbooks and garden calendars, or the staging of special events. Many garden projects investigated other funding sources, including in-kind donations, private donations, and enterprises that would generate income. Various sources of federal funding were also available to community garden organizations. Quite a few organizations relied on federal funds from the Comprehensive Employment and Training Act (CETA) to pay salaries to employees. When CETA ended in 1979, they had to scramble for other sources. The Community Development Block Grant Program (CDBG) established by Congress in 1974 provided federal funds for neighborhood facilities and public services in low- and moderate-income communities. Energetic young staff were often brought in through local, educational, and national service programs, such as Volunteers in Service to America (VISTA).[20]

SUPPORT FROM THE USDA: THE COOPERATIVE EXTENSION URBAN GARDEN PROGRAM

Urban garden promotion received a big boost in 1976 when the federal government announced the Urban Garden Program, to be administered by the Department of Agriculture's cooperative extension service. The program was initiated by Congressmen Frederick Richmond, a Democrat from Brooklyn, New York, who wanted to secure funding support for existing community gardens, and James Whitten, a Democrat from Mississippi and chairman of the Subcommittee on Rural Development, Agriculture, and Related Agencies and soon-to-be chairman of the House Appropriations Committee, who saw this program as an opportunity to woo an urban constituency for the Department of Agriculture.[21] In his testimony before Chairman Whitten's subcommittee, Congressman Richmond cited many reasons for urban gardens, including the need to reconnect urban citizens with agriculture, the increasing interest in gardening, the lack of information and resources available to urban residents, the opportunity to encourage home production and home economic practices, and the possibility of offering jobs to urban youth. Initially, $1.5 million was provided to establish urban offices and set up garden projects in six cities—Chicago, Detroit, Houston, Los Angeles, New York, and Philadelphia. In 1978, funding increased to $3 million and ten new cities were added: Atlanta, Baltimore, Boston, Cleveland, Jacksonville, Memphis, Milwaukee, Newark, New Orleans, and St. Louis. The program selected the largest cities in the nation, with only one per state and excluding the District of Columbia. In 1985, the program grew to twenty-one cities with a budget of $3.3 million, and by 1988 it had grown to twenty-three cities. Results in terms of participation and estimated value of the food produced for 1979 and 1985 are provided in table 9.

According to federal guidelines, program money was to be used "for employing people having general qualifications of Extension Service agents to assist in teaching and demonstrating gardening and 4-H type work, as well as nutrition assistance in our large cities."[22] Federal money could only be used for educational programs, to organize new programs, and to teach gardening and food preservation techniques. Except for demonstration gardens, the federally sponsored offices could not provide community gardens. Materials such as seeds, fertilizer, and topsoil had to be acquired through local sources or in-kind donations. Beyond the funding limitations, however, the federal mandate allowed for a variety of approaches, depending on the local situation and the priorities of the state USDA cooperative extension service office. In some cases, a staff person devoted to urban gardening was added to an existing office, while in others a new department in a new urban office was established. In cities where garden programs already existed, such as New York and Boston, the extension service provided technical and educational support. While the Baltimore program coordinated garden education with a farmers' market program, Philadelphia's Penn State Urban Garden Program established demonstration gardens. A common service was to disseminate findings from the extension service's

TABLE 9

Results of the USDA Cooperative Extension Urban Garden Program, 1979 and 1985

	1979		*1985*	
	Total Participants	Estimated Value	Total Participants	Estimated Value
FIRST SIX CITIES				
Chicago	1,767	$193,890	2,531	$275,000
Detroit	6,995	484,500	1,145	172,298
Houston	1,250	192,275	2,793	850,000
Los Angeles	2,160	77,405	15,175	2,893,414
New York	27,982	300,000	33,752	1,863,975
Philadelphia	4,205	337,100	82,996	2,238,712
TEN ADDED IN 1978				
Atlanta	922	28,360	14,331	498,223
Baltimore	4,551	31,045	6,000	723,630
Boston	16,468	227,615	9,700	2,250,000
Cleveland	14,389	360,000	17,225	1,907,080
Jacksonville	289	11,400	5,543	1,330,320
Memphis	1,048	194,000	3,192	1,408,077
Milwaukee	3,357	115,710	5,639	647,000
Newark	1,261	16,930	3,200	461,268
New Orleans	487	12,155	1,201	151,326
St. Louis	1,107	53,000	2,530	271,866
FIVE ADDED IN 1985				
Bridgeport	—	—	1,707	135,500
Denver	—	—	12,500	2,455,389
Indianapolis	—	—	891	1,660
Phoenix	—	—	426	27,690
Seattle	—	—	300	2,205
TOTALS	88,238	$2,635,385	222,777	$20,564,633

SOURCES: Letter from C. S. Oliver, assistant dean and assistant director, to extension directors in urban gardening states, May 9, 1979. Letter from Mary Nell Greenwood, administrator, to state extension directors in the Urban Gardening Program, December 13, 1985. Both sources from archived materials at Bancroft Library, University of California.

research departments to the urban public, through flyers, pamphlets, and classes.

In Los Angeles, the urban garden program was called LA Common Ground. The Los Angeles County University of California Cooperative Extension provided information on local growing conditions and connected the new program with the community outreach already occurring through the federal Expanded Food and Nutrition Program. Common Ground provided services to low-income families and communities in Los Angeles, Lynwood, Compton, Inglewood, and Long Beach. It facilitated community gardens in neighborhoods, schools, and homes, and coordinated volunteer networks, including master gardeners, master composters, master food preservers, and Gardening Angels, who worked with school garden programs. To address Los Angeles's multiethnic population, Common Ground held workshops and produced materials in English, Spanish, and other languages as appropriate. Education topics extended beyond gardening to related issues such as nutrition, conflict resolution, and leadership management. In the early 1990s, extension service staff and supporters developed a nonprofit organization called LA Harvest to supplement their work.[23] Between 1977 and 1992, Common Ground helped to start over one hundred community gardens, two-thirds of which were operational for over four years, six of them for over ten years. Over 25,000 people participated in the workshops, school programs, and gardens.

By 1980, the federal Urban Garden Program had served nearly 200,000 urban residents, including approximately 65,000 youth. In 1982 alone, an estimated $17 million worth of food was produced. While some hoped to expand the program to all fifty states, political support for this was weak. A ceiling of $3.6 million was placed on the budget even as the number of cities increased, which meant reduced staff and strained program resources. The program's funding was a line item on the Department of Agriculture budget, yet department officials repeatedly omitted it from their budget requests and Congressman Whitten found that he had to keep reinstating the money. In 1992, Whitten relinquished leadership of the House Appropriations Committee and the Agricultural Subcommittee. In 1993, the Urban Garden Program was removed as a line item from the Department of Agriculture's budget, with the rationalization that the program would continue through allocations from the cooperative extension's general funds. Each city dealt with the change differently, with some programs ending and others downsizing.

Besides the urban garden programs in selected cities, some state extension service directors supported community gardens in other ways. For example, in California, the cooperative extension provided a home and a community-garden public service program in most counties. A 1976 survey of thirty-six counties revealed ninety-four community gardens that served 1,533 individuals and 1,625 families. The programs were staffed by extension services, or through funds from CETA or the county.[24] Many cooperative extension offices also assisted urban gardening through a master gardener program. This scheme was started in 1973 in the state of Washington and eventually expanded to other states. Participants were volunteers who received thirty to sixty hours of intensive horticultural training. They then volunteered from ten to two hundred hours in extension-related activities such as tending educational booths at festivals and fairs, setting up demonstration gardens, speaking to garden clubs and classes, answering phone questions at the extension service office, and volunteering at community gardens.

STAYING CONNECTED TO THE COMMUNITY

While community gardens were arguably recreational and open space activities, they differed fundamentally from other kinds of parks or recreation programs because of the participants' active role. Citywide organizations could promote and support a garden, but local involvement was essential for its survival. Hard lessons were learned by city agencies and others who jumped on the community garden bandwagon without local support or commitment to follow through on maintenance. Likewise, advocates voiced concern about the heavy-handedness of some public agencies and bureaucracies that formed around community gardening. "Handed-over" gardens, although developed with the best intentions, were often abandoned because the communities were not involved in their development.

In New York, for example, the Housing, Preservation, and Development Agency established a garden program that built ninety-six "interim" gardens between 1976 and 1982 at a cost of $3.6 million. Sites were designed and built by agency staff without citizen involvement, and it was assumed that community members would maintain the sites once the work was complete. In the end, the program was considered a failure. It was eventually revised to make better use of existing organizations with neighborhood connections and to encourage sustained participation. Another well-intentioned but ultimately inefficient effort was the Revival program in Boston, established in 1975. With the intention of expanding the existing

Park and Recreation Department's victory garden program, the number of gardens grew from fourteen to over twice that number in two years. Although the program provided money for grading, irrigation, fencing, and topsoil, the lack of community input on design, plus poor supervision of the contractors, often left the garden sites in worse condition than before.

Learning from these and similar experiences, many garden organizations set up procedures to encourage community participation and build leadership as part of the process of garden development. While directed by a mission to promote new gardens in under-served areas, many garden organizations focused as much on community development as on garden development. Typically, they required that the resident group reach a certain level of internal organization and be able to define its needs before proceeding with the garden. The sponsoring organization might provide materials and possibly trained design and work crews, but control and decision-making rested ultimately with the community group. Tessa Huxley, former director of Green Guerillas, was adamant that the gardeners should be responsible for design and development:

> The important point to remember is to get gardeners involved in creating their own space with their own hands. For many people this is the first opportunity they have ever had to control the creation and development of a project. Our job as technical assistance coordinators is to assist them in doing what they have dreamed about. Community gardeners should build their own gardens because this building process will give them the skills to maintain and repair their garden fixtures.[25]

THE GARDEN ACTIVIST

Many staff members in the garden organizations of the early 1970s and 1980s came to community gardening through horticulture. Yet while the work required gardening skills, it also called for an ability to work closely with diverse groups in the physical and social context of urban neighborhoods. For example, Sally McCabe, trained in agriculture and a staff member of the Penn State Urban Garden Program from 1977 to 1985, soon realized that, although she was employed to teach gardening techniques, this was secondary to social activities and organizing: "At the beginning we spent a whole week preparing; we thought we'd save the world by teaching people to garden. But then we learned that if we'd listen we'd learn how to do this. Gardening was the means, not the end."[26]

Other garden organization staff became involved through community activism or education. For example, Charlotte Kahn, cofounder of Boston Urban Gardeners, discovered that gardening was a useful means to help Boston youth caught in the middle of a political and racially charged new school busing program. "I had come to Boston with a social goal. I hadn't planned on getting involved in gardening. I was thinking of it as a hobby, but the kids were so attracted to the garden. Faced with all the violence, it just seemed good to have more gardens."[27] Terry Mushovic, former director of the Penn State Urban Garden Program, reported that getting residents to see the potential in their neighborhood vacant lots was only the first step in a process of neighborhood renewal:

> Most people had no vision that it was possible to take these vacant lots and do anything with them. And so it was an interesting time to convince people that it was possible. Then as I saw the commitment on the side of the different community people, and what it did to bring them together, then I could see the vision of where it could go. It was very personally satisfying to work with individuals who were up against tremendous odds and still believed they could make a difference in their neighborhood.[28]

While some of the early organizers have moved on to other careers, many are still active in garden organizations. Several of these longtime community garden staff members comment that working for a garden organization is more than a job; it is a passion and a ministry. Days might be spent rounding up materials and working with city officials, while evenings and weekends are the times to meet with community residents. Frustration over finding funds and securing gardens is more than counteracted by the rich personal relationships they develop with the gardeners.

THE AMERICAN COMMUNITY GARDENING ASSOCIATION

To facilitate an exchange of information among community garden organizers around the country, the Chicago Department of Human Services convened the first national conference of garden organizers in 1978. Approximately 150 organizers and activists attended, including staff from the USDA cooperative extension urban garden program in various cities. It was at this first conference that forming a national community garden organization was suggested, and a small group of organizers volunteered to develop the idea.[29] At a second conference in 1979, again hosted by Chicago, the American Community Gardening Association (ACGA) was officially founded. The mission of the organization was to promote community gardening, to provide opportunities for networking between garden organizations, to develop an information clearinghouse, and to help establish new programs. The forty-nine charter members from twenty-two states represented garden organizations, cooperative extension service programs, public agencies, horticultural societies, and universities. By 1983, the ACGA had 150 members—some organizations and some individuals—comprising a network of 1,200 community garden organizers who in turn represented thousands of gardeners.

One of the early projects of the ACGA was to publicize community gardens. In 1983, the ACGA partnered with the Glad Bag Company to establish a national community garden contest. Through this arrangement, the ACGA was able to bring in revenue to cover some staff and publishing costs while also increasing public awareness of community gardening through Glad Bag's public relations department. Contest categories and prizes changed over time. In the first year, the contest offered $20,000 in cash prizes, including a grand prize of $1,500. The nation was divided into eight regions to address climatic considerations and coordinate site visits by members of the ACGA board, who served as judges. Prizes were awarded in each region in the categories of large site, small site, and new site. The grand prize winner was the Edgement Solar Garden in Dayton, Ohio. The next year the contest was expanded to include twelve categories, with both regional and national winners. In 1986—the final year of the contest, due to a corporate takeover of the Glad Bag Company—there were 678 entries that represented all states plus the District of Columbia. Table 10 shows the results of the national competition for 1985 and 1986. Though short-lived, the contest raised public awareness of community gardening through media coverage in newspapers, on radio, and in television spots. It also provided the ACGA an opportunity to profile the types of gardens entered in the contest, most of which were urban and on public land.

Through its annual conferences, the ACGA has for over twenty-five years provided a means for garden activists across the country to exchange information and establish relationships. Conferences' presentation topics are wide-ranging. The 1979 conference, for example, included presentations on garden design, site preparation and maintenance, children's gardens, food delivery systems, horticultural therapy, comprehensive citywide community garden programs, container gardens, and other topics. The ACGA also started a quarterly journal in 1982 that continues to be a major information resource. Based on requests voiced at the annual conferences, the first four

TABLE 10

National Winners of the ACGA/Glad Bag Community Garden Contests

Categories	*1985*	*1986*
Large sites	Fort Mason Community Garden, San Francisco, CA	Manhattan Community Garden, Manhattan, KS
Small sites	Pleasant Village Community Garden, New York, NY	Pinehurst P-Patch, Seattle, WA
New sites	Gardening Opportunities, Greenville, KY	Mt. Auburn Food Park, Cincinnati, OH
Special populations	Hampton Victory Garden, Hampton, NH	Cheyenne Community Botanical Garden, CO
Food for others category	—	Community Service Harvest Garden, Niles, MI

SOURCE: Files of the ACGA, Philadelphia Green, Pennsylvania Horticultural Society.

issues addressed funding, land acquisition, special population groups, and community development. The journal often includes profiles of programs and the latest tips with the aim of enthusing readers about the progress of the community garden movement.

EVOLUTION AND EXPANSION INTO OTHER PROGRAMS

While neighborhood community gardens were the main focus of garden activities in the early 1970s, other types of garden programs complemented these efforts and helped to broaden the audience. Company-sponsored community gardens, children's gardens, and job-training gardens were linked to the community garden movement through literature and advocacy.

Company Gardens

A small but well-publicized percentage of the community gardens of the 1970s and 1980s were company gardens. These gardens, considered a form of recreation for employees and a good public relations scheme, were usually located at corporate headquarters where land was available. For example, in 1972, employees at the Hewlett-Packard headquarters in Palo Alto, California, requested a garden that ultimately resulted in a six-acre community garden used by 450 employees. The National Gardening Association promoted this kind of community garden through publications and a technical assistance program. Its 1985 book, *Theory G: The Employee Gardening Book,* cited multiple advantages of company gardens that included not only benefits to the employee (fitness, social contact, and household income subsidy) but also benefits to the business in terms of community relations, employee productivity, and economy. The book highlighted examples from over fifty company gardens, including Reader's Digest (New York), AT&T Bell Labs (New Jersey, Massachusetts, and Colorado), Bethlehem Steel Company (Indiana), Eastman Kodak (New York), Mitre Corp (Massachusetts), and Whitten Machine Works (Massachusetts). Although these gardens received public attention, they were largely a fad, and many either ended or evolved to become community gardens with little connection to the company.

Children's Gardens

As was true of garden promotion in earlier eras, children were considered an important group to reach with gardening opportunities. A commonly voiced concern was that children were disconnected from natural systems—they were growing up thinking tomatoes came from a can, not from the earth. Whether in a neighborhood garden or a school garden, children were encouraged by educators, activists, and others to get their hands dirty and learn firsthand about ecology. For example, in 1970, the president of the General Federation of Women's Clubs promoted children's gardens as her personal project; as a result, information and encouragement were sent out to over 14,000 women's clubs. Past projects were profiled in education and gardening journals to show the possibilities. On one serendipitous occasion, Philadelphia Green assisted a garden project at Vare Middle School, which had had a model school-garden program back in the 1910s. A few new projects developed in the early 1980s, like California's Life Lab Science Program, which advocated garden-based curricula to teach scientific processes.[30] In 1982, Gardens for All established an annual youth garden grant program that awarded $500 to $750 worth of tools, seeds, and educational materials. Publications such as the *National Gardening Association Guide to Kids' Gardening* provided technical advice and suggestions for community outreach and children's garden activities.[31] However, interest in children's gardens needed time to ferment. While projects did develop in various cities, national promotion was weak. As we will see in chapter 9, school gardening did not really become a national movement until the 1990s.

Job-Training and Entrepreneurial Gardens

Just as the growing number of underemployed people in the 1890s led to the founding of vacant-lot cultivation associations, so the economic needs of neighborhoods in the 1970s and 1980s prompted an array of garden programs that incorporated job training and income generation. The hope was that such projects would both increase the employment opportunities of local residents and ease the financial burden on garden organizations and municipalities alike. The garden organizations hoped to offset the insecurities of funding through grants and donations by developing earned-income ventures that also advanced social and environmental goals. To this end, in the 1980s, several groups experimented with greenhouse production programs. One of the early success stories was GLIE Farms, an herb nursery project in the South Bronx.[32] Citing research that showed the profitability of urban agriculture and direct marketing, coupled with the readily available land and labor in the area, the project started in 1981 as a for-profit business that would sustain its non-profit parent agency, the GLIE Community Youth Program. Initial funds came from the youth program and Bronx 2000 Local Development Corporation. By 1983, the project was running eight nurseries on approximately 2.5 acres of city-owned land and earned a reported $15,000 per month. The program was administered by five staff, who trained and employed eight nursery workers as well as summer youth employees. Another well-publicized program was the Landscape Skills Training Program developed by the Boston Urban Gardeners in 1984, which combined job preparation courses with technical training in botany and landscaping.[33] It also provided job placement services, so that all graduates from the first four years were offered jobs with the Boston Department of Parks and Recreation, local housing authorities, or private landscape firms and greenhouses. While praised as models to be replicated elsewhere, programs like these tended to succeed as short-term efforts that relied on particular alliances between the organization and local political leaders and agencies. Susceptible to problems as a result of changes in the political and social climate of supporting city agencies, these programs ultimately ended with little public notice. As will be seen in chapter 9, interest in economic development has continued to inspire entrepreneurial garden programs.

GARDENS AS COMMUNITY OPEN SPACE AND THE QUESTION OF PERMANENCE

When the various community-based efforts in a given city were considered together—a community garden on one block, a locally built playground on another, a minipark down the street, and so on—it became clear that an alternative open space system had emerged in response to local need. Community open spaces such as these differed from the traditional pastoral or recreational parks in appearance and function. Often developed on small sites that had been otherwise overlooked, community open spaces were often proposed and incrementally designed by citizens to satisfy new recreational or social roles in a neighborhood. In their book *Community Open Space,* authors Mark Francis, Lisa Cashdan, and Lynn Paxson defined community open space as "any green place designed, developed, or managed by local residents for the use and enjoyment of those in the community."[34] This interdisciplinary team's research in New York City highlighted multiple examples of community-built spaces, described the impetus behind and the techniques used by local groups, and raised awareness of the potential cumulative impact on people and neighborhoods. Community gardens, along with miniparks, sitting gardens, and playgrounds, were typical manifestations of this new kind of open space. This research catalyzed the formation of the Neighborhood Open Space Coalition in 1980 as a network of organizations with a mission to share information and protect community open space in New York.[35]

To understand the extent of this alternative community open space in New York City, the Neighborhood Open Space Coalition conducted an inventory in 1983 that documented 410 sites on 143 acres that had been created and maintained by over 10,000 volunteers. Compared to the 25,000-acre city park system, the acreage of community open space might have seemed insignificant until considered in light of the number of sites and their integration into neighborhood recreation. While the city operated 1,490 parks, community residents maintained 410 other sites, which represented 23 percent of all park space in the city. These sites were scattered throughout the city, with the majority built on 2,500- to 5,000-square-foot vacant lots. These spaces were built largely through volunteer labor and in-kind contributions. The coalition estimated that whereas it cost $50 per square foot to build a park, community gardens cost an average of $5 per square foot. The inventory found that about 75 percent of community open spaces were on city-owned land, made vacant because of abandonment by the original owners. However, there was still a concern about permanence.

Between 1980 and 1983, nearly 10 percent of the gardens in Manhattan were sold for development.[36] As a result, coalition members started to look for alternative forms of control, including longer lease arrangements, the formation of land trusts, and purchasing land.

Even though urban residents were taking action to create places to garden, the opportunistic means of acquiring land continued to threaten community gardens. In its 1982 poll, Gardens for All reported that 7 million people nationwide were interested in gardening but lacked space. Seventy-six percent of respondents—gardeners and nongardeners alike—wanted permanent gardens in their communities. However, as the economic situation improved in the 1980s, some sites were lost to development, particularly housing construction. The issue was not universal and depended on local political and economic conditions. In some cities, elected officials and city workers were sympathetic to garden groups, while in others the relationship was almost antagonistic, especially when gardens were pitted against housing proposals. For example, gardeners in Boston's South End redevelopment area organized in 1986 after the first garden site was taken over for a housing project. They started working with the redevelopment agency and housing interests to save gardens through a land trust and other means. A more ambivalent atmosphere existed in New York City. Site permanence was—and is—a critical issue to community garden organizers. The May 1982 and Fall 1987 issues of *Journal of Community Gardening* were devoted to site permanence, urging garden organizers to secure their sites through ownership by land trusts, long-term leases, lobbying for the inclusion of gardens in recreation master plans, and advocacy for garden space within housing developments. These issues have continued to galvanize organizers, as we will see in chapter 8.

EIGHT

Community Greening

Urban Garden Programs from 1990 to the Present

MANY COMMUNITY GARDENS BUILT IN the 1970s and 1980s continued to evolve in the 1990s. Neighborhoods saw gardens change, new ones created, and some lost, but overall the movement grew. More organizations formed with the goal of encouraging garden projects. As of 1999, members of the American Community Gardening Association hailed from over seven hundred organizations that represented over half a million people engaged in garden activities in nearly every state. Its membership included individual gardeners, neighborhood groups, citywide gardening organizations, extension agents, educators, designers, and social-service providers. Under the umbrella of community gardening were to be found neighborhood gardens, school gardens, job-training programs, institutional gardens that provided horticultural therapy, and other types of garden programs whose goals were determined by their context and participants.

Along with these developments came the need to reevaluate norms and assumptions within the movement. Some activists began to expand the movement's borders, seeking broader influence in matters of community development, social justice, education, and environmentalism. In 1990, a proposal was made to change the name of the American Community Gardening Association to the American Community Greening Organization. What exactly "greening" involved was open to debate. In an article in the ACGA journal, Marc Breslav described greening as "the intentional, beneficial activity by which plants, through assistance or benign neglect by people, are encouraged to return or thrive within or near any area of human settlement, and thereby encourage feelings of empowerment, connectedness, and common concern among the settlement's human residents and visitors."[1] To meet this broader responsibility, a broad professional discipline equipped with technical as well as organizational skills would be needed.

The debate revealed some conflicts. Some practitioners felt they should keep to the main agenda, which was to assist people wanting to grow food in cities. Others sought to delve into a broader range of activities. The debate also exposed the disparity developing among organizations within the ACGA, which included small volunteer groups, large nonprofit organizations with million-dollar budgets, and public agencies that included gardening as one of their many responsibilities. Were the issues faced by a volunteer garden group the same as those faced by an organization with a paid staff of twenty? What was the relationship between a neighborhood garden and a highly programmed job-training garden program? While everyone agreed that gardening provided many different benefits, bigger questions were emerging: What is the priority of the community gardening movement? Is it gardening or something larger? In the end, the ACGA did not change its name, but its journal was renamed *Community Greening Review*. The ACGA's recently developed vision statement reveals a holistic agenda that uses gardening as a catalyst for a broad range of social benefits:

> Our vision is that community gardening is a resource used to build community, foster social and environmental justice, eliminate hunger, empower communities, break down racial and ethnic barriers, provide adequate health and nutrition, reduce crime, improve housing, promote and enhance education, and otherwise create sustainable communities.[2]

THE ACGA SURVEYS OF 1990 AND 1996

If it was difficult to get a clear picture of the national movement's goals, it was just as challenging to describe what urban gardening meant within one city. With so many people involved—institutions, public agencies, nonprofit organizations, and individual gardeners—and so many different kinds of gardens and programs, the cumulative effect could only be guessed. For example, while a community garden might have lost its land, just around the corner a school might have started to build a new garden as part of its curriculum. Were these groups working together or in competition for limited resources? In some cities, gardens were anchored in the public planning of open spaces, recreation, or institutions. In others, gardening was barely recognized and organizers faced the possibility of losing land or other resources.

To better understand the community it was serving, the ACGA conducted two surveys of its membership.[3] The first survey, in 1990, gathered results from thirty-six organizations in twenty-four cities. The second, in 1996, received responses from thirty-eight cities, of which fifteen had been represented in the 1990 survey. Both surveys reflected a broad range of organizations, from large to small, from long established to relatively new, and both nonprofit and public agencies. The results, while not inclusive of all cities and organizations, provided important information on the number and types of gardens, the functions of the organizations, and the sustainability of urban gardening in cities across the nation.[4]

The 1990 data represented 2,329 gardens, while the 1996 data represented 6,020 (see table 11). In both surveys, the most common garden type was the neighborhood garden—a place that supplied households with land to grow food and flowers. Besides offering individual garden plots, many of these gardens also provided play areas for children, picnic and barbecue areas, demonstration areas for native habitats, and compost and recycling facilities. Another type of garden cited in the surveys was the public housing garden, largely represented by the New York City Housing Authority Tenant Garden Program. In addition, respondents listed gardens at schools, mental health and rehabilitation facilities, and senior centers, among other locations. The 1996 survey paid particular attention to new types of garden projects that had received publicity, such as job-training and entrepreneurial projects. However, this category comprised less than 1 percent of the gardens included in the results.

The 1996 survey found that the cities with the most gardens were New York (1,906 gardens), Newark (1,318), Philadelphia (1,135), Minneapolis (536), and Boston (148). While larger cities generally had more gardens, population did not always correlate with the number of gardens. For example, Portland, Oregon, with a population of 445,000, listed 23 gardens, while Minneapolis, with a population of 363,000, had 536. Comparing the number of gardens with the population, the cities with the most gardens per ten thousand residents were Newark (49), Minneapolis (14.8), Philadelphia (7.3), Trenton (6.8), and Pittsburgh (2.9).

Losses and gains in the five years between the surveys provided a means to evaluate garden stability. The 1996 survey showed that more than 2,123

TABLE 11

American Community Garden Association Survey Results, 1990 and 1996

	1990 Results		*1996 Results*	
Number of gardens reporting	2,329		6,020	
Number of cities represented	24		38	
	Number of Gardens	Percentage of Total Gardens	Number of Gardens	Percentage of Total Gardens
Gardens by type:				
Neighborhood garden	1,538	66.0%	4,055	67.4%
Public housing[a]	269	11.5	983	16.3
School	203	8.7	496	8.2
Mental health/rehabilitation	—	—	87	1.4
Senior center/senior housing	105	4.5	85	1.4
Job/entrepreneurial	—	—	36	0.6
Other	214	9.2	278	4.6
Gardens older than ten years	—	—	2,047	34.0
New gardens built	523 in last two years	22.5	2,123+ in last five years	35.3
Gardens lost in last five years	74	—	542	—
Garden sites considered permanent	154	6.6	318	5.3

SOURCES: American Community Gardening Association, *Findings from the National Community Gardening Survey,* ACGA Monograph (1992); American Community Gardening Association, *National Community Gardening Survey,* ACGA Monograph (1998).

NOTE: Some of the numbers differ from those in original sources due to errors in the summary tables there.

[a] Most of the public-housing gardens belonged to the New York City Housing Authority Tenant Garden Program. In 1990, 237 of the 269 were from the New York program. In 1996, New York represented 834 of the 983 public-housing gardens.

new gardens, or 35.3 percent of the total number, had been created. However, 542 gardens had been lost in the same five-year interval. Some cities showed an increase, some maintained relatively stable numbers, and others experienced a decline in the number of gardens. Among Newark's 1,318 gardens in 1996, 516 were new and only 14 had been closed. Some cities that showed a relatively steady number of gardens had actually lost gardens and built approximately the same number of new ones. For example, Pittsburgh reported 108 gardens, 48 of which had been built between the surveys while 50 had been discontinued.

Of the gardens that closed down in the early 1990s, the primary reason was gardeners' lack of interest in maintaining the garden (49 percent), followed by loss to a public agency (20 percent) and loss to a private developer (15 percent). How a garden had been initiated—whether it was started by an outside organization or was community-sponsored—did not seem to relate to its longevity. Efforts on the part of garden organizations to establish land ownership or assure longer lease agreements seemed to be on the rise. Ten of the fifteen cities surveyed in both years reported an increase in their land holdings, in land protected through land trusts, or in longer-termed contracts, such as ten-year leases. However, only 5.3 percent of all gardens reported in 1996 were protected through land trusts or other ownership, and just fourteen cities reported any significant municipal policies in place to secure gardens in land use planning. While 34.0 percent of the gardens in the 1996 survey were over ten years old, many were still on short-term leases and their permanence remained unsure.

EXTENDING THE CAPACITIES OF GARDEN PROGRAMS

Of the activities and services provided by responding organizations in the 1990 ACGA survey, the top five were community organizing, assisting with land acquisition, horticultural assistance, educational programs and resources, and assistance in garden design and layout. These represent the core activities of most programs today as well; however, some programs provide other services, such as assistance in marketing and business development, job training and placement, political lobbying, and horticultural therapy. To meet such a variety of objectives, garden organizations often rely on staff and volunteers with a broad range of expertise—horticulturalists, community organizers, educators, and landscape architects. While many organizations maintain longstanding staff who have made a career of community garden facilitation, another resource for many organizations is recent college graduates who decide to work for a garden organization for a few years before moving on to new careers or pursuing advanced degrees. Young staff members have frequently been hired through national service programs, such as VISTA or Americorps. Terry Mushovic, former director of the Penn State Urban Garden Program, has described garden staff as "people who want flexibility, creativity, and that kind of passion and commitment to working on making where they live and where they are a little better."[5]

As gardening organizations have grown more sophisticated, they have begun to work proactively to protect gardens by extending their activities into community development. They have had to become politically savvy in order to be included in planning decisions and funding pools, whether public or private. While municipally run organizations may appear to have a better connection to public policy, this is not necessarily the case. Both nonprofit organizations and city agencies have actively lobbied to raise awareness among policy makers for garden funding and site permanence. To raise attention and prove the legitimacy of urban gardens as part of effective municipal planning, organizations have conducted citywide inventories, quantified participation and outcomes, and facilitated the drafting of public documents. Some groups have worked to change the zoning classification of garden sites to an open-space designation to reduce pressure from those who want to develop the land. Others have been successful in getting gardens included in municipal open space and recreational planning, as in the cases of San Francisco and Seattle described later in this chapter.[6]

Another way garden organizations have expanded their influence is through collaboration. Some cities have more than one gardening organization, so rather than competing for the same resources, it makes sense for them to work together and gain strength in numbers. In Boston, for instance, Garden Futures was developed in 1994 as a collaborative group representing Boston's four main community garden organizations—Boston Urban Gardeners, the Boston Natural Areas Fund, Dorchester Gardenlands Preserve, and the South End Lower Roxbury Open Space Land Trust. The collaboration's initial objective was to inventory the resources, needs, and capabilities of Boston's nonprofit-owned community gardens. A grant allowed Garden Futures to conduct an in-depth study in four areas: capital and maintenance; organization and governance; public awareness; and training, education, and leadership development. The resulting report provided a first step toward declaring gardens a holistic benefit to the public and a beneficial use of land.[7] The study also recommended establishing minimum standards for garden conditions and criteria on which to evaluate the leadership of gardens. Given the range of interests that apply to gardening, garden organizations can easily link with environmental organizations, teaching institutions, neighborhood associations, and others. Such collaborations have helped to form land trusts, procure large grants, and pass supportive legislation, besides expanding the influence of garden organizations into community development, urban design, education, and environmental policy.

While working to promote and support urban gardening at a citywide scale, garden organizations have also continued to encourage self-reliance on the part of the gardeners. Most garden organization staff have insisted that a garden should not be "given" to neighborhood residents, but instead assistance should be provided so they can build the garden themselves and have a sense of ownership over it. One long-time New York community garden organizer emphasized that the community garden movement has to be more about *community* than about gardening.[8] To this end, discussions in community garden circles have increasingly emphasized that the process of building a garden must also nurture community investment and local leadership.

In 1995, the ACGA initiated a mentorship program for new garden organizations called From the Roots Up. This one-year intensive training program provided workshops and mentors for fledgling city gardening organizations that worked with low-income communities. Mentors were assigned to help individual projects with organizational development, community organizing, coalition building, leadership development, fund raising, and publicity. Between 1996 and 1999, seventeen groups participated in From the Roots Up. Based on experiences gained in the program, the ACGA published a handbook, *Growing Communities Curriculum,* and developed a two-day intensive workshop. The curriculum is grounded in five core beliefs developed by community gardeners and organizers:

1. There are many ways to start or manage a community garden.
2. In order for a garden to be sustainable as a true community resource, it must grow from local conditions and reflect the strengths, needs and desires of the local community.
3. Diverse participation and leadership, at all phases of garden operation, enrich and strengthen a community garden.
4. Each community member has something to contribute.
5. Gardens are communities in themselves, as well as part of a larger community.[9]

The handbook rarely mentions gardening itself, addressing instead the broad range of skills needed to run a garden organization and empower local garden groups. It provides an information base on four issues: principles to guide community gardening, participatory methods, organizational development, and how to run workshops and meetings.

Given the two main reasons that gardens are discontinued—lack of sustained interest and loss of land—it behooves garden organizations to simultaneously address the community's commitment to the project and the arrangements for procuring land. Much of the lobbying, collaboration, and community organizing that garden organizations do is to secure garden sites. One approach that combines the building of community leadership and the pursuit of land security is the development of neighborhood land trusts. The protection of gardens through land trust is not a new idea. In the mid-1970s, Trust for Public Land (TPL) began to assist garden groups in Vermont, New York, Newark, and Oakland with site preservation.[10] Currently, the effort to establish a land trust is itself used to strengthen the garden participants' commitment to the responsibilities inherent in permanent land stewardship. Following this principle, Philadelphia activists formed the Neighborhood Gardens Association/A Philadelphia Land Trust in 1986, and Chicago's Neighborspace land trust was organized in 1996. Boston is furthest along in protecting its gardens; of its 140 gardens, sixty are held in land trusts and twenty are owned by nonprofit organizations.[11]

URBAN GARDEN PROGRAMS IN FOUR CITIES

The four garden programs presented here provide a sampling of the range and kinds of activities, staffing, funding, and so on that shape urban garden programs. The three organizations—Seattle's P-Patch, the San Francisco League of Urban Gardeners (SLUG), and the Pennsylvania Horticultural Society's Philadelphia Green—reflect different organizational structures and different urban contexts. Seattle P-Patch is a municipally run program with a nonprofit partner, SLUG is a nonprofit organization, and Philadelphia Green is a program within a larger nonprofit organization. The fourth "program" is in fact the ongoing story of organizing for urban gardens in New York City, an effort that has involved many organizations. Although context and organization vary, each case reveals the ever-changing nature of urban garden organizing as it adapts to local needs, types of participation, demands for site security, and new program directions. In each story we can see the effort involved in justifying and expanding urban gardening programs as well as the hurdles that challenge their ultimate survival and sustainability.

Seattle's P-Patch

When Rainie Picardo retired from farming in 1971, he decided to let local residents and students garden on his old three-acre truck farm site in Northeast Seattle at no cost.[12] But faced with annual property taxes of $688 per year, Mr. Picardo decided that he could not continue this arrangement and started the procedure to sell the property in 1973. Supported by popular interest in continuing the garden, an alternative proposal surfaced for the city to lease the site. The city council approved the lease on April 16, 1973, with a $950 appropriation for a ten-month "experiment" in community gardening. Under the direction of the parks department, the site was prepared for 195 community plots and fourteen youth-service plots. Gardeners were to pay $10 per year toward water for a plot of up to four hundred square feet. By May 20, all the community plots had been rented. At the end of the season, only ten gardeners had quit. The season closed with a harvest luncheon attended by the mayor, the city council, and at least sixty gardeners.

Given the success of the "experiment," the city council appointed a task force of citizens and city agency representatives to organize the "P-Patch program," a publicly administered program named after Picardo.[13] Three years after the initial garden was laid out, the city had developed ten more garden sites, totaling 725 plots and serving 2,500 gardeners. Permanency was obtained in 1975 when the city purchased the original P-Patch for $78,000. Since then, P-Patch has continued to develop gardens in neighborhoods, at senior centers and schools, in public housing, and at the city arboretum. Some are permanent gardens, while others are intended to be an interim use of land that may be developed later.

Over time, the program has evolved to accommodate changes in budget, urban development policies, and social concerns about education, the environment, and economic opportunity. Services that had initially been provided, such as rototilling and staking garden plots each spring, ended in the 1980s due to budget cuts. Fees have slowly risen, although there has always been an exemption for families below the poverty level. New programs have been added such as Lettuce-Link, established in the early 1990s to encourage gardeners to cultivate extra food for the needy. Lettuce-Link provides drop-off boxes and volunteers to deliver produce to the food bank daily. At different times in its history, P-Patch has supported various children's and youth programs. While P-Patch enjoys a certain level of security as a municipally run program, the addition of the Friends of P-Patch,

Figure 56. Over twenty-five years after its creation, the original P-Patch site exudes a sense of fertility and careful tending. Photograph by author, 2001.

a nonprofit advisory and support organization, has allowed gardening to advance in new directions not possible under a public agency. Established in 1979 and incorporated as a nonprofit organization in 1994, Friends of P-Patch connects P-Patch with an array of other organizations and individuals through its membership. It is effective at lobbying and promotion and is able to start new programs, conduct fundraising, apply for grants, and own property.

The City of Seattle has continued to support community gardening. In 1992, the city council passed resolution 28610, which declared general support for community gardening and specific support for making surplus land available for gardening. The city's recent comprehensive plan, Toward a Sustainable Seattle, includes the goal of one community garden per 2,500 households. Gardens are represented on planning maps to insure their consideration in development proposals and planning. As of 2004, P-Patch served over 4,600 gardeners by providing over 1,900 plots in 52 garden sites around the city. Friends of P-Patch has helped to secure five of these gardens through ownership arrangements.

A relatively recent addition to urban gardening in Seattle is the Cultivating Communities program, which started in 1995 as a cooperative effort by Friends of P-Patch and the Seattle Housing Authority. It provides land, training, and technical support so families living in public housing can grow food for consumption or for sale. Participating families are predominantly Southeast Asian immigrants from agrarian backgrounds. The program provides an opportunity for them to continue using their traditional farming practices and to grow culturally specific foods. In addition, it has helped to establish economic opportunities for immigrant families through community-supported agriculture. Families in the program grow produce that is available to subscribing consumers for weekly pick-up. For an annual fee, subscribers receive a weekly box of produce for twenty-four weeks of the year. Because the program is still developing, its economic viability has not yet been determined.

San Francisco League of Urban Gardeners

When community gardens were in demand in the 1970s, the City of San Francisco established a municipal garden program through the Department of Public Works. By 1979, this program was managing seventy-five gardens and had a staff of thirty-five garden coordinators and leaders, paid primarily through CETA funds. When CETA was discontinued and funds from taxes were reduced due to passage of property-tax reform bill Proposition 13, the program lost its funding and ultimately closed in 1980. Concerned about maintaining the existing garden projects, a group of gardeners and activists formed the San Francisco League of Urban Gardeners in 1981, which was incorporated as a nonprofit organization in 1983. SLUG's main purpose was to promote community and school gardens and provide horticultural education. By 1986, its staff of six oversaw forty-seven sites that totaled ten acres, with an operating budget of $156,000. Over two thousand people were members, and there was a waiting list of at least four hundred people who wanted garden plots. That same year, as city planners were drafting a new open space plan, SLUG staff wrote a community garden master plan and submitted it to the city. The document not only provided information on existing gardens, but it also suggested ways for gardens to work better with city agencies, identified new sites in underrepresented neighborhoods, and listed resources available to promote community gardening. This master plan was eventually incorporated into the open space element of the city's master plan.

In addition to facilitating existing and promoting new community gar-

Figure 57. A Laotian woman at a community garden in Seattle, Washington, using a traditional method to separate seeds from chaff. Photograph by author, 2001.

dens, SLUG began to promote environmental education through SLUG-sponsored classes and training programs related to gardening and composting. At the Garden for the Environment, SLUG's demonstration garden, San Franciscans learned about gardening, drought-tolerant planting, erosion control, composting, and garden designs that increase accessibility for the physically disabled. Volunteers at the garden provided additional advice or suggested further resources for visitors. Over time, SLUG expanded the educational opportunities it offered by initiating a native plants nursery, a habitat conservation department, and an adult environmental education program.

In the 1990s, under the leadership of director Mohammed Nuru, SLUG focused more of its programming on low-income and underserved communities. Through education programs, job training, and entrepreneurial businesses, SLUG provided much-needed resources to many struggling families. With an adult work crew, youth employment, and other job-training programs, SLUG served as a local employment opportunity in Bayview/Hunter's Point, a distressed neighborhood with one of the high-

Figure 58. The San Francisco League of Urban Gardeners' Garden for the Environment is a resource for information on composting methods, native plants, handicapped-accessible gardening beds, erosion control, irrigation, and more. Photograph by author, 1992.

est unemployment rates in the city and also the location of the SLUG office.[14] SLUG's design and construction work crew, 70 percent of whom came from the neighborhood, contracted with the city to build and maintain gardens and other kinds of spaces. SLUG also developed programs for transitional employment in the welfare-to-work process, providing street sweeping and landscaping jobs while also teaching literacy and computer skills to participants.

The organization expanded further as it developed several youth programs, including the Youth Garden Internship Program and the Environmental Justice Program. Both programs employed youth to provide gardening and outreach services while receiving training in leadership, conflict resolution, and other personal development skills. Youth engaged with the Environmental Justice Program were trained as community organizers who distributed information on lead poisoning, nutrition, smoke abatement, and more. The Youth Garden Internship Program, which engaged young people in gardening, landscaping, and community education at public housing projects and surrounding communities, started in 1995 at the Alemany Public Housing Project. Funds were provided by the mayor's office, among other sources, and the program had a contract with the housing authority to landscape the grounds. In the first year, twenty young people and two supervisors were hired to develop a garden on a 4.5-acre site leased from the Department of Parks and Recreation and adjacent to the public housing. Over time, the site, which had been subjected to illegal dumping, was transformed into St. Mary's Urban Farm, a vibrant oasis of vegetable gardens, flower beds, orchards, and native habitat areas, with a pond, greenhouse, and windmill. The participating youth and volunteers grew food and flowers that were distributed to tenants and to local senior centers and other institutions. They also maintained a recycling/compost education area, managed a beehive, and gave tours to school children and other visitors. The garden included an area where local residents could garden on their own. The success at the Alemany Housing Project led to similar programs at other housing projects in San Francisco. When SLUG sent some of the youth interns to national ACGA conferences, other garden activists were awed by the interns' confidence and the program's success, which inspired other groups to start similar efforts in their cities.

Building on these youth programs, SLUG created an entrepreneurial job-training program that provided more jobs and educational opportunities—including teaching transferable business skills—to young adults who were often graduates of the SLUG youth programs. Under the catchy brand name Urban Herbals, SLUG began to produce and market jams, vinegars, and salsa. Using a kitchen rented from the Hunter's Point Naval Shipyard, employees developed the recipes, began production, and assisted with marketing and recordkeeping. The products were sold directly to consumers as well as in some local and national grocery stores. The program was initially subsidized through various funding sources; however, SLUG staff and board members hoped that the program would ultimately achieve economic self-sufficiency through sales. Approximately one-quarter of the ingredients came from SLUG gardens. This program, too, received national attention and spurred similar efforts by other garden organizations.

As of 2002, SLUG was one of the largest community-garden organizations in the nation, with an operating budget of $3.5 million and employing 150 to 200 people, with more employed during the summer months. The organization received funds from Community Development Block Grants, contracts with city agencies, and other sources. Its activities

Figures 59 and 60. *Top:* St. Mary's Urban Farm at Alemany Public Housing in its second year. A great deal of labor was required to return the degraded site to health. Photograph by author, 1996. *Bottom:* Five years later, St. Mary's Urban Farm has been transformed into a thriving garden, thanks to the efforts of participating youth and volunteers. Photograph by author, 2001.

continued to expand, with new programs like a tool lending center where residents could borrow a range of tools for garden and home repair. With its well-publicized training and outreach programs, SLUG was pushing the boundaries of what most people considered urban gardening. Cory Calandra, director in 2002, summarized SLUG's mission as the use of gardening for personal growth and community empowerment. "The garden is a vehicle," she said, "but we need to look at the whole individual."[15]

However, as successful as SLUG appeared to be, underlying management concerns soon came to the surface. Heavy reliance on city contracts for many of the training and education programs meant that SLUG was vulnerable to political and economic changes that affected city policies, such as reduced funding as a result of the economic downturn of the "dot-com" crash, or political pressure from unions that felt that such programs took jobs away from city employees. Meanwhile, discontented community gardeners felt that SLUG had strayed from its mission of promoting gardening and was not paying attention to their needs. Serious questions arose about how SLUG managed its finances and performed in city-contracted work, ultimately leading to investigation by the city attorney. Publicity over the investigation, as well as civil suits over debts and other scandals, began to tarnish SLUG's reputation. As of 2003, the budget dropped to $1.2 million, which meant that several programs had to be closed, including St. Mary's Urban Farm and Urban Herbals. The situation worsened, and in July 2003, SLUG laid off all its employees and suspended most operations. The city removed SLUG as managing agent for the forty-two community gardens on city- and county-owned properties. The organization was also debarred from city contracts for two years. As of the summer of 2004, with no budget, the SLUG office consisted of two volunteers who helped to answer phone calls and manage the few sites still under SLUG's leadership, including the Garden for the Environment. Director Cletis Young expressed the hope, however, that once SLUG addresses these problems, its supporters will build on its rich history and successes to re-create SLUG with a revived mission and leadership.

Philadelphia Green

Philadelphia has a rich history of urban garden programs. The Philadelphia Vacant Lots Cultivation Association, started in 1897, was the longest-running program of its kind, continuing until at least 1927. Philadelphia school gardens, war gardens, and victory gardens, as well as the Neighborhood Garden Association described in the introductory essay to part III, compose a nearly continuous presence of urban garden programs in the city. In recent times this trend has continued. The Pennsylvania Horticultural Society, established in 1826, expanded its activities into community outreach projects in 1974. These projects included working with schools, establishing community gardens, a garden contest, a Junior Flower Show, and a prison landscape and nursery training program. When financial support decreased for the city's vest-pocket park program and community gardens, the Pennsylvania Horticultural Society filled the need and further expanded its community projects, starting with a community vegetable garden program in 1973–74 and leading up to the official founding of the Philadelphia Green program in 1978.[16] Collaboration between the Penn State Urban Garden Program, established in 1977, and Philadelphia Green resulted in a strong urban gardening presence in the 1980s through the 1990s. The loss of several large community garden sites to housing and commercial development led to the establishment of the Neighborhood Gardens Association/A Philadelphia Land Trust in 1986. Altogether, Philadelphia has had one of the strongest and most comprehensive community gardening and greening efforts in the nation.

Philadelphia Green, one of the largest organizations affiliated with the ACGA, illustrates the cutting edge in community greening efforts. In 1978, Philadelphia Green had ten full-time horticulturists for four program areas: vegetable gardens, sitting gardens, street-tree planting, and window boxes and tire-urn plantings. The vegetable garden program worked with community groups of at least ten neighbors who had obtained permission to garden on a site. Philadelphia Green provided each neighborhood group with fencing, soil, tools, plants, and seeds as well as help with planning and technical advice. The participants provided the labor of putting up the fence, spreading soil and fertilizer, planting, and maintaining the gardens. The sitting garden program developed small neighborhood spaces for flowers and benches. The street tree program provided technical assistance and a crew to help remove concrete and debris so street trees could be planted. Lastly, the window-box and urn program was a street beautification scheme that evolved from the defunct Neighborhood Garden Association. In 1978, Philadelphia Green worked with fifty-five groups that started vegetable gardens averaging three thousand square feet each; forty blocks installed and planted window boxes and tire urns; ten neighborhoods began sitting gardens; and eighteen blocks planted street trees. By 1982, Philadelphia Green had worked with community groups on 300 vegetable gardens, 220 garden blocks, 100 street-tree and curbside-planter blocks, and 58 sitting gardens.[17]

From 1980 to 1993, Philadelphia Green developed eight community projects under its Greene Countrie Towne program, whose name was inspired by city founder William Penn's plan for one acre of cultivated land for every ten acres of development.[18] The program intended to work with community-based organizations for a three-to-five-year period to beautify low-income neighborhoods by adding window boxes, gardens, trees, and gateways. The program was based on the philosophy that "a concentration of community gardening activities in a neighborhood inspires significant changes in its appearance, strengthens community organizations, engenders pride, and often empowers other community development activities."[19] Philadelphia Green provided organizational assistance, technical and design advice, all the gardening materials, and a work crew to make deliveries and do some of the construction. In some cases, neighborhood groups approached Philadelphia Green to become Greene Countrie Townes, while at other times staff members identified target areas. Criteria for selection included the following: the median family income per block was at or below the city's median income, at least 60 percent of the housing was owner-occupied, there was evidence of commitment to gardening, and neighborhood organizations were in place to provide support and leadership. If all these standards were met, the Greene Countrie Towne process began, with the staff working with community groups to collect additional data on community leadership, attitudes, open space, demographics, and a vision for the community. Acknowledging the existing social networks, staff worked closely with local groups, city officials, and neighborhood residents. In turn, the residents formed a community club and worked with Philadelphia Green to develop, plant, and maintain the greening projects. Projects included street tree planting, reclamation of vacant lots for gardening and sitting areas, and the creation of Garden Blocks, which transformed bare sidewalks by installing and planting flowers in window boxes, tire urns, wine barrels, and other sidewalk containers. Other activities included educational programs, contests, events, and working with schools. J. Blaine Bonham, executive vice president of the Pennsylvania Horticultural Society, who has led Philadelphia Green through most of its existence, described the lessons learned through the Greene Countrie Towne program:

> One of the compelling lessons here was the enormous dedication of people in these communities to try to turn their communities around. And not

until you do these things can you refute suburbanites' or exurbanites' views of people in cities as being lazy and all addicted to drugs and not working. Because these folks, I mean hundreds, thousands, over all the years in all those communities, pulled together to develop 20, 30, 60, 80, 100 projects, planting trees, doing vacant lots. Now some didn't last. We learned a lot about what kind of infrastructure has to support it. But, in almost all of those communities, if you went back there, you'd see that some are still spectacular. Others have disappeared through no fault of anyone, but the economic issues overwhelmed them.[20]

In 1987, Philadelphia Green expanded its mission from neighborhood projects to working on downtown areas and city gateways. This program, initially called City Center Green before changing its name to The Public Landscape Project, worked with corporations, government agencies, community groups, and individuals to improve and maintain highly visible landscapes. Projects included the Azalea Garden at the Philadelphia Museum of Art, planting along John F. Kennedy Boulevard near the main train station, and the Twenty-Sixth Street corridor from the airport, as well as other street tree planting and ornamental planting projects. More recently, the program has started to manage landscapes, including Penn's Landing, for a fee. Philadelphia Green has several landscape architects on staff to oversee this program.

With over 15,000 vacant lots and 27,000 vacant structures in the city, in 1995 Philadelphia Green became increasingly involved with addressing the vacant land issue. In the 1990s, they started promoting collaborations with community development corporations to include open space planning in community development strategies. Philadelphia Green is trying to convince policy makers and development corporations that gardening creates social networks that in turn get neighborhood residents to interact with each other and become involved in community improvements. Gardening is also faster and cheaper than other interventions, such as housing. As Bonham commented, "It may be crass fact, but gardening is logical because it works and it is cheap."[21] To this end, the Pennsylvania Horticultural Society initiated a study of urban vacant land to complement a similar report conducted by the city's planning commission. The report recommended that the city streamline land acquisition and title consolidation processes, establish policies and procedures to make it easier for community groups to utilize land, draw federal and state attention to the reclamation of contaminated sites, and develop a neighborhood open space management program. As a result of the report, Philadelphia Green proposed establishing a model of such a program that would work with community development corporations and other community groups to strategize vacant land management and open space planning.

In 1996, Philadelphia Green initiated the New Kensington Project, an open space management plan that incorporated several elements of vacant land management. It undertook the project in collaboration with New Kensington Community Development Corporation, with funds from the city's Office of Housing and Community Development plus additional support from the William Penn Foundation, the Pew Charitable Trusts, the Local Initiatives Support Corporation, the Philadelphia Urban Resources Partnership, and other small foundation grants. The neighborhood of South Kensington faced many of the issues common in low-income areas of Philadelphia. The area had experienced a drop in population from 32,000 to 15,600 in forty years, resulting in 1,100 vacant lots, of which 70 percent were private, tax-delinquent properties. By November 1999, $2.5 million had been spent on the project, and community residents had taken part in cleaning 370 parcels, building sixty-two community gardens, improving 158 side yards, developing a demonstration garden, and initiating various education programs. To further this area of work, Philadelphia Green sponsored a report on the feasibility of nonprofit as well as for-profit urban agriculture as a land-use alternative for Philadelphia neighborhoods experiencing declining population and increasing vacant land.[22]

Philadelphia Green currently is one of the largest citywide garden advocacy programs in the nation. It is funded through proceeds from the Philadelphia Flower Show, grants, contracts, and donations. In addition to Philadelphia Green's many projects, the city's gardening future is being secured through the Neighborhood Gardens Association/A Philadelphia Land Trust, which currently owns twenty-four properties representing twenty-one gardens and holds two garden leases.[23]

New York City's Community Gardens

The history of New York City's urban gardens is as complex as the city itself. In the 1970s, the first group that organized to advocate community gardens in New York City was the Green Guerillas, founded in 1973. In a city with densely built neighborhoods and few parks, urban gardens satisfied needs for more open space, for recreation, and for social outlets.

The Bowery/Houston Garden (later renamed after Liz Christy, founder of the Green Guerillas) received the first lease for a community garden from New York City. Ongoing efforts to promote community gardens in other places in the city eventually led Green Guerillas to form into a nonprofit organization in 1976. In addition, the Council on the Environment, established in 1970 as part of the mayor's office to coordinate various public agencies, developed an Open Space Greening Program that included assistance to gardens.[24] Through this program, the city provides office space, telephones, and vehicles while their nonprofit wing funds salaries, program expenses, and materials. In 1978, Operation GreenThumb was established as part of the New York City Department of General Services to administer community garden programs and issue interim leases for gardens on city-owned lots. Through Operation GreenThumb, community groups could lease city-owned properties for one dollar per year for establishing neighborhood-sponsored pocket parks, community gardens, and children's play areas. A variety of other organizations have also promoted urban gardening, including Bronx Frontier Development Corporation, Cornell University Cooperative Extension, the Trust for Public Land, the New York Horticultural Society, and the Brooklyn and Bronx Botanical Gardens. To coordinate activities, the Neighborhood Open Space Coalition, discussed in the previous chapter, formed in 1980 as a network for information, resources, and advocacy.[25] However, even with the presence of these organizations, New York City community gardens struggled for survival in a city where open land is scarce.

A 1983 inventory by the New York Neighborhood Open Space Coalition provides a snapshot of conditions at that time. The study analyzed 410 community open spaces, including parks, playgrounds, community gardens, and sitting parks, covering approximately 150 acres. Of these sites, 80 percent were gardens and parks on vacant lots and 69 percent included vegetable gardens as a land use. The reasons these gardens and parks were started included sanitation (73 percent), recreation (36 percent), education (28 percent), social benefits (25 percent), economic benefits (22 percent), and other reasons associated with nutrition, community organizing, and memorials. Seventy-nine percent were sponsored by civic groups such as nonprofit organizations, schools, youth groups, and religious institutions. Approximately 75 percent were leased from the City of New York and only 24 of the 410 were owned by community organizations. The average age of a garden or park was three and a half years, and the average size was 5,225 square feet. The average cost was $46,756, of which $16,365 was initial capital, $734 annual capital, and $7,949 sweat equity (the estimated worth of the voluntary labor invested). Ninety-two percent of total maintenance was accounted for through sweat equity.

Through the 1980s to the present, Operation GreenThumb has continued to oversee leases for city-owned land. Until 1984, the typical lease was a one-year renewable for $1 per year, with a thirty-day notice if revoked. In 1984, Operation GreenThumb developed a long-term open space lease that provided five-to-ten-year leases for garden sites. This program was available to garden sites appraised at less than $20,000 that carried liability insurance. The long-term lease agreements required a higher rent—$120 to $360 per year, according to the appraised value of the property. In 1987, eligibility was extended to lots appraised at $35,000 or less and rent was stabilized at $120.[26] By 2000, GreenThumb had helped over 800 community groups lease over 1,000 lots that totaled more than 125 acres.

However, development pressure has always been an issue for New York's community gardens. One of the first gardens to receive public attention for its destruction was Adam Purple's Garden of Eden on the Lower East Side in 1986. A squatter garden with little legal standing, the garden lost its site to a subsidized housing development. To avoid similar situations that pitted much-needed housing against gardens, activists strategized ways to develop inclusive plans that blended housing and gardens. In the mid-1980s, for instance, one developer whose project was going to demolish an existing garden worked with the gardeners to develop a new garden nearby, now called the Westside Community Garden. While there were some successful saves, gardens elsewhere continued to be threatened. In 1996, six gardens in Bushwick, Brooklyn, and seventeen in Harlem had their leases canceled to make way for subsidized housing. This pattern has continued, causing greater concern and even more organizing for garden protection.

The story of the Clinton Community Garden in New York City illustrates an inventive if unusual strategy for preservation that resulted in ownership. The garden was started in 1978 on a site at Forty-Eighth Street between Ninth and Tenth Avenues in Manhattan, leased from the city through Operation GreenThumb for $1 per year. In 1980, there were ninety plots plus communal space. Although the garden had a locked gate, the key was easily available to neighbor residents—over four hundred copies had been distributed. In 1981, the city decided to sell the parcel at auction for $375,000, though city officials and neighborhood residents noted it was worth closer to $800,000. The gardeners, the Trust for Pub-

Figure 61. The Westside Community Garden in New York City. When an established community garden had to be destroyed for new housing, the developer worked with the gardeners to create this smaller but more public community garden nearby. Photograph by author, 1992.

lic Land, Green Guerillas, and Housing Conservation Coordinators (a local neighborhood organization) organized into the Committee to Save Clinton Community Garden. Their plan was to acquire the necessary money by selling square inches for $5 contributions. The city accepted this plan and held off the auction. The first inch was sold to Mayor Edward Koch. Celebrities, such as community members Kevin Kline, Joanne Woodward, and Pete Seeger, joined in the effort, and the garden group successfully purchased the site.

The status of New York's community gardens became even more vulnerable in the 1990s. In 1995, GreenThumb's long-term lease program ended, and leases were replaced with license agreements. In May 1998, Mayor Rudolph Giuliani placed an "emergency hold" on all GreenThumb garden properties, transferring them from their previous home in the Parks Department to the Department of Housing Preservation and Development (HPD). No new gardens were permitted on HPD property after this. Then, to raise revenue, New York City's Office of Management and Budget man-

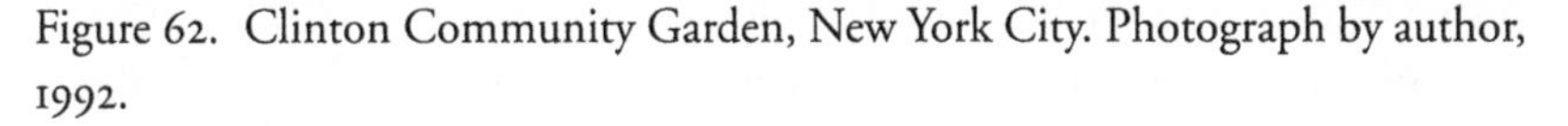

Figure 62. Clinton Community Garden, New York City. Photograph by author, 1992.

dated that the HPD and the Department of Citywide Administrative Services' Division of Real Estate Services dispose of properties in their inventories, through either development or auction. Given that almost half of the 750 GreenThumb gardens were now located on HPD properties, this policy threatened the future of many gardens. The Giuliani administration emphasized that the gardens, including the GreenThumb sites, were never meant to be permanent. However, the HPD agreed to transfer thirty-six of the gardens back to the city's Parks Department, pending local community board approval. Some community boards passed resolutions to preserve other garden sites, pending HPD acceptance. However, for the 113 gardens not protected or released, some of which were more than ten years old, an auction was scheduled for May 1999. The city planned to offer the sites to the highest bidder for unrestricted use. Sale of all the gardens together was predicted to earn a minimum of $3.5 million.

As the auction deadline approached, organizing to protect the gardens intensified. Rallies were held outside the mayor's office, with some protesters dressed as fruits and vegetables and singing songs such as "We Shall Till the Soil," to the tune of "We Shall Overcome." Meanwhile, various community and environmental organizations—Green Guerillas, New York State Attorney General Eliot Spitzer, the New York Environmental Justice Alliance, the Puerto Rican Legal Defense Fund, the Natural Resources Defense Council, and others—initiated at least four lawsuits. The Trust for Public Land proposed to buy the garden sites for $2 million, but the city refused the offer. Giuliani was reported to have taunted the gardeners: "This is a free market economy. Welcome to the era after communism."[27] Newspaper coverage of the battle was heavy and slanted in favor of the gardens.

The day before the auction, the New York Supreme Court stopped the action, stating that the city needed to prove that there would be no environmental harm from the sale. Whereas before the city would not deal with the Trust for Public Land, last-minute negotiations resulted in the purchase of 113 gardens. The Trust for Public Land bought sixty-two gardens and the remainder were purchased by New York Restoration, an organization spearheaded by actress Bette Midler, for a total of $4.2 million. Under the new arrangement, TPL acts as an interim land trust and the sites continue to be managed by community residents. The goal is to develop multineighborhood land trusts to own the sites permanently. Currently, TPL is working with gardeners to establish new nonprofit organizations and land trusts to maintain the gardens into the future.[28]

Protecting the 113 community gardens from auction did not end the instability of community gardening in New York. In July 1999, for example, seven Harlem gardens were bulldozed the day before a court-ordered stay of demolition went into effect. The lines were drawn much as in past conflicts over garden sites: the city council and development proponents argued that the sites were needed for housing, while garden advocates countered that there were other, equally developable vacant sites nearby. Thus, success in saving some gardens does little to assure the protection of others because no policy or procedure is in place to either protect gardening as a form of land use or establish clear distinctions between garden sites and other vacant lots. In a special report published after the successful purchase of the 113 gardens, the Green Guerillas called for further action:

> Clearly, this was a victory in the short run. The loss of these gardens—including some of the oldest gardens in New York City's network—would have been devastating. A precedent would have been set allowing the City to sell community gardens to the highest bidder without current public reviews and with no plans to replace the services lost. But forcing supporters

of community gardens to pay the City millions of dollars to secure a future for community gardeners is bad public policy. We simply cannot allow this to become the model for garden preservation in New York City. Community gardeners and garden supporters must continue to press on with demands for public policies that preserve and protect community gardens without private money. Gardeners deserve nothing less.[29]

Tessa Huxley, director of Green Guerillas from 1980 to 1985, made the following observation:

> I think in the end New York is very pro-developer, and they hate the idea to take potential development out of the realm of potential. I don't understand why, since it seems so motherhood and apple-pie to me. If you put a city planning hat on and look at some of the neighborhoods like the Lower East Side, which has had forty-odd community gardens in a square mile, there was a very good reason to have so many even if it seems like a really big number. Because when it had been fully built out, before all the fires and so forth, if it had been a city, it would have been the densest city in the world. And that was too dense, we all know that is too dense, so the community gardens help to make it impossible to make it become that again. So you'd think from a city planning point of view that at least the city planning department would get it and help us. But in the end it seems like something the mayors don't want to get into. It is such a contrast to other cities like Seattle. Why, I still don't understand.[30]

NINE

A Look at Gardens Today

THIS HISTORICAL REVIEW OF URBAN garden programs would not be complete without a look at some current gardens and their participants. Selecting a few illustrative examples is difficult due to the variety of program objectives plus the influences of region and socioeconomic context. While the garden projects cited in previous chapters could be chosen with the benefit of hindsight, current projects are still evolving. As with most of the projects already described, the main sources of written information on current gardens maintain an encouraging, optimistic outlook that tends to downplay any obstacles or criticism. That said, such positive descriptions illustrate the kind of promotion that the garden programs are using to garner public support.

The projects included in this chapter demonstrate four of the ongoing roles played by garden projects today: neighborhood garden, community food security source, job-training and entrepreneurial garden, and school garden. The gardens profiled—the Wattles Farm and Neighborhood Garden, the Urban Garden Program of the Los Angeles Regional Food Bank, the Berkeley Youth Alternatives Community Garden Patch, and the Edible Schoolyard—can all be considered exemplary projects. Some might say that they are models for others to emulate; however, the particularities that shape each case cannot be replicated. To begin with, each of these projects is located in California and therefore benefits from a nearly year-round growing season that might be the envy of project organizers in the Midwest or the East. Besides climate, other factors such as political connections, land availability, and funding sources determine the nature and scope of each project.

THE NEIGHBORHOOD GARDEN

Neighborhood community gardens remain the most common representation of urban garden programs today. Such gardens give individuals in cities, suburbs, and small towns alike the opportunity to sign up for a plot of land to garden. Some gardens are efficiently subdivided to yield the maximum number of plots. Many include shared spaces, such as sitting gardens, areas of native habitat restoration, butterfly gardens, communal orchards or herb gardens, and play areas. Just like the perennial plants that the gardeners grow, the neighborhood garden itself goes through cycles of use. During the wet and cold seasons, it may appear unoccupied and the debris of the previous season—tomato cages, buckets, hoses, and so forth—may give the garden a forlorn, abandoned quality. However, on spring days—especially in the evening hours after work or on the weekends—the garden comes alive with people reclaiming their plots for the upcoming season. On a sign usually posted at the gate, passersby can see who to call for information about getting a plot as well as announcements for public events, such as an upcoming spring workday or a harvest fair in the fall. Even though gardens often are locked and so available for use only by the gardeners, many gardens arrange public events, open houses, or other forms of community outreach.

For many of the gardeners and their families, the neighborhood garden serves as a semipublic park. Time spent in the garden is considered recreation and restoration by most gardeners. Removed from both the workplace and duties at home, the garden provides a place where one can focus on a gardening task or simply putter. This appreciation of neighborhood gardens as open space was validated in a study by landscape architect Mark Francis.[1] A community garden adjacent to a park in Sacramento, California, provided the ideal context for comparison. Through interviews with gardeners, city staff, and nearby residents, Francis found that the park cost twenty times more to develop and was twenty-seven times more expensive to maintain than the community garden, yet received less than one-quarter of the use the garden did. The report showed that city residents—gardeners and nongardeners—valued community gardens as open space more than city officials.

The people who use neighborhood gardens vary as much as the plants that are grown in them. Some gardeners come from the geographical area around the garden, often reflecting the demographics of that neighborhood. Others seek out neighborhood gardens and may travel greater distances for the opportunity to tend a plot. Researchers have conducted studies to understand who gardens and why, but the small sample sizes and the particularities of most gardens due to location and social context make it difficult to generalize to gardeners nationally.[2] Popular journalism and publications about gardening tend to describe a kaleidoscope of gardeners in terms of ethnicity, gender, and age. There is a general assumption that most community gardens are in low-income communities, but they also appear in middle-class and gentrifying communities. A neighborhood garden is just as likely to be located on a vacant lot surrounded by single-family homes in Los Angeles as between the apartment towers of Battery Park City in New York.

Even though hard data on neighborhood gardens are difficult to obtain, the literature is full of anecdotal evidence from the gardeners describing the personal benefits gardening has brought them. Such stories highlight the experiences of individuals who reconnect with their rural background, first-time gardeners who are thrilled with their newfound connection to natural processes, and others who, in the process of building their garden, were empowered to take on other community problems. Some people are growing food to save money, while others grow gourmet varieties or ethnic vegetables not readily available in stores.

Especially in urban settings, neighborhood gardens express the cultural diversity of their participants. In cities such as New York, Seattle, or Los Angeles, immigration patterns are reflected in the design and planting of community gardens. The *casita* gardens in the Lower East Side of New York display the Caribbean agricultural traditions of the primarily Puerto Rican gardeners as well as other symbols of Puerto Rico—religious iconography, masks, and murals. Besides providing a space for families to grow vegetables, fruits, and medicinal and culinary herbs, the casita gardens are social spaces that often include a small structure surrounded by a clean-swept yard known as the *batey*.[3] In Los Angeles, the gardens at the Watts Growing Project are reminiscent of gardens and farms in Mexico and Central America. Cultivation practices and crop selection here create a space distinctly different from the gardens cultivated by Laotian and Hmong immigrants in Seattle.

The Wattles Farm and Neighborhood Garden

Driving west on Hollywood Boulevard, about three-quarters of a mile from the famous Mann's Chinese Theater, one begins to see the dense vegetation of a thick avocado orchard along one side of the street. A remnant from a different era in Los Angeles's history, the orchard now forms a screen along one side of the Wattles Farm and Neighborhood Garden, hiding the busy activities taking place there.[4] To the north of the 4.2-acre garden, the old Wattles Mansion, built in 1907, now houses a nonprofit historic preservation organization, Hollywood Heritage. Viewed from above, the garden is the southern tip of a finger of green that starts at Hollywood Boulevard and extends up into the mountain park of Runyan Canyon, an important open space for the surrounding communities.

The origin of the Wattles Farm and Neighborhood Garden dates back to 1972, when Mark Casady of Mayor Tom Bradley's office initiated the Los Angeles Gardens and Farms Program. With funds from CETA, this program started over twenty gardens, including the Wattles Farm. The garden site was leased from the Department of Recreation and Parks. Groundbreaking for the garden began in 1975, when the initial thirty members started clearing brush and weeds and tried to rejuvenate the neglected avocado orchard. To manage garden affairs with greater autonomy, the gardeners formed a nonprofit organization in 1978. By the mid-1980s, the garden had 165 members.

Since Wattles Farm is on a hillside, the 169 garden plots, each measuring approximately fifteen feet by fifteen feet, are terraced to maximize sun exposure and minimize erosion. The layout leaves many irregular areas along the edges of the garden and at other odd spots that members can claim as a "community plot" to grow plants for general use and enjoyment, such as flowers, native plants, herbs, and so on. For instance, one long-time gardener who grew up in the Philippines has developed a community plot devoted to exotic tropical plants such as coffee, bananas, and special herbs. There are also communal orchards where plums, apricots, peaches, and other fruit are grown. At the transition from the avocado orchard to the gardens there is a small picnic area, outdoor classroom, tool shed, and compost area. A bulletin board announces meetings, concerns, and other information. Given the large site and its urban context, the entire garden is surrounded by a tall chain-link fence, with two locked entrance gates.

To receive a plot, gardeners must sign an agreement and pay yearly dues. Rules and regulations cover cultivation practices as well as responsibilities

Figure 63. The Wattles Farm and Neighborhood Garden in the Hollywood hills, Los Angeles. Photograph © by Lewis Watts, 2001.

that members bear as part of the group. The gardeners must agree to keep their plots and the surrounding paths free of overhanging plants, weeds, and debris. To avoid casting shade on other plots, gardeners are not allowed to plant tall trees, shrubs, or vines, or build obtrusive structures. Organic gardening is mandated, and use of chemical herbicides and pesticides is prohibited. Each gardener is expected to contribute at least one hour per month to community cleanup. The second weekend of the month is designated as cleanup time. There are also occasional general meetings to elect officers and discuss matters pertaining to the garden. A fourteen-member elected board of directors runs the garden. A subcommittee, called the Gardenmasters, oversees various responsibilities, including new member orientation and enforcement of rules and regulations. If a gardener is not maintaining his or her site, a Gardenmaster has the authority to first warn the gardener and then give a termination notice.

While it is hard to see any of the garden from the street, visiting during a weekend workday may provide the opportunity to ask permission to take a stroll through the garden. Just such an occasion arose for me a few summers ago when I visited the garden during the annual garden meeting and potluck. The gardeners present ranged in age from children to seniors.

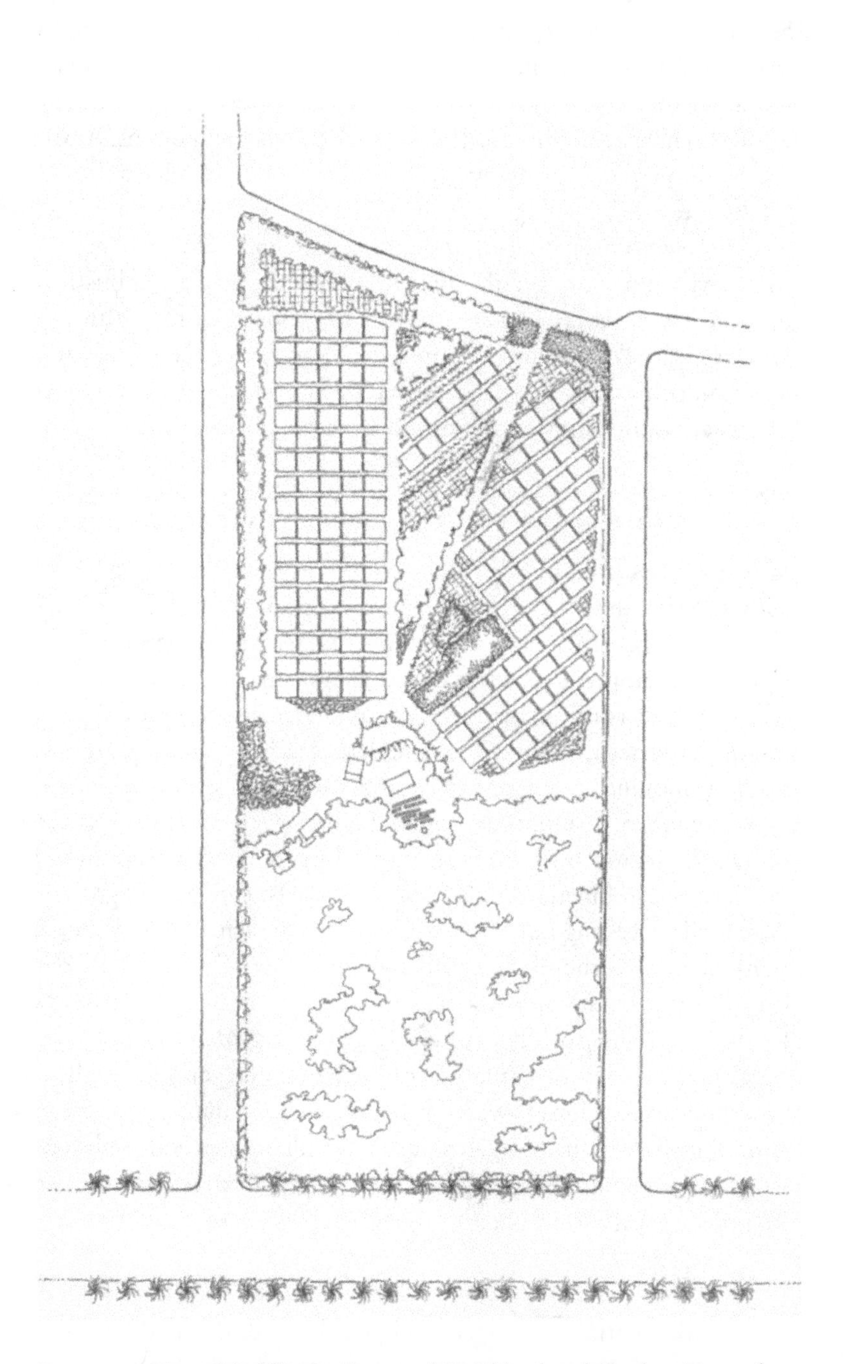

Figure 64. Plan of the Wattles Farm and Neighborhood Garden. Drawn by Kristofer Johnson.

Several older women spoke with strong Russian accents and had brought traditional Russian foods to the potluck. After the meeting, two of the founding members—now in their seventies—went to work in the communal orchard, climbing ladders to pluck peaches and plums. At the other end of the garden, a group of young women were enjoying a picnic under shady trees after cleaning up their shared plot and the surrounding paths. As I spoke with the gardeners and sampled their tomatoes and greens, I was struck with how removed this place felt from the hustle and bustle of Hollywood. However, as I looked out of the garden to the palm trees and rooftops of the city below, I appreciated how appropriate this garden is to its urban context. In a city where views determine housing prices, the gardeners at Wattles enjoy a million-dollar view.

COMMUNITY FOOD SECURITY AND GARDENING

In recent years, community food security has emerged as a new concept in addressing urban food systems in light of nutritional needs, economic factors, and cultural expectations. Community food security is the ability to acquire culturally appropriate food through local, nonemergency sources.[5] Because it offers a way to consider the problems of food access comprehensively, community food security links the concerns of urban planners, farmers, food banks, community development corporations, social service agencies, and environmentalists. Community gardens figure into the discussion as a resource for people to grow fresh fruits and vegetables for their own consumption or for sale. In combination with farmers' markets, cooperative groceries, and local food processing centers, the community garden adds resiliency and options to neighborhoods, particularly inner-city neighborhoods that have been deserted by the large supermarkets and, as a result, have fewer resources for fresh food and pay higher prices. In addition, gardening allows greater variety of food choice, particularly access to ethnic and gourmet foods. This has economic implications as well. The USDA estimated in 1993 that urban gardeners involved in its programs grew $16 million worth of fresh food. One study of gardens in Newark, for example, found that an average 720-square-foot plot earned $500 from a $25 investment.[6]

The Urban Garden Program of the Los Angeles Regional Food Bank

In a South Central Los Angeles neighborhood dominated by warehouses, auto salvage yards, and train tracks, the sight of a block of greenery punctuated by banana trees and twelve-foot-high corn takes a visitor by surprise.

Figure 65. Founding members of the Wattles Farm and Neighborhood Garden are still involved. Photograph © by Lewis Watts, 2001.

Surrounded by a tall chain-link fence topped with barbed wire, the garden may not initially strike the viewer as beautiful, but if the banana fronds and corn tassels entice you to go through the gates, you will find yourself transported into a very different world. For here at the Urban Garden Program of the Los Angeles Regional Food Bank, low-income families are making use of an opportunity to grow culturally appropriate food and subsidize their household food expenses.[7]

In the aftermath of the 1992 civil disturbances in Los Angeles, the idea of starting a garden emerged as a way to heal the community. Spearheaded by the Los Angeles Regional Food Bank, the development of the garden complemented the food bank's mission to provide food for the city's needy. The initial site was a 7.5-acre vacant parcel owned by the City of Los Angeles and adjacent to the food bank. It had been empty following failed attempts to develop first an incinerator and then housing on the site. Setting up the garden cost just over $8,000, with funding and support coming from the USDA, the City of Los Angeles, thirty Los Angeles restaurateurs, and others.[8] Initial site preparation was provided by the city and the Los Angeles Conservation Corps. The original garden consisted of sixty individual plots of four feet by sixteen feet, and one hundred family plots of forty-five feet by fifty-five feet. In 1995, the food bank received a $19,500 USDA Urban Resources Partnership grant to expand the project to include another seven-acre site, where an additional 150 plots were added.

Anyone qualified to receive food from the food bank is eligible to sign up for a garden plot. With all of the approximately 315 plots already assigned, there is a waiting list of potential gardeners. Most participants are immigrant families from Mexico, El Salvador, and other Central American countries, as well as Caribbean and African American families. The Food Bank provides water and some seed and supplies. It also employs an on-site garden coordinator who gives technical assistance and oversees general maintenance. LA Harvest, the nonprofit organization that evolved out of the Los Angeles Cooperative Extension Urban Garden Program, provides education as necessary. Besides the family gardeners, a local elementary school has a garden here. This garden has a distinct look. The sides of the tall chain-link fence surrounding the entire site are punctuated by central gateways that are locked in the evening. The garden is bisected by a wide dirt driveway where gardeners park their cars. Some have set up small enclosures that serve as gathering places and stores. The foot traffic makes the garden an ideal marketplace for selling garden produce as well as other wares, such as compact discs, toys, and ice cream. The plots—averaging 1,500 square feet—are larger than an average community garden plot and provide room for a range of crops. With boundaries marked by fences, cactus, and tall plants, the garden plots are more like enclosures or rooms than is typical for neighborhood gardens that have smaller plots. Many gardeners have built small shelters in their gardens for food preparation, eating, and socializing. Families grow foods from their homelands: tomatillos, jicama, *chile de arbol,* chayote, *verdolaga, epazote,* bananas, and other crops. In addition, people have personalized their gardens: one garden includes a shrine to the Virgin of Guadalupe, and another displays the flag of Mexico. However, the garden's future is now tenuous. The gardeners are embroiled in a site-ownership lawsuit that has also raised concerns over ethnic and racial representation in the garden and local district. Participants still garden, but now their gardening is also a political act.

JOB-TRAINING AND ENTREPRENEURIAL GARDENS

Given the ready market of urbanites hungry for fresh, high-quality produce and the need for new economic opportunities in communities that often have unused plots of land, it is easy to understand why social service institutions, philanthropic organizations, and garden organizations have started

Figure 66. At the Urban Garden Program of the Los Angeles Regional Food Bank, families from Central and South America grow many of the staples and specialty crops necessary for traditional cooking. Some entrepreneurial gardeners grow extra to sell or barter. Photograph © by Lewis Watts, 2001.

entrepreneurial and training programs that link agricultural production and sales with education and job training. This was particularly the trend in the 1990s. The intention was to spark economic development in local communities while also providing a social outlet and educational services. Most of these programs serve a constituency that has limited access to jobs, such as at-risk youth and adults who are disabled, homeless, or recently out of incarceration. Participants are typically paid, and they work fifteen to twenty hours per week. Their time is divided between garden upkeep and training programs that teach basic job, technical, and leadership skills. They may also learn marketing skills through the sale of produce at the garden, at farmers' markets, and to local restaurants. Some programs have developed community-supported agriculture arrangements, in which neighbors are shareholders who pay a set amount and receive a box of produce weekly or monthly. Several gardens with an entrepreneurial bent have forayed into processing value-added products, such as salsa, vinegar, or crafts. These efforts tend to be structured as small income-generating businesses within the organization. The proceeds help fund the employment program and are supplemented by grants, donations, and other sources.[9] Given their multi-

Figure 67. Altar with statue of the Virgin of Guadalupe at the Urban Garden Program of the Los Angeles Regional Food Bank. Photograph © by Lewis Watts, 2001.

dimensional approach, these projects attract the interest of educators, environmentalists, activists, and business people and invite them to literally get their hands dirty as they contribute to community building and local economic development.

The Berkeley Youth Alternatives Community Garden Patch

On a Tuesday afternoon in the summer of 2002, the Berkeley Youth Alternatives (BYA) Community Garden Patch was a busy place.[10] Two of the youth gardeners had just left for the Berkeley Farmers' Market laden with produce, flowers, and plant starts to sell. Meanwhile, at the garden, the community-supported agriculture manager, also a high school senior, was busy checking to see that each box of food had its share of produce and was ready for pick-up or delivery. Already one of the neighbors had arrived to pick up her box of produce. The Ped-Ex delivery person was loading boxes onto his bicycle for delivery to people who could not pick up their produce and who had paid an additional charge for this service. In the back of the garden, the garden supervisor turned a compost pile while another youth gardener watered a newly planted arugula bed. The next day was just as busy, as children from the BYA Team Nutrition program swarmed to the children's garden to see if any new strawberries had appeared.

The idea for the Garden Patch started in 1993 when Berkeley Youth Alternatives Executive Director Niculia Williams, tired of seeing children arrive at the youth center with fast-food breakfasts, decided to start a small garden. Very quickly, the idea blossomed into a much grander plan to serve BYA's mission in multiple ways. BYA is a long-standing nonprofit youth center that provides a variety of services, including after-school and summer programs, sports leagues, crisis counseling, academic tutoring, and youth employment, to predominantly low-income, at-risk children and their families. The envisioned garden would expand existing programs by providing a children's garden, community activities, and a youth market garden. As part of the budget-conscious nonprofit sector, BYA envisioned the youth market garden as an income-generating venture that would pay for all the programs of the Garden Patch.

To garner support and better serve the neighborhood, BYA held a series of community meetings to develop garden goals and site design. The initial site—behind the old bread factory where BYA was housed—was too shady for a garden, so Williams and her board arranged to lease a nearby half-acre site from the city for one dollar per year on a three-year renewable lease. In September 1993, BYA youth and staff, neighborhood residents, city officials, and about one hundred Summer-of-Service volunteers celebrated the groundbreaking of the Berkeley Youth Alternatives' Community Garden Patch, which was so named by children at the BYA center. And a groundbreaking it was, since the site, previously a railroad right-of-way and then a storage area for city equipment, had compacted and degraded soil that required considerable reclamation work. The program started with almost no funding and relied on volunteers and in-kind contributions of soil, wood, gravel, and plants. Workdays were held every month for the first year, and volunteers were sought from the neighborhood, service organizations, environmental groups, and people working off traffic and other civic violations through community-service hours. Over time, the BYA staff acquired grants to hire youth to work in the garden after school and on weekends. Over the five years it took to develop the garden, the initial pickaxes and sledgehammers gave way to spades and hoses as the compacted earth became a lush garden.

Based on community input, the garden was designed to be attractive

Figure 68. This photograph shows the condition of the site for the BYA Community Garden Patch in 1993, before volunteer workdays began. Photograph by author.

from the street. A see-through wood-and-wire fence is set back to provide a fifteen-foot-wide flower garden along the sidewalk that is always accessible to the public even when the rest of the garden is closed. A small bosk of fruit trees and flowerbeds aligns with the street. An office located in a shed, decorated with paintings by BYA children, is near the front of the garden and can be opened as a market stand. Next to it is the greenhouse, built with rammed-earth walls and recycled window sashes. An existing willow tree provides shade for an outdoor classroom. Adjacent to this space are fourteen community garden plots. Away from street traffic is a children's garden, designed and built by a UC Berkeley landscape architecture student working with BYA youth employees. At the rear of the garden is a composting area. The rest of the garden is devoted to beds for the market garden.

The Garden Patch is intended to be a community resource. The community garden plots are actively used, with a waiting list of people who want plots. During the spring and summer, BYA's Team Nutrition program brings children from BYA's after-school and summer programs here to

Figure 69. By 1996, the Garden Patch was nearing completion. Crops are grown in raised beds as well as in the ground after much soil amending. Photograph by author.

grow food that is then used to make snacks, and discuss nutrition. Occasionally, arrangements are made with local schools and daycare centers that want to use the garden; for instance, one year, children from a nearby school for developmentally disabled youth had several plots they maintained weekly. A sign at the fence encourages the public to visit the garden when it is open after school and on Saturdays. Social events, such as the annual harvest fair and the Bay Area Open Garden Weekend, provide opportunities for extended outreach. Occasionally, when interest is shown and funding is available, the BYA garden supervisor and youth provide educational workshops on composting and gardening.

The heart of the Garden Patch program is the youth market garden, which employs between two and six young people, depending on funding. They work fifteen hours per week under the direction of a garden supervisor. Since the first sale of lettuce at a farmers' market in July 1994, which earned sixty-seven dollars, the market garden has expanded into a wider range of organic produce, flowers, plant starts, and a few value-added products such as wreaths and garlic braids. Youth sell their products at the

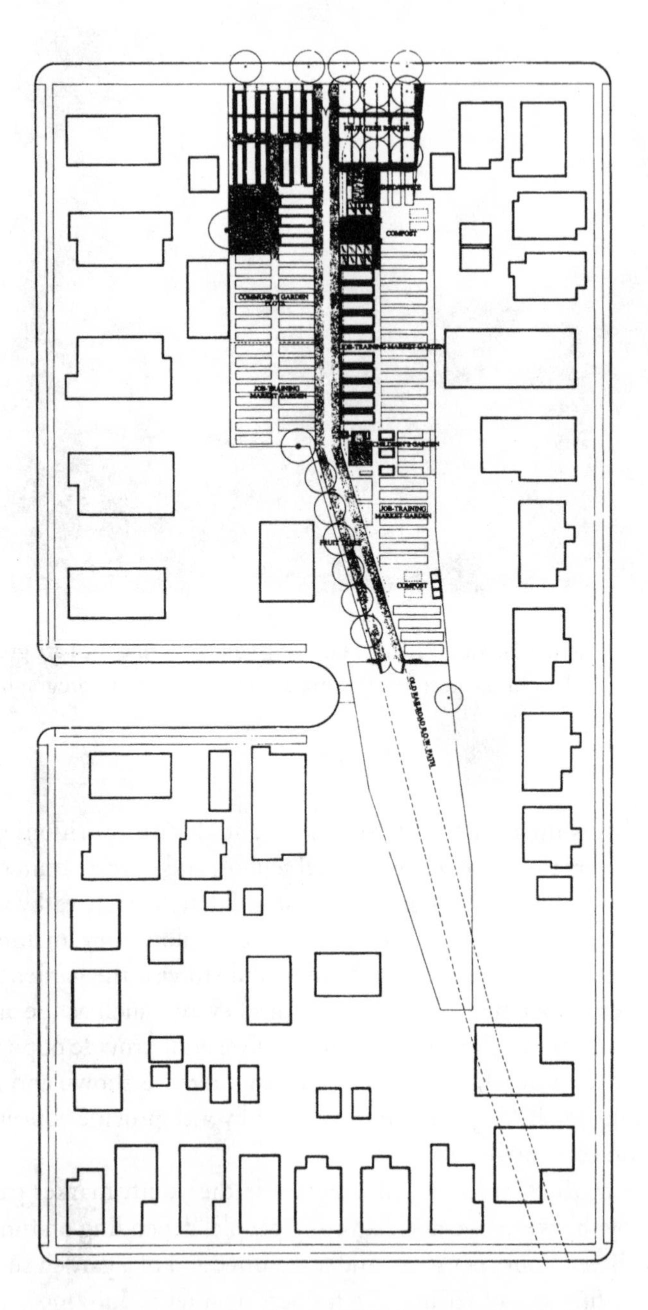

Figure 70. As the plan for the BYA Community Garden Patch shows, the garden fits within a block on what used to be a railroad corridor. The immediate neighborhood includes single-family houses, apartments, and a park. Drawn by author.

Figure 71. A Garden Patch youth employee preparing produce for the farmers' market. Photograph © by Lewis Watts, 2001.

garden, at the farmers' market, to local restaurants, and occasionally to a small local grocery store. In 1998, the garden earned $10,223 from sales. In 1999, BYA established a community-supported agriculture program and a flower delivery business. In addition to gardening, marketing, and record-keeping, the youth employees spend a few hours per week in training, which includes landscaping techniques, resume writing, computer skills, counseling, and field trips.

Concerned that the youth market garden was not meeting the goal of making the Garden Patch financially self-sufficient as expected, in 1998 BYA engaged researchers to explore the Garden Patch's market potential. The process of internal audits, interviews with experts, and surveys of existing and potential client markets revealed that the market garden was earning approximately one-seventh of its budget, with the rest of the funds coming from grants, donations, and in-kind contributions. Factors such as the small site, untrained workers, and time spent on training and community outreach, as well as crop selection choices based on educational benefit rather than profit, all contributed to the poor financial returns. As a semi-autonomous program within the BYA organization, the direction of the

Figure 72. A Garden Patch youth employee sells produce and flowers at the Berkeley Farmers' Market. Photograph © by Lewis Watts, 2001.

project was largely determined by the Garden Patch supervisor. Changes in personnel led to shifts in the emphases given to community outreach, education, gardening practices, and marketing. At the same time, data collected from interviews, focus groups, and surveys showed tremendous community support for the program as a youth training opportunity and community desire for locally grown fresh produce. Wages were the largest expense associated with the program, but about one-third of employees' time was devoted to training and other nongarden activities. Together, these findings helped BYA develop a realistic vision for the youth market garden. As of 1999, the youth market garden was expected to earn 15 percent of its budget and to expand the public exposure of its training and youth services.

The Garden Patch has continued to evolve. The project has expanded to three sites with varying degrees of use. The original Garden Patch remains the focal point for the project. A fire that destroyed part of the BYA center in 1995 left a site adjacent to the center that was made into the BYA Flower Garden, though its use is now sporadic. The BYA/Chaparral House Intergenerational Garden, located on the grounds of a nearby senior assisted-living center, has been used by the BYA youth market garden, seniors, and local residents. The garden still provides a training and employment opportunity for BYA youth as well as other young people assigned to the project by other youth service agencies. In 2002, garden supervisor Jason Uribe estimated that garden sales provided $18,000 of the $65,000 budget, and most of this income had been generated by the community-supported agriculture project. At that time, there were twenty-five members—six full members who paid $75 per month for a weekly box of produce, seven who paid $45 for half subscriptions, seven who paid $25 for quarter subscriptions, and five who were low-income and paid $12 per box. Although the flower bouquet delivery business was closed in September 2001, BYA youth continue to sell flowers and bouquets at the farmers' market and to community-supported agriculture participants. Then, in 2004, BYA staff decided to end the community-supported agriculture program because of its time-consuming organizational needs. Produce continues to be sold at the farmers' market and to BYA families and neighbors. The program keeps seeking to balance earned income and funding through grants and other sources. The BYA staff is always looking for new funding opportunities that will allow for more youth employment and community involvement.

SCHOOL GARDENS

Children have always been a special audience for urban garden programs. Today, a rebirth in interest has led to new programs that use gardens to teach natural science, nutrition, interpersonal skills, and self-esteem. With more people subsisting on diets of processed foods and the sedentary activities that fill children's days, gardens also address concerns about the increasing number of overweight children and the need for teaching better eating habits.

Currently, California leads the effort to establish gardens in schools. In 1995, California's Superintendent of Public Instruction Delaine Eastin established a garden agenda within the Nutrition Education and Training Program, with the stated goal of "a garden in every school" by 2000. A survey in 1996 found that at least one thousand schools had instructional gardens; estimates are that it will take a few more years to reach all eight thousand eligible schools. The goal of the program is to increase appreciation of fresh vegetables and fruits and to make connections between food and agriculture. Another program that has received good publicity in national magazines and newspapers is the school garden program in San Antonio, Texas, where 192 school gardens were started between 1990 and 1998 with the help of Bexar County Master Gardeners and Cooperative Extension Services.

The National Gardening Association has also initiated a national campaign to promote "a garden in every school."

The Edible Schoolyard

> The mission of the Edible Schoolyard at Martin Luther King Middle School is to create and sustain an organic garden and landscape which is wholly integrated into the school's curriculum and lunch program. It involves the students in all aspects of farming the garden—along with preparing, serving and eating the food—as a means of awakening their senses and encouraging awareness and appreciation of the transformative values of nourishment, community and stewardship of the land.[11]

Where asphalt once provided a monochrome landscape for children to play on, a one-acre garden now catches the eye with a multitude of textures and colors. Here, children receive thoughtful and holistic lessons about the natural world, including their own bodies, through gardening, cooking, and eating. The Edible Schoolyard at Martin Luther King Jr. Middle School in Berkeley, California, has been put forward as a model program to inspire school gardens elsewhere, although its unique context and leadership have created rare opportunities that make this garden exceptional.[12]

The project started in 1994 with the wishful thinking of a school principal and a local restaurateur. During a radio interview, renowned restaurateur Alice Waters of Chez Panisse, in Berkeley, commented on the Martin Luther King Jr. Middle School grounds, which she passed daily on her way to work. Mostly composed of asphalt, dried-up grass, and leggy shrubs, the grounds could not inspire children, she thought. Having heard the interview, the principal, Neil Smith, called Alice Waters and invited her to tour the school to see some recent improvements, and during the tour he asked for her suggestions and help. Ms. Waters's vision of an entirely edible, organic landscape provided the inspiration that eventually led—after much organizing and fundraising—to the Edible Schoolyard. The official groundbreaking ceremony was held October 8, 1995.

The garden has a slightly wild appearance, with intermixed plantings, winding paths, and a seasonal creek meandering through the site. Pro bono advice from architects and landscape architects, volunteer work by a stonemason, and the efforts of the on-site garden manager have combined to create a garden rich in detail and visual delight. The garden also includes a circular willow gazebo for class discussions, a bread oven, and a composting

Figure 73. The Edible Schoolyard has a wild look even while it produces the ingredients for creative food and craft projects as well as science lessons. Photograph by author, 1999.

area. The garden is open to the public whenever the Schoolyard is open. Beside the garden there is an office to support organizational activities and a kitchen and dining room for cooking classes and meals.

The Edible Schoolyard is managed as a nonprofit organization that works in partnership with the school, with its offices located on the school property. With its own full-time staff, Americorps workers, and many volunteers, the Schoolyard actually adds to the staffing resources of the school. Having a celebrated chef as the founder has helped with publicity and funding for the program. The costs associated with staffing and materials are funded through the Chez Panisse Foundation, private donations, grants, and corporate donations. However, curriculum development and student participation require teacher involvement. Together, staff and teachers develop study plans that correlate garden activities with classroom work. For example, a history class studying a particular culture might cook a meal reflective of that culture in the Schoolyard kitchen. Teachers are encouraged, not required, to use the garden. A core group of teachers utilize the garden as much as possible in their teaching while other teachers do not use it at all.

Figure 74. In a temporary building, students at the Edible Schoolyard learn how to use the garden's produce to prepare healthy and tasty meals. Photograph © by Lewis Watts, 2001.

As this book goes to press, the school serves over 800 students in the sixth, seventh, and eighth grades. Sixth-graders work in the garden in the fall as part of their math and science curriculum, visiting the garden 1½ hours per week for eight to ten weeks. In the spring they move to the kitchen. The garden serves seventh-grade classes in humanities/social science studies. They start in the kitchen in the fall and then move to the garden in the spring. Eighth-grade science classes frequently use the garden for intensive study or a specific project, such as creek restoration or soil studies. A summer program is also run by Edible Schoolyard staff and volunteers.

Although the Edible Schoolyard is described as a model of school gardening, critics suggest that its celebrity sponsorship and its location in progressive Berkeley are special circumstances that have assured its success. Only the future will prove whether similar programs can develop elsewhere. Indeed, Berkeley is unique in its support of school gardens and local food systems. Such gardens have a long history in Berkeley: the Garden City (described in chapter 2) was built on the University of California's campus in the 1910s. The LeConte Elementary School garden began in the early twentieth century and

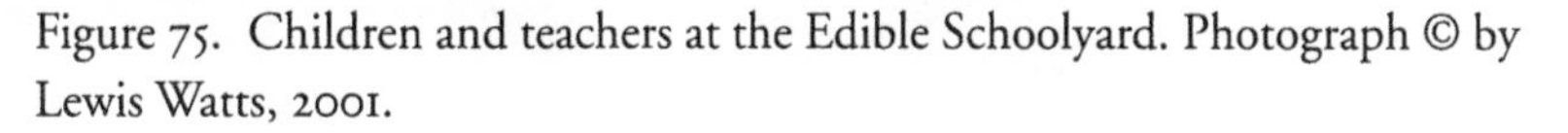

Figure 75. Children and teachers at the Edible Schoolyard. Photograph © by Lewis Watts, 2001.

was reinstated in the 1980s. The Berkeley Garden Collaborative connects a network of garden groups, environmentalists, and others to promote community gardens, job-training gardens, school gardens, and similar projects. The Center for Ecoliteracy, a nonprofit organization founded in 1995, established the Food Systems Project to create a school food policy, establish gardens, integrate gardens into classroom curricula, and challenge the typical approach to school food service with options for fresh produce and garden-grown foods. In 1999, Berkeley's school board passed a measure intended not only to encourage students to garden but also to incorporate school-grown food and other locally grown, organic food into school kitchens.

The projects described in this chapter reveal the kind of nurturing that is required to take a garden project from initial idea to completed site and then sustain it. Each one has evolved in its own way in response to the changing needs of its constituents, swings in funding and support, and unforeseen opportunities. Thus although these projects have been put forward as models, each is unique enough with regard to location, constituency, and other programmatic factors that they cannot really be duplicated; however, they can serve as sources of inspiration and information. Each of these projects offers lessons in taking advantage of opportunities, working with obstacles and setbacks, and coming up with creative solutions. Revisiting and evaluating them again and again over time are the way to understand what it takes to keep a successful garden program growing.

CONCLUSION

Sustaining a City Bountiful

PROVIDING A PLACE FOR PEOPLE to garden in the city seems to be a simple idea. And as this historical overview has shown, programs have surfaced time and again to facilitate this need. Each phase of urban garden promotion—vacant-lot cultivation associations, school gardens, civic beautification gardens, war gardens, relief gardens, victory gardens, and community gardens—has been shaped by its own social, political, economic, and environmental context. Yet when taken together, these phases are expressions of a consistent thread through American social history: the impulse to provide communal land so that people can garden. This overview has also shown that the promoters of urban gardens have rarely considered them simply as places to grow food and flowers; rather, they have viewed them as a means to address much larger social concerns, such as economic relief, education reform, and civic accord. Whether praised by the president or an anonymous gardener, gardens have been lauded as symbols of hope for a better, more cooperative, and more beautiful and healthy world.

Yet appealing as the idea of urban gardens may be, it has been difficult to sustain the gardens themselves. In the past, programs were developed to be temporary, so that with each new crisis, new organizations and procedures had to be invented, even though similar appeals were used each time to obtain public support and land. Today, as advocates assert that gardens should be permanent community resources, we are poised to change this legacy. It is a good moment to consider the purpose of urban gardens, why they have been conceived as temporary, and what actions are needed to validate them as permanent community resources. At issue is the contradiction between the general appeal that the *idea* of urban gardens garners and the historical ambivalence toward sustaining gardens and programs. The public acknowledges the benefits from urban gardening during crises, so why does support for the gardens dwindle in times of peace and plenty? The litany of benefits associated with gardening has effectively obscured the real issue: Is the urban garden intended as a means to other ambitions or an end in itself? If it is a means to other ends, the garden is only useful until those other goals are met. If, however, the garden is the desired product, the issue of permanence can be argued with greater authority. To answer this question, we must step away from an intuitive faith in gardening and examine the imbedded cultural assumptions that make gardening a solution to so many different issues and concerns. To do so uncovers the conscious and unconscious goals that influence urban garden programs.

GARDENING AS A MEANS TO OTHER ENDS

The reasons given to justify urban gardens have remained surprisingly consistent over the past one hundred years. When faced with economic depression, inflation, or disenfranchised communities, urban garden programs have been opportunities to subsidize household food expenses, provide job training, and help enterprising individuals earn income. In times of civic unrest, the garden has been a nucleus for building morale, developing relationships with neighbors, and restoring the soul. Faced with environmental degradation, gardens brighten up declining neighborhoods, instill a sense of pride, and encourage other kinds of community improvements. If one primary concern instigated the campaign, other benefits were quickly listed as further justifications. For instance, the primary objective of vacant-lot cultivation associations was to provide economic relief, but they also advanced certain social agendas: encouraging a work ethic, providing healthy recreation, and inspiring migration back to the country. In this manner, urban garden programs have dovetailed with many different agendas, including the desire to disperse urban populations, Americanize immigrants, improve education, ease labor relations, and counter social anomie. As an action with a satisfyingly tangible outcome, gardening has been an almost knee-jerk response to crisis.

To understand why gardens address so many different concerns requires looking to the core cultural values that find expression in urban garden programs: the value of nature in the city, and the value of individualism. These values influence program development—whether impelling a philanthropic group to support a garden for its educational potential or a neighborhood group to start a community garden as an act of self-reliance—and the participatory nature and tangibility of gardens express and promote these values. In this capacity, garden programs serve to further a vision of what *should be* in times when society is unclear about where the future is heading.

Gardens as Nature in the City

As "nature in the city," the urban garden gives its urban constituency the opportunity to express values associated with countryside and wilderness. In truth, of course, the city is as "natural" as the garden—both express the relationships people have with natural systems that shape their habitat—but the need to juxtapose nature and the city is deeply embedded in American culture. Often serving as a foil to urban conditions, the urban garden is proposed as a counterbalance to urban stresses. When the gardener is on his or her knees planting in the soil, the paved streets and tall buildings are blocked from view by vegetation while the sounds of the city are slightly buffered by trickling water from a hose or the buzzing of insects around the gardener's ear. The garden has been a refuge to the nineteenth-century immigrant living in a cramped tenement district, the unemployed laborer tired of pounding the streets and needing a distraction, the harried war-industry worker during the world wars, and the disillusioned urbanite needing a break from a city reeling from civic unrest and injustice.

Besides providing a physical and emotional oasis, the garden also represents an opportunity to convey certain social values associated with nature. In particular, it provides a bridge to virtues attributed to country living and the agrarian lifestyle. One way to consider urban gardens is as an urban rendition of a traditionally rural occupation—agriculture. The nation's agricultural heritage endures as a significant influence on American cultural identity. Probably the most quoted appeal for an agrarian ethic comes from Thomas Jefferson, who stated, "Those who labor on the earth are the chosen people of God, if ever He had a chosen people, whose breasts He has made his peculiar deposit for substantial and genuine virtue."[1] Jefferson's yeoman-farmer image personified the values of self-reliance, democracy, and virtuous citizenship, and the connection between these social virtues and the act of cultivation found expression in urban gardening. Promoters of early urban garden programs hoped that urban participants would realize these virtues and thereby improve their economic and social situation, and perhaps even migrate from the city into the country. Today, when most Americans have little contact with agriculture, we still want the garden to remind us of our cultural history and create sympathy for farmers and love of the country. Thus, it is common to see farming imagery in urban gardens—scarecrows, hay bales, gazebos, and so forth. Community gardens are often called urban farms and the gardeners "city farmers."[2]

The garden has been envisioned as a place where nature can teach all her lessons. The specific lessons hoped for have often been linked to the social concerns of the time. Advocates at the turn of the twentieth century hoped that school gardens would teach civic responsibility and aesthetics along with natural science. Today, school gardens are intended to teach human connectedness to the environment and natural systems. Sometimes the lessons are didactic, conveyed through workshops and classes. But quite often the lessons are subtle and rely on personal enlightenment through the process of working in the garden—digging in the soil, witnessing worms, seeing detritus become rich soil, watching plants grow, and marveling at the production of fruit and flower.

Individualism and Self-Help

The hands-on, participatory quality of gardening has made the garden a self-help resource. The garden is a low-cost and direct means to obtain one of life's basic needs—food. An individual or a community can make use of an underutilized resource such as a vacant lot and, with a little sweat and nurturing, grow food with minimal outside assistance. Beyond feeding the gardeners, the act of gardening provides evidence of self-effort—that the gardeners are taking matters into their own hands and doing everything possible to improve their situation. To social reformers at the turn of the century, the garden was a means to separate the worthy from the unworthy among the poor. The programs were directed to the "self-respecting man" who did not want charity, but wanted work "that he may honestly earn food and shelter for his family."[3] Similarly, during the world wars and the 1930s depression, gardening was encouraged as an act of national self-help. Rather than sit idly and let the mind turn to dangerous topics, the individual could show his or her initiative and contribution to solving the national crisis. In a 1942 *House and Garden* article, editor Richardson Wright argued that gardening provided an expression of successful effort:

> This presumption of victory is in line with the blind faith that leads gardeners to sow seed—dead-looking seed that they tuck into the brown earth—with certainty and assurance that it will spring to life. Every garden ever made was a Victory Garden. In expansive eras of peace and the grim, restricted years of war alike, the purpose of gardening is to win a victory. Come wind and lashing rain and scorching sun, come plant pests and diseases innumerable, come merciless drought and unheralded frost, still, somehow these are surmounted until victory is attained. The crop is victory.[4]

Whether countering economic, political, or social constraints, the philosophy of personal responsibility for improvement was intended to apply to the neighborhood as well as to the individual. In both the past and the present, the act of building a garden on a vacant lot was intended to yield not simply a prettier lot but an inspired and activated community. Having started with such a satisfying product as a garden, the community would hopefully move on to other civic improvements as well.

That gardening satisfies deeply ingrained values associated with nature and individualism helps to explain why gardens are promoted during times of crisis, and why they have frequently been targeted to particular groups that society sees as needing acculturation. In times of social upheaval, we seek reassurance of order and purpose. When global wars and economic systems wreak havoc on our daily life, the tangibility of the garden keeps us focused on what we can do to improve our situation. We may not be able to stop war, global warming, or a depersonalized global economy, but we can grow our own food and build ourselves a refuge and oasis.

The situation becomes more complex when values such as these are structured into a program. Normative values—what should be—have infused the justifications for garden programs in both the past and the present. In the early twentieth century, environmental determinism—the belief that the environment determined people's behavior—found resonance in garden programs intended to teach the poor and immigrants to be better citizens and workers. Nature and self-help were pervasive themes in the language of the promotional literature. And—at the risk of sounding overly critical—there are still instances in which people propose gardening as a means to teach a particular worldview, environmental ethic, or organic diet. The cultural assumptions underlying such schemes are often unspoken. As a result, urban garden programs run the risk of carrying cultural biases and pursuing unclear goals that cannot be achieved.

The "Gardens as Good" Dilemma

Another way that cultural values associated with gardens find expression is in the general goodwill that urban garden programs garner. This goodwill has favored garden programs with leadership, participants, land, and other resources. Very few people will stand up and say that a children's garden is a bad thing. Few frown when they see a group of gardeners volunteer their time to make a garden on an otherwise vacant lot. Most people view urban garden programs as symbols of hope and positive action toward individual and social betterment. However, relying on vague benevolent approval avoids the exercise of clarifying and articulating a garden's real objectives. The attitude is that gardens are nice without being necessary. As a result, urban gardens hold little power when pitted against development or serious funding concerns. The assumption that gardens can garner support through their inherent goodness permits a lack of accountability about the good that gardens actually do.

The belief that urban gardens ameliorate a range of problems has produced righteous justifications of gardening without attention to actual results. In cases where food production is the core reason for an urban garden, the results can be measured by the monetary value of the food produced. But other less tangible benefits—restoration, health, civic-mindedness, and skills—have rarely been evaluated. Most historical and contemporary promotional material relies on anecdotal accounts to confirm these benefits. Furthermore, some of these broader goals are tied to economic, political, social, or environmental systems in complex ways. Thus it is difficult to assess the garden's impact separate from that of other community processes and activities. However, unsubstantiated claims for gardening's benefits do not help to sustain a purpose for gardens when the crisis they were originally meant to address has subsided.

Gardens as a Panacea

A potentially more serious by-product of an unarticulated faith in gardening as a cure for crisis is the tendency to propose urban gardens as treatments for largely unsolvable issues. The credibility of urban gardening as a change agent is compromised when the gardens serve as opportunistic, stopgap measures that in fact mask the real issue at stake. In times of crisis, the neighborhood garden becomes a place to go, to get active, to meet neighbors, and to make daily life more palatable. The tangible nature of the results satisfies political leaders and donors looking for a "photo op," while the larger issues that prompted the gardens in the first place, such as environmental injustice, educational disparity, or lack of economic opportunity, are more or less ignored. For example, turn-of-the-century reformers, unable or unwilling to address the flaws of the industrial capitalist system, justified vacant-lot cultivation as a self-help subsidy so unemployed laborers could eat until the economy improved. The secondary agenda that gardening would get participants out of urban industrial lifestyles and into the country largely failed because of inattention to how the urban poor could possibly afford to purchase land. Today, while a community garden may serve as a rallying point for community organizing, it cannot by itself solve the bigger problems facing urban communities—it cannot single-handedly stop drug sales on the adjacent street or the lack of public services to maintain vacant land. By promoting gardening as a solution to complex social concerns, not only is the investment burden for what turns out to be only a stopgap measure put on the individual and the neighborhood, but the unlikelihood of meeting these larger goals also devalues the benefits that gardens do provide. A similar shortsightedness is evident when gardens are promoted as environmental restoration without attention to protecting garden sites from development. Hours of volunteer labor and money for materials may improve a degraded vacant lot, but these resources are ultimately lost because the site was never conserved as open space with the long-term goal of ecological restoration.

Gardens satisfy cultural values and help people feel that they are doing something to improve their situation. Difficulties come when gardens become a tool for paternalistic goals or when the goals are too broad to be achievable by gardens alone. This analysis suggests the importance of articulating the objectives of an urban garden program in terms that can be validated. It also suggests that urban gardens need to be looked at as part of a larger set of actions that pertain to local, regional, national, or even global issues. Highly participatory, physically tangible, and generally liked, urban garden programs have many important community functions; they deserve greater scrutiny so they can be made more rigorous and the gardens more sustainable.

IMPLICATIONS FOR GARDENS AS COMMUNITY DEVELOPMENT

Considering urban gardens as an element in community development may clarify some of the vagueness concerning purpose and outcome that characterized earlier urban garden campaigns. Community development is a broad term that encapsulates a variety of social, economic, and physical improvements meant to empower a neighborhood or group so it can advance itself. To be successful, community development must not only address current conditions but also commit to dispelling larger economic and social forces that inhibit a community's self-actualization. While there are many neighborhood improvement projects that make urban living more fulfilling, an effective community development strategy requires attention to increasing the community's capacity to meet its social, economic, and physical needs. This definition of community development poses a challenge to urban garden programs. Many advocates already describe urban gardens as community development because they serve so many community functions, such as social interaction, beautification, economic opportunity, and education. However, do these programs actually empower individuals and communities over the long run? Clearly, some urban garden projects, particularly in the past thirty years, have done so. However, as we have seen, in the case of other garden projects agendas of personal development or civic improvement were touted but not necessarily structured into the programs in a concrete, effective way.

If an urban garden program is intended as a community development tool, it needs to be evaluated in that light. A more realistic approach for a garden, perhaps, is to identify the specific resources it can provide and why these may benefit the neighborhood. Some possible objectives include food, recreation, education, income, local activism, and restoration of the environment. Each objective has its own set of considerations that affect plan, design, and program.

Food

The most obvious resource that urban gardens provide is a place to grow food. Whether to subsidize household food expenses or to grow foods that are part of a cultural tradition or cuisine preference, the urban garden allows people without private land to use shared land for food production. Thus gardens fit into an agenda to increase community food security. However, it should be stressed that gardens are not the ultimate solution to food security but instead provide one piece of a more comprehensive strategy. Not everyone wants to garden or is able to garden, so gardening should not be offered as the only means or mandated in any way.

For a garden to produce healthy and abundant food, certain basic physical considerations must be addressed. First, the garden needs good, rich, tillable earth. Nurturing good soil is an investment that assures that the land will be available for future growing seasons. In fact, with proper care and as the gardener gains experience, the soil should be able to produce richer harvests in years to come. To build good soil takes time, labor, and resources that need to be valued along with the seasonal food yields. Regarding gardens in this way may increase the acknowledgment of investment in the site and so counter possible development proposals that seem to assume that a garden is just unbuilt earth. Another issue is site security, to insure that the gardener will be able to reap his or her harvest without fear of it being stolen or vandalized.

Recreation

Gardening is an important recreational opportunity that many urban people would like to access. However, just as people go to parks for different reasons, so too people may seek different experiences in the garden. While one person might want solitude and an opportunity to enjoy nature, another may seek an aerobic workout, and yet another may want to socialize. Each of these purposes has a different design and planning implication. Balancing such different interests requires community participation in the design process.

Gardens differ significantly from other types of nature-based recreation, such as parks and zoos, in that the users nurture the site and actively help in its maintenance. Unlike visitors to the park, the participants are invested in the garden space, thereby making the garden seem semiprivate. In a garden there is probably far less institutional control over what the space looks like than in a park. On the other hand, there may be conflicts of interest regarding how the space is used or maintained that must be negotiated among the gardeners.

Education

Urban gardening has served educational agendas throughout its history. While past efforts to instill moral lessons or Americanize immigrants sound patronizing today, we continue to see the garden as an educational tool. Many citywide garden organizations have child- and adult-education programs to teach gardening, recycling, composting, and nutrition. The key issue in garden education is awareness of implied cultural assumptions. For instance, gardening can be a tool to teach American agricultural traditions; however, participants in that heritage were not only yeoman farmers but also slaves and tenant farmers, and the scale has ranged from small family farms to agribusiness. Many ethnic groups have traditions that can be included in gardening and nutrition courses. Thus, sensitivity to cultural context is a prerequisite to a successful urban garden education program.

Economic Opportunity

Both past and present programs have used market gardening for generating income and for job training. Today's job-training and entrepreneurial garden projects illustrate the difficulties in justifying gardening solely in economic terms. These limitations include the facts that gardens employ only a few people, that they teach technical skills that may not be directly applicable to the urban job market, and that they often struggle to earn income from sales. Yet while gardens cannot substitute for a comprehensive community job-training and employment program, they can supplement and complement such efforts. One way to consider the urban garden is as a sheltered workplace that blends training in practical gardening skills with horticultural therapy, general education, community outreach, and environmental restoration. Such a project requires additional funds and support for its nonmarket aspects. Another option is to simply allow the enterprising gardener to develop his or her own informal cottage industry. The gardeners and the nature of the informal economy that exists in a neighborhood largely determine whether the gardening program needs to be involved in orchestrating trade.

Community Activism

While local activism is not an inevitable outcome of urban gardening, there are many examples of how the process of building and maintaining a community garden can inspire it. Gardens can bring people together who might not otherwise interact, and they provide a responsive medium where people can see results from their efforts. Several organizers share the belief, expressed by Tessa Huxley, that "once people discover they can make a garden, they realize they can do anything."[5] Successful urban garden programs activate neighbors to personally commit to making positive community change. If the objective of a given program is to initiate community activism, more attention might be placed on group processes, mechanisms for inclusion, and leadership development than on the garden itself. The process of neighborhood organizing may lead to community objectives related to issues other than gardening, such as housing, job training, or safety. In situations such as these, although the garden serves an important function in the ritual forging of community bonds, its permanency may not be a long-term goal or high priority. If this is the case, it should be stated so from the beginning and the garden developed in a way that does not waste resources or result in burnout or disappointment of volunteers.

Environmental Restoration

Urban gardens add biological diversity to urban areas. Butterfly gardens, native plant gardens, creek restoration, and other features are often included in gardens as a means for people to interact with natural systems that are otherwise largely hidden in the urban context. The garden creates an opportunity for people to see their own impact on ecological systems through such activities as composting garbage or measuring the inputs and outputs of food production. Environmental restoration is still a design and planning process. Different garden designs allow differing expressions of "nature." This can be seen in gardens inspired by nostalgia for the country that include Victorian gazebos and scarecrows, artistically inspired gardens where paths and planting beds take fanciful shapes such as hearts or peace signs, and very structured gardens in which rectilinear paths and beds contrast with organic growth patterns of plants. Restoration also requires the commitment of attention and time. Building diversity and habitat is a long-term investment.

KEY STRATEGIES FOR SUSTAINING URBAN GARDENS

Shifting away from normative agendas for urban garden programs and toward articulating the actual resources that gardens provide clarifies their real effectiveness in meeting the needs of individuals and communities. Yet, however useful a garden project may be in fulfilling such needs, that project will have minimum impact if it is implemented as a stopgap measure. For a garden to sustain itself as a community resource, three critical areas need to be addressed: balancing local and interest-based leadership, land security, and public support.

Balancing Local and Interest-Based Leadership

In each era, urban garden promotion has been inspired by a slightly different combination of interests. While today we think of community gardens as the epitome of grassroots activism, garden projects both current and past have often been developed from the top down. Earlier phases of garden programs were of a philanthropic nature, often imposed on the needy who had little recourse to other means of support. The national campaigns during the world wars relied on middle-class women's clubs and volunteer organizations to coordinate local efforts. A shift toward more local leadership occurred in the 1970s when gardens became rallying points for neighborhood activism. As neighbors reclaimed vacant land, citywide garden organizations formed to provide technical assistance and to promote gardening in other communities. Today, groups with special interests in environmentalism, education, or certain population sectors, such as the homeless, public-housing residents, or youth, also start garden projects. Thus garden programs continue to navigate a course between local interests and special interests aligned to more general issues.

This distinction between interest-based advocacy and local participation has implications for the sustainability of urban gardens. To increase local resources and expand social networks that support the garden, both top-down and bottom-up investment are necessary. Urban gardens can be venues to attract new resources—ideas, skills, money, and social connections—from outside the community. However, giving too much emphasis and control to outside groups can undermine the community leadership necessary to sustain the garden. In past cases, urban garden projects drew on external leadership during crises but found that the leadership dissipated after each crisis was resolved. Even if local interest continued, the neighborhood was often left without the authority or networking capacity to secure the site and program permanently. Given that gardens must be cultivated and cared for by participants, urban garden programs have a unique and critical need to sustain community interest. To counter the possibility of later disinterest, urban garden projects must stay in touch with the gardeners and the community at large to evaluate whether the garden satisfies their personal reasons for involvement. Garden programs need to be flexible so participants can change them to better fit their evolving needs.

By defining urban garden programs as community resources that are contributing to community development, the role of interest-based garden organizations becomes one of facilitation and collaboration. An interest-based garden organization benefits from working with community-based organizations that have local experience and knowledge of existing social conditions, such as local service providers, community development corporations, or other neighborhood institutions such as churches, schools, and clubs. Through this kind of collaboration, garden organizations can provide critical information and technical assistance while not forcing gardening as a panacea to local problems.

Securing Land

Historically, a crisis such as war or depression provided justification for requesting land for gardens, usually on a temporary basis. Once the crisis had passed, the land often became valuable for development and the garden site was ultimately lost. Past programs did not intend to permanently serve urban residents. Instead, garden advocates often hoped to inculcate in participants the desire to acquire their own property. Today, most gardens make opportunistic use of vacant or underutilized land, but changes in the economy or local conditions may mean that, sooner or later, the site becomes valuable for other development. The battle to protect community gardens in New York City in the late 1990s highlights the fragility of unprotected gardens in cities that experience development booms. No one wants to see the product of hours of hard work leveled by a bulldozer.

The opportunism inherent in nonpermanent gardens has had some important benefits. In some ways, the tenuous status of some gardens has kept the participants prepared to exploit opportunities, work together, and fight to protect their efforts. As neighbors collectively work to establish a garden out of nothing, they develop social networks and learn organizing skills. Likewise, having to lobby or fight for garden permanency has served as a catalyst for community activism. However, it is heartbreaking to see a community fail to protect its garden because it does not have the power to stop development. The enthusiasm and energy that go into urban gardens can be lost if city officials do not consider the garden a legitimate resource in the community.

For these reasons, as well as to retain the increased land values that might result from the community's investment in the garden, organizations need to pay attention to site procurement. There is no single, foolproof mechanism to secure land for gardens. In cases where an active and organized group assumes full responsibility and there is ongoing demand for gardening space, outright ownership of the site and long-term leases are preferable solutions. A land trust or community organization might hold the lease and oversee property issues such as taxes and insurance. In other cases, where a community is in transition or is not able to take ownership, one strategy is to pressure the public sector to support urban gardens as recreational and open space: perhaps the city should dedicate a certain percentage of land to gardening, similar to park or playground ratios. Gardens can be included as a valid form of land use in city plans and zoning. However, given the unique need for personal responsibility in gardening, such a top-down strategy cannot guarantee that local interest would sustain the garden program in all contexts. A better strategy is to structure a public decision-making process that responds to community-driven initiatives. The efforts of neighborhood groups that organize enough to petition for a garden and have the wherewithal to maintain it should be legitimized by public officials, and such groups should be provided with public resources to sustain their efforts.

Public Outreach

Public support for gardens is necessary for acquiring land, funding programs, and increasing community involvement. City gardening organizations and national organizations, such as the American Community Gardening Association, can lobby at federal, state, and local levels to raise awareness of the many important roles urban gardens play in communities. The message must be carefully crafted to address the public's perceptions of garden projects and the reasons the public might be willing to support them as public resources. Unfortunately, many advocates "preach to the converted," ignoring the fact that not everyone wants to garden or agrees that the public should provide gardening space for individuals to use. Therefore, it is important to convey the role of gardens as local resources that also serve the public interest. Instead of relying solely on anecdotal accounts, concrete proof of garden outcomes—such as the impact of environmental restoration, the reduction of vandalism, or the increase in property values—needs to be made public. This sort of presentation requires stronger evaluative processes for garden programs, plus commitment by the national movement to support and encourage a research agenda that can be shared by many organizations.

Because gardens are a physical presence in the city, the way they appear to the general public influences perceptions of how they serve their community. Gardeners should be concerned with how their gardens look from the street as well as how inviting the project seems to be to newcomers. While gardens look full and green in spring and summer, they can appear unkempt and lifeless in the winter. Effort may be necessary to cultivate the public's appreciation of a garden's seasonal appearance and activity. While fences and gates may be required for safety and security reasons, they need not be designed to make the garden seem closed off from the public. To counteract criticism that urban gardens are semiprivate spaces, garden groups should encourage broader participation through harvest fairs, children's gardens, social activities, and classes. Outreach should not be limited to garden-specific interests but should be expanded to address larger community needs through involvement with neighborhood associations and other organizations.

THE OUTLOOK FOR THE FUTURE

For many participants, urban gardening is a labor of love that combines the best of environmental ethics, social activism, and personal expression. Even when faced with social turmoil, degraded sites, and ambivalent policy makers, urban gardeners maintain a faith that what they do not only helps the individual but strengthens the community. This book, a historical overview of urban garden programs, celebrates the efforts of thousands of gardeners and reveals a consistent desire for land to garden. However, it also reveals structural problems that inhibit the sustainability of the gardens. Urban garden programs have been consistently enacted as expressions of local resiliency in times of crisis, but with a lack of rigor that undermines their permanence. While garden promoters are proud of their grassroots activism and the self-reliance that gardens make visible to their community, garden programs also have a legacy of conservatism and paternalism. Hopefully, by exposing the environmental determinism inherent in past promotional efforts, we can consciously avoid making some of the same assumptions today. Faith in gardening should not be blind faith but instead an inspiration that is then proven through self-reflection and critical evaluation of gardening's actual returns. Constructive criticism, while sometimes difficult to hear, is often the best way to see what areas need improvement, as well as to recognize accomplishments and celebrate progress.

Through the food and flowers they grow, the people they involve, and the physical environment they create, urban gardens are community resources at multiple levels. Urban garden projects are uniquely capable of providing a food source, a hobby, a place to socialize, and a place to express urban ecology all in one. They also represent the participants' investment of time and effort to make these opportunities possible. Garden projects are most likely to serve community development objectives when the participants themselves have control of program development and the resources that go into it—particularly the land and the value of their labor. While not a substitute for other community resource needs, garden programs can inspire volunteerism and activism and forge new links between community members, local and national organizations, and federal agencies. As they build on many social values related to nature and individualism, urban garden projects bridge public interests with local community development.

Every gardener hopes for a bountiful harvest. As tassels form on corn stalks and tomato blossoms transform into fruit, the gardener anticipates the tasty meals soon to be created from the harvested foods. But even after the last tomato is eaten, the gardener knows that the garden must still be tended—old plants taken out, compost turned, new crops planted—if the next harvest is to be even better. Those who support and nurture urban garden programs feel the same way. The reward of seeing an active urban garden—the busy gardeners, the children exploring for insects, the exuberant growth of plants—compensates for the hard work of planning and organizing the garden. The thought that the garden will continue to be useful to its participants inspires proactive efforts to nurture community leadership and site permanence. A patchwork of such projects in neighborhoods throughout the city suggests another scale of bountiful harvest. The notion of a City Bountiful suggests a city abounding with vegetables, fruits, and flowers that people grow for themselves and their community. Such a city provides opportunities for people to engage with their environment through the process of gardening and with their community through the social interaction of organizing and maintaining the gardens. Like the garden itself, the City Bountiful is a vision that needs to be nurtured.

NOTES

PREFACE AND ACKNOWLEDGMENTS

1. Louise Klein Miller, *Children's Gardens for Home and School* (New York: D. Appleton and Company, 1904); M. Louise Greene, *Among School Gardens* (New York: Russell Sage Foundation, 1910); Charles Lathrop Pack, *The War Garden Victorious* (Philadelphia: J. B. Lippincott Company, 1919); M. G. Kains, *The Original Victory Garden Book* (New York: Stein and Day, 1942); Boston Urban Gardeners, *Handbook of Community Gardening* (New York: Charles Scribner's Sons, 1982); Patricia Hynes, *A Patch of Eden* (Junction, VT: Chelsea Green Publishing Company, 1996); Dianne Balmori and Margaret Morton, *Transitory Gardens, Uprooted Lives* (New Haven: Yale University Press, 1993).

INTRODUCTION. GARDEN PATCHES IN AMERICAN CITIES

1. Jacob Riis, "What Ails Our Boys," *The Craftsman* 21, 1 (October 1911): 3–10, quote p. 8.

2. Laurie Belton, "Youth Update," *San Francisco League of Urban Gardeners Update* (Summer 1997): 11–12, quote p. 12.

3. In his 1981 article "Reaping the Margins: A Century of Community Gardening in America," *Landscape Journal* 25, 2: 1–8, Thomas J. Bassett outlines seven phases of community gardens and hypothesizes that the replication of the idea stems from its success as a supportive institution during periods of social crisis. I am indebted to his scholarship.

4. American Community Gardening Association, *Findings from the National Community Gardening Survey,* ACGA Monograph, 1998. While this survey was conducted in 1996–97, it is still the most recent attempt to summarize participation.

PART I. INTRODUCTION

1. See Kenneth Jackson, *Crabgrass Frontier: The Suburbanization of the United States* (New York: Oxford University Press, 1985); Sam Bass Warner, *The Urban Wilderness: A History of the American City* (New York: Harper and Row, 1972); and Gwendolyn Wright, *Building the Dream: A Social History of Housing in America* (Cambridge: MIT Press, 1981).

2. Quoted in May Vida Clark, "Conference on Agricultural Depression," *Charities Review* 5, 2 (December 1895): 105–10, quote p. 105.

3. See Bolton Hall, *Three Acres and Liberty* (New York: Grosset and Dunlap, 1907); and Peter Schmitt, *Back to Nature: The Arcadian Myth in Urban America* (New York: Oxford University Press, 1969).

4. See Liberty Hyde Bailey, *The Country-Life Movement in the United States* (New York: Macmillan, 1911); and William Bowers, *The Country Life Movement in America, 1900–1920* (Port Washington, NY: National University Publications/Keenikat Press, 1974).

5. See Roy Lubove, *The Progressives and the Slums: Tenement Housing Reform in New York City: 1890–1917* (Pittsburgh: University of Pittsburgh Press, 1962); and Paul Boyer, *Urban Masses and Moral Order in America, 1820–1920* (Cambridge: Harvard University Press, 1978).

6. Richard Hofstadter, *The Age of Reform: From Bryan to FDR* (New York: Alfred A. Knopf, 1959), 5.

7. Quoted in Albert Shaw, ed., "City Gardens versus Hoodlumism," *The American Review of Reviews* (November 1911): 622; quotation marks in original.

CHAPTER ONE. AN ALTERNATIVE TO CHARITY

1. Leah Fedder, *Unemployment Relief in Periods of Depression: A Study of Measures Adopted in Certain Cities, 1857–1922* (New York: Arno Press, 1936 [reprint 1971]).

2. G. J. H. Crespi, "Allotments," in *Good Words,* edited by Donald Macleod (London: Isbister and Co., 1898), 779–83. For information on English allotment gardens, see David Crouch and Colin Ward, *The Allotment: Its Landscape and Culture* (London: Faber and Faber, 1988). Americans were informed about European allotment gardens in journal articles at the time when vacant-lot garden-

ing was being proposed. See Rev. J. Frome Wilkinson, "Pages in the History of Allotments," *The Contemporary Review,* 65 (January–June 1894): 532–44.

3. New York Association for Improving the Condition of the Poor (AICP), "Cultivation of Vacant Lots by the Unemployed," *AICP Notes* 1, 1 (December 1898): 13.

4. Quoted in Melvin G. Holli, *Reform in Detroit: Hazen S. Pingree and Urban Politics* (New York: Oxford University Press, 1969), 71.

5. Ibid., 72.

6. Cornelius Gardener, "An Experiment in Relief Work," *Charities Review* 4, 5 (March 1895): 225–28, quote p. 226.

7. Quoted in Frederick W. Speirs, Samuel McCune Lindsay, and Franklin B. Kirkbride, "Vacant-Lot Cultivation," *Charities Review* 8, 1 (March 1898): 74–107, quote p. 78.

8. Vacant Lot Gardening Association, *Vacant Lot Gardening Association Season of 1907* (New York, 1907), 5.

9. E.g., see "An Experiment in Relief by Work," *Charities Review* 4, 5 (March 1895): 225–28.

10. When Bolton Hall began organizing for a New York City vacant-lot farming program, he secured the cooperation of the United Hebrew Charities, the Charity Organization Society, the Federation of East-Side Workers, and the Association for the Improvement of the Condition of the Poor (AICP). The result was a committee of ten members that included representatives of several philanthropic organizations. See Michael Mikkelsen, "Cultivation of Vacant City Lots," *Forum* 21 (March–August 1896): 313–17.

11. AICP, "Cultivation," 27–28.

12. Ibid., 10.

13. Ibid., 28.

14. Ibid., 9.

15. Ibid., 14.

16. Ibid., 44.

17. Philadelphia Vacant Lots Cultivation Association (PVLCA), *Tenth Annual Report, 1906,* 6.

18. PVLCA, *Eighth Annual Report,* 1904, 10.

19. May Vida Clark, "Conference on Agricultural Depression," *Charities Review* 5, 2 (December 1895): 105–10, quote p. 109.

20. AICP, "Cultivation," 29.

21. Ibid., 20.

22. PVLCA, *Fifteenth Annual Report,* 1911, 10.

23. PVLCA, *Third Annual Report,* 1899, 4.

24. Speirs, Lindsay, and Kirkbride, "Vacant-Lot Cultivation," 87–88.

25. Ibid., 94.

26. AICP, "Cultivation," 11.

27. Ibid., 10.

28. PVLCA, *Seventh Annual Report,* 1903, 10.

29. Speirs, Lindsay, and Kirkbride, "Vacant-Lot Cultivation," 89.

30. AICP, "Cultivation," 33.

31. PVLCA, *Second Annual Report,* 1898, 3.

32. Jane Stewart, "Market Gardening on Vacant Lots for the Unemployed," *Country Life in America* (April 1903): cci–ccii, quote p. cci.

33. PVLCA, *Twelfth Annual Report,* 1908, 9.

34. "Farming on Vacant City Lots," *Garden and Forest* 9 (March 4, 1896): 91–92, 100. *Garden and Forest* was a journal dedicated to horticulture, landscape design, and forestry managed by Charles Sargent of the Arnold Arboretum.

35. PVLCA, *Eighth Annual Report,* 1904, 9.

36. PVLCA, *Second Annual Report,* 1898, 10.

37. PVLCA, *Seventh Annual Report,* 1903, 10.

38. AICP, "Cultivation," 9.

39. PVLCA, *Thirty-First Annual Report,* 1927, 1–2.

40. Gregory Smith, "To the Editor of Garden and Forest," *Garden and Forest* 9 (April 1, 1896): 139.

41. Michael Mikkelsen, "Cultivation of Vacant City Lots," *Forum* 21 (March–August 1896): 313–17.

42. AICP, "Cultivation," 17.

43. PVLCA, *Third Annual Report,* 1899, 9.

44. PVLCA, *Second Annual Report,* 1898, 8.

45. AICP, "Cultivation," 12.

46. Ibid., 11.

47. Ibid., 18.

48. Ibid., 11.

49. PVLCA, *Seventh Annual Report,* 1903, 6.

50. PVLCA, *Fifth Annual Report,* 1901, 8.

51. PVLCA, *Sixteenth Annual Report,* 1912, 7.

52. Mapping Philadelphia vacant-lot gardens is complicated since some annual reports list locations while others list the names of landowners or the name of the farm.

53. PVLCA, *Fifth Annual Report,* 1901, 1.

54. PVLCA, *Eighth Annual Report,* 1904, 18.

55. One person who continued to advocate for vacant-lot cultivation was Bolton Hall (1854–1938). As treasurer of the Vacant Lot Gardens Association, Hall published an article in *The Survey* in 1910 (in two parts, in the February 19 and March 19 issues) that summarized a range of vacant-lot cultivation projects. These included the Philadelphia Vacant Lots Cultivation Association and several projects at sanitariums, schools, prisons, and hospitals. A reprint of his "Garden Plots for Institution Inmates" from this article was distributed along with a cover

essay, "Vacant Lot Garden Work for 1910." In this essay he mentions that the New York Vacant Lot Gardens Association had secured a piece of land in Berkeley Heights, New Jersey, for New Yorkers "to work away from their present homes." See Bolton Hall, "Vacant Lot Garden Work for 1910" (New York Vacant Lot Gardens Association, n.d.). Bolton also wrote two books to promote back-to-the-land cultivation: *Three Acres and Liberty* (New York: Grosset and Dunlap, 1907) and *A Little Land and a Living* (New York: Arcadia Press, 1908). Hall remained a vacant-lot-cultivation and back-to-the-land advocate throughout his life, and his work included advocacy for garden programs during World War I, as evidenced by various letters included in his archives at the New York Public Library.

56. Fedder, *Unemployment Relief.*

57. Quoted in Hall, "Vacant Lot Garden Work for 1910" (March 19, 1910): 9.

58. "Gardens and the Unemployed," *The Craftsman* 27, 6 (March 1915): 708–10, quote p. 710.

59. PVLCA, *Eighth Annual Report,* 1904, 17.

60. Speirs, Lindsay, and Kirkbride, "Vacant-Lot Cultivation," 85.

CHAPTER TWO. THE SCHOOL GARDEN MOVEMENT

1. Attributed to George Fox, quoted in James Ralph Jewell, *Agricultural Education Including Nature Study and School Gardens,* Bulletin 2, Department of Interior, Bureau of Education (Washington, DC: GPO, 1907), 30.

2. Ibid., 37.

3. M. Louise Greene, *Among School Gardens* (New York: Russell Sage Foundation, 1910), 3.

4. The earliest source is E. Gang's 1899 account in *School Gardens* (Washington, DC: GPO, 1900; reprinted from U.S. Bureau of Education, Report of the Commissioner of Education, 1898–99, chapter 20), which Gang credits as a translation from Rein's *Pedagogical Cyclopedia.* Similar reviews are found in Helen Putnam's "School Gardens in Cities" (a lecture given before the Rhode Island Normal School, April 1, 1902); Otis W. Caldwell, *The School Garden* (Charleston, IL: Eastern Illinois State Normal School, 1903); B. M. Davis, *School Gardens for California Schools* (Sacramento: Superintendent State Printing, 1905); Jewell, *Agricultural Education;* and Greene, *Among School Gardens.* These authors do not provide references for their historical statements. For a recent account, see Aarti Subramaniam, "Garden-Based Learning in Basic Education: A Historical Review," monograph, 4-H Center for Youth Development, University of California, summer 2002, available online: http://fourhcyd.ucdavis.edu.

5. See Frederick Froebel, *The Education of Man,* trans. Josephine Jarvis (New York: A. Lovell and Company, 1886).

6. Michigan State Superintendent of Public Instruction, *A Study of School Gardens and Elementary Agriculture for the Schools of Michigan,* Bulletin 10 (1904), 46.

7. See a description of the Canadian system in B. M. Davis's 1905 publication *School Gardens for California Schools.* Also see James W. Robertson, "The Macdonald Manual Training Schools," *Canadian Magazine* (April 1901); *Memorandum re: Rural Schools and Household Science* (Ottawa, Can.: Macdonald Printing Fund, 1902).

8. Fannie Griscom Parsons (Mrs. Henry Parsons), *The First Children's Farm School in New York City, 1902, 1903, 1904* (New York: DeWitt Clinton Farm School, 1904), n.p.

9. Margaret Knox, "The Function of a School Garden in a Crowded City District," in School Garden Association of New York (SGANY), *Second Annual Report* (New York: School Garden Association of New York, 1910), 32–33, quote p. 33.

10. Ellen Eddy Shaw, "The Place of Children's Gardens," *Nature-Study Review* 6, 2 (February 1910): 43–45, quote p. 43.

11. Jewell, *Agricultural Education,* 40.

12. Jacob Riis, "What Ails Our Boys?" *The Craftsman* 21, 1 (October 1911): 3–10, quote p. 8.

13. For a history of education reform, see Diane Ravitch, *The Great School Wars: New York City, 1905–1973: A History of Public Schools as Battlefields of Social Change* (New York: Basic Books, Inc., 1974); and Samuel Bowles and Herbert Gintis, *Schooling in Capitalist America: Educational Reform and the Contradictions of Economic Life* (New York: Basic Books, 1976).

14. The *Nature-Study Review,* published from 1905 to 1923, was an important resource for discussions related to all phases of nature study in education and frequently included topics related to school and civic gardening. See Van Evrie Kilpatrick, ed., *Nature Education in the Cities of the United States* (New York: School Garden Association of America, 1923). Also see Peter J. Schmitt, *Back to Nature: The Arcadian Myth in Urban America* (New York: Oxford University Press, 1969).

15. Louise Klein Miller, *Children's Gardens for School and Home: A Manual of Cooperative Gardening* (New York: D. Appleton and Company, 1904), 116.

16. Caldwell, *The School Garden,* 10–11.

17. Jewell, *Agricultural Education,* 46.

18. Country Life Commission, *Report of the Country Life Commission* (Washington, DC: GPO, 1909), 125. For a history of agricultural education, see A. C. True, *A History of Agricultural Education in the United States, 1785–1925* (Washington, DC: GPO, 1929).

19. Lydia Southard, "The School Garden as an Educational Factor," *New England Magazine* 26, 6 (August 1902): 675–78.

20. Ernest B. Babcock and Cyril A. Stebbins, *Elementary School Agriculture* (New York: MacMillan Company, 1911), 1.

21. Agrarianism encompasses not only the rural lifestyle but also agriculture as a socioeconomic consideration. However, as a social identity, agrarianism is often conflated with an idealized agricultural tradition, that of the yeoman farmer. For discussion of the impact of agrarian ideals on agricultural policy, see William P. Browne et al., *Sacred Cows and Hot Potatoes: Agrarian Myths in American Policy* (Boulder: Westview Press, 1992). A related concept is pastoralism, which Leo Marx describes as the literary interpretation of the American landscape's effect on morality and character. See Leo Marx, "Pastoral Ideals and City Troubles," in *Western Man and Environmental Ethics,* edited by Ian G. Barbour (Reading, MA: Addison-Wesley Publishing Company, 1973). Other writers have described the agrarian/pastoral ideal's influence on America's urban development. See James L. Machor, *Pastoral Cities: Urban Ideals and the Symbolic Landscape of America* (Madison: University of Wisconsin Press, 1987); and Morton and Lucia White, *The Intellectual versus the City: From Thomas Jefferson to Frank Lloyd Wright* (Cambridge: Harvard University Press and MIT Press, 1962).

22. H. D. Hemenway, *How to Make School Gardens: A Manual for Teachers and Pupils* (New York: Doubleday, Page, and Company, 1903), xiv.

23. Quoted in R. L. Templin, ed., *Information and Suggestions on School Gardens, Children's Home Gardens, Junior Clean-Up Work, and How to Make Your Home and Community a More Desirable Place to Live* (Cleveland: Children's Flower Mission, 1915), 25.

24. H. D. Hemenway, "School-Gardens at the Hartford School of Horticulture, Hartford, Connecticut," *Nature-Study Review* 1, 1 (January 1905): 29–36, quote p. 36.

25. Beverly Thomas Galloway, *School Gardens: A Report upon Some Cooperative Work in Normal Schools of Washington, with Notes on School Garden Methods in Other American Cities,* U.S. Department of Agriculture, Office of Experiment Stations Bulletin 160 (Washington, DC: GPO, 1905), 7. Lee Cleveland Corbett, *The School Garden,* United States Department of Agriculture Farmers' Bulletin 218 (Washington, DC: GPO, 1905).

26. According to Greene in *Among School Gardens,* the Massachusetts Horticultural Society had initially sent Henry Lincoln Clapp to study gardens in Europe, but this is not mentioned in other sources or by Clapp himself. See Henry Lincoln Clapp, "A Public School Garden," *New England Magazine* 26, 4 (June 1902): 417–27; Clapp, "School Gardens," *Education* (May 1901); Massachusetts Horticultural Society, *Report of the Committee on School Gardens and Children's Herbariums of the Massachusetts Horticultural Society for the Year 1901* (Boston: Massachusetts Horticultural Society, 1902); and Dick J. Crosby, "The School Garden Movement," *American Park and Outdoor Art Association* 6 (1902): 9–16.

27. George H. Martin, *School Gardens in the Public Schools of Massachusetts* (n.p., 1906), reprinted from the Sixty-Ninth Report of the State Board of Education, January 1906. Also see Boston School Garden Committee, *The Annual Report of the Boston School Garden Committee Nineteen Hundred and Five.*

28. Articles and bulletins on school gardens often included descriptions of the same model programs, often using the same text, anecdotal stories, and facts. See Miller, *Children's Gardens for School and Home;* Susan Sipe, *Some Types of Children's School Garden Work,* U.S. Department of Agriculture Office of Experiment Stations Bulletin 252 (Washington, DC: GPO, 1912); and Templin, *Information and Suggestions.*

29. Fannie Griscom Parsons (Mrs. Henry Parsons), "The Second Children's Farm School in New York City: DeWitt Clinton Park, 53rd Street and 11th Avenue," *The Charities* 11, 10 (September 5, 1903): 220–23, n. 1, p. 220. Also see Parsons, *The First Children's Farm School in New York City;* and Galloway, *School Gardens,* 34–36.

30. Parsons, "The Second Children's Farm School in New York City": 223.

31. Fannie Griscom Parsons (Mrs. Henry Parsons), "A Day at the Children's Farm School in New York City," *Nature-Study Review* 1, 6 (November 1905): 255–59. The last report of the DeWitt Clinton Farm School was in 1908; however, it is unclear exactly when the program ended. Today, DeWitt Clinton Park is a recreational facility with no trace of the farm school.

32. The School Garden Association of America (SGAA) grew out of the New York School Garden Association, established in 1908. It held annual meetings in conjunction with the annual meetings of the National Education Association. At its 1939 meeting, it became the Department of Garden Education of the National Education Association. In 1945, this department merged with the National Science Teachers Association. *Garden Magazine* served as the official organ of the SGAA until 1939; *Garden Digest* was the official organ of the Department of Garden Education from 1939 to 1943, after which mailed mimeographs served for communication.

33. International Farm School League, *Annual Report* 1911, 1. Information on the league is limited to reports from 1907, 1908, and 1911. The first two accounts report that the DeWitt Clinton Farm School's founder, Mrs. Fannie Griscom Parsons, served as president, with Grover Cleveland as honorary vice president. In 1911, Henry Griscom Parsons, son of Fannie Parsons, was secretary to the league. It is unclear when the league disbanded.

34. SGANY, *Second Annual Report,* 10.

35. Greene, *Among School Gardens,* 35.

36. C. D. Jarvis, *Gardening in Elementary City Schools,* Department of Interior, Bureau of Education Bulletin 40 (Washington, DC: GPO, 1916), 5.

37. Ibid., 7.

38. Ibid., 29.

39. Louise Klein Miller, "The Civic Aspects of School Gardens," *Nature-Study Review* 8, 2 (February 1912): 74–76. It is unclear if she was paid by the school board or the Home Gardening Association. For more information on Cleveland's program, see the Art Education Society and the Home Gardening Association of the Cleveland Public Schools, *Fifth Annual Report* (1904), *Sixth Annual Report* (1905), and *Seventh Annual Report* (1906).

40. Public Education Association of Philadelphia, *Twenty-Fourth Annual Report* (Philadelphia: Public Education Association, 1905), quote p. 16; italics appear in bold in original. A report on Philadelphia school gardens was jointly published by the Civic Club, Civic Betterment Association, Public Education Association, and City Parks Association, entitled *Philadelphia School Gardens* (n.d., although this seems to be from 1905, given the data it includes). Also see Philadelphia Board of Public Education, *Municipal School Gardens* (Philadelphia: Board of Education, 1905, 1906); and Philadelphia Vacant Lots Cultivation Association, *Eighth Annual Report* (1904). The latter is particularly interesting in light of the collaboration between a vacant-lot cultivation association and school gardening. In its 1904 annual report, the superintendent of the Philadelphia Vacant Lots Cultivation Association reported that its participants installed seven children's gardens that served approximately 550 children. With the Civic Club of Philadelphia, it managed two gardens, one for seventy girls at the Church Home for Children and the other as part of a garden and playground complex at Water View Park in Germantown. In the 1905 report, a letter from the Civic Club's Garden Committee reported on its collaboration to develop three garden projects that accommodated twelve hundred children. The Vacant Lots Cultivation Association also worked with the Colored Boys Club to start a school garden in Germantown. However, in the 1908 annual report, the supervisor stated that the vacant-lot cultivation association was no longer involved in school gardens.

41. Detroit Bureau of Governmental Research, "Report on the Home and School Garden Movement of the Recreation Commission," April 1918.

42. Elizabeth Rafter, "Home and Club Gardens," *Charities* 11, 10 (September 5, 1903): 210–18, quote pp. 210–11. For information on D.C.'s school gardens, see Board of Education of the District of Columbia, *Outline of Work: Home Gardening: Graded Schools* (Washington, DC: Press of Gibson Brothers, 1906); Washington Branch of the National Plant, Fruit, and Flower Guild, *Seventh Annual Report* (1902) and *Eighth Annual Report* (1903); Susan Sipe, "Practical Aid to the School Garden Movement by the United States Department of Agriculture," *Nature-Study Review* 8, 2 (February 1912); and Galloway, *School Gardens,* 8–15.

43. Galloway, *School Gardens,* 14.

44. Davis, *School Gardens for California Schools* (Sacramento: Superintendent State Printing, 1905).

45. For information on the Los Angeles garden programs, see B. M. Davis,

"Nature-Study in the Los Angeles State Normal," *Los Angeles Normal Exponent* (June 1901); Clayton Palmer, *Elementary Horticulture for California School* (Los Angeles: Los Angeles State Normal School, 1910); Sipe, *Some Types of Children's School Garden Work;* Los Angeles City School District, *An Outline of Instruction for School Gardening and Agriculture,* School Publication 9 (Los Angeles: Los Angeles City School District, 1918); Clayton F. Palmer, "Agriculture and Gardening in the Public Schools," *National Education Association Annual Report,* 1913; and Margaret Dolan, "Beautifying Work as Nature-Study," *Nature-Study Review* 11, 2 (February 1915): 52–57.

46. Ellen Eddy Shaw, "Prize Winners in Children's Garden Contest," *Garden Magazine* 17, 2 (March 1913): 102.

47. Mary Richards Gray, "Putting Your Civic House in Order: How the Young Members of the Family Help," *The Craftsman* 30, 3 (June 1916): 283–323; Clayton Palmer, "Agriculture in the Elementary Schools of Los Angeles City," *Nature-Study Review* 17, 5 (May 1920): 217–20.

48. Prior to 1901, the University of California in Berkeley was the main institution in the state that taught agriculture. In 1905, appropriations were made for the University Farm, which later became the University of California at Davis. By 1911, agricultural education existed at normal schools, polytechnic schools, public high schools, public elementary schools, state industrial schools, and private schools and colleges. For a description of work under way in school gardening and the university's outreach activities in 1911, see Ernest B. Babcock, "Cooperation between the Schools and the College of Agriculture," *The University of California Chronicle* 13 (Berkeley: University Press, 1911, 1912). See also E. B. Babcock, C. J. Booth, H. Lee, and F. H. Bolster, *Development of Secondary School Agriculture in California,* University of California College of Agriculture Circular 67 (Berkeley: University of California, 1911).

49. Ernest B. Babcock, *Suggestions for Garden Work in California Schools,* University of California College of Agriculture Circular 46 (Berkeley: University Press, 1909).

50. *Junior Agricultural Supplement* 1, 2 (January 1917): 4.

51. Cyril A. Stebbins, "Growing Children in California Gardens," *Nature-Study Review* 8, 2 (February 1912): 67–74, quote p. 70. See also Stebbins, "California 'Garden City,'" *Garden Magazine* 15, 1 (February 1912): 25. More information on the university's programs is available at the Bancroft Library, University of California, Berkeley.

52. A gardening program similar to the Berkeley Garden City was the Worcester Good Citizens' Factory in Worcester, Massachusetts, started by the Worcester Social Settlement. The settlement worked in an area called the Island District, which housed twenty-two nationalities and over twenty thousand children. The garden reportedly reduced juvenile crime by 50 percent in the immediate community, produced $2,341 worth of vegetables, raised the health rate by

72 percent in the district over three years, and enhanced property values by $50,000. According to a report by the National Recreation Association, the Worcester garden was still in existence in 1940. See R. J. Floody, "Worcester Garden City Plan: or the Good Citizens' Factory," *Nature-Study Review* 8, 4 (April 1912): 145–50.

53. Jarvis, *Gardening in Elementary Schools,* 7.

54. Roland W. Guss, "A Graded Course of Garden Work and Nature-Study," *Nature-Study Review* 12, 5 (May 1916): 213–25.

55. Arthur Dean, "The New Education," *The Craftsman* 24, 5 (August 1913): 463–71.

56. Frederick Harvey Bolster, "The High School Garden," M.A. thesis, University of California, Berkeley, 1918, 6.

57. Ethel Gowans, *A Vegetable Gardening Syllabus for Teachers* (U.S. Bureau of Education, April 1915).

58. "School Gardens," *Nature-Study Review* 1, 1 (January 1905): 28–29, quote p. 29.

59. George Washington Carver, *Nature Study and Children's Gardens,* Teacher's Leaflet 2, Extension Division (Tuskegee, AL: Tuskegee Normal and Industrial Institute, 1906), quote n.p.

60. Mary Leland Butler, "A New Kind of School Garden," in Society of American Florists and Ornamental Horticulturists, *Report, Compilations and Suggestions on the Methods of Teaching Horticulture in Public Schools,* presented by Committee of the Society of American Florists and Ornamental Horticulturists at the convention in Dayton, Ohio, August 21, 1906: 10–12, quote p. 12.

61. Greene, *Among School Gardens,* 131.

62. "Agriculture in Public Schools," *Garden and Forest* 6, 257 (January 25, 1893): 37–38.

63. The School of Horticulture was established as part of the Handicraft School of Hartford in 1901. It initially served the pupils of the Watkinson Farm School for Homeless Boys, with the goal of "turning boys toward the country." Under the direction of H. D. Hemenway, the program expanded to involve other schools and to provide a broad range of activities: training for teachers, an apprentice program, nature-study programs, and more. For more information see Trustees of the Handicraft Schools of Hartford, *Report of the Director of the School of Horticulture* (1903), and *Report of the Director of the School of Horticulture* (1904) (Hartford, CT: Handicraft Schools of Hartford, 1903, 1904); Margaret Klein, "A School of Horticulture for Young People," in Society of American Florists and Ornamental Horticulturists, *Report, Compilations and Suggestions:* 14–15. The 1912 annual report of the School Garden Association of America mentioned that several universities offered summer courses in school gardening and related agricultural work, including the University of California, Cornell, Harvard, and Chicago.

64. Circulars produced by the Bureau of Education included *Instruction for School-Supervised Home Gardens, Course in Vegetable Gardening for Teachers, Winter Vegetable Gardens, Organic Matter in Home Gardens, Hotbeds and Cold Frames for Home Gardens, Raising Vegetable Plants from Seed, How to Make the Garden Soil More Productive, Planting the Garden, The Part Played by the Leaf in the Production of a Crop, A Suggestive Schedule for Home-Garden Work in the South, List of Publications for Use of School Home-Garden Teachers,* and *School Home Garden Results of 1916.* These are available at the National Agricultural Library in Greenbelt, Maryland.

65. Greene, *Among School Gardens,* 194.

66. Ibid., 198.

67. Michigan State Superintendent of Public Instruction, *A Study of School Gardens,* 56.

68. Greene, *Among School Gardens,* 270.

69. Quite often, justifications for school gardens cited the success of African American agrarian education in the South. The most famous debates about the intent of African American education—for intellectual development or for practical skills—were occurring at this time between W. E. B. DuBois and Booker T. Washington. See W. E. B. DuBois, *The Souls of Black Folk* (New York: Penguin Books, 1989; originally published by A. C. McClurg and Co., 1903); and Booker T. Washington, *Up from Slavery* (New York: Bantam Books, 1901). For discussion of their debate, see Thomas Harris, *Analysis of the Clash over Issues between Booker and DuBois* (New York: Garland, 1993).

70. J. E. Davis, "The Whittier School Garden," *Southern Workman* 31, 11 (November 1902): 598–603, quote p. 602.

71. See "Nature-Study and Gardening for Indian Schools," *Nature-Study Review* 2, 4 (April 1906): 141–43. For more information on debate about Native American education as assimilation, see David Wallace Adams, *Education for Extinction: American Indians and the Boarding School Experience, 1875–1928* (Lawrence: University Press of Kansas, 1995).

72. See Arthur Dean, "Educating the Institutional Child: Right Labor as the Great Factor in Developing Youth," *The Craftsman* 24, 5 (August 1913): 509–17; Henry Griscom Parsons, *Children's Gardens for Pleasure, Health, and Education* (New York: Sturgis and Walton Company, 1910); Chester Tether, *Hints for Special Class Gardens* (Oswego, NY: State Normal and Training School, 1919); and Greene, *Among School Gardens.*

73. Home Gardening Association of Cleveland, *Sixth Annual Report* (1905), 28.

74. Charles Harcourt, "Reform for the Truant Boy in Industrial Training and Farming: An Effort to Improve Existing Laws and Lessen the Evil," *The Craftsman* 15, 4 (January 1909): 436–46.

75. Sipe, *Some Types of Children's Garden Work,* 17.

76. According to a report in the *San Francisco Call,* as of April 21, 1906, approximately two hundred thousand people were encamped in Golden Gate Park. By April 25, this number had dropped to forty thousand, many of whom lived in tents provided by the army. The last refugees were moved out of the park in January 1907.

77. Bertha Chapman, "School Gardens in the Refugee Camps of San Francisco," *Nature-Study Review* 2, 7 (October 1906): 225–29, quote p. 229.

78. Ibid.

79. Ibid., 226.

80. Corbett, *The School Garden,* 33.

81. Van Evrie Kilpatrick, "Editorial: School Gardening in America," *Nature-Study Review* 11, 2 (February 1915): 79–80, quote p. 80.

82. School Garden Association of New York, "Summary of Principal's Reports on Vacant Lots and Lands Owned and Tilled Which Are Available for Garden Purposes within the City Limits," *Ninth Annual Report 1917* (New York: School Garden Association of New York, 1917), 10–11.

83. School Garden Association of New York (SGANY), *Annual Report 1911* (New York: School Garden Association of New York, 1911), 13; also the report on p. 38. See also Anna Hill, "St. Mary's Park School Garden," *Garden Magazine* 15, 4 (May 1912): 253.

84. For a brief description of Pittsburgh's school garden in a park, see Greene, *Among School Gardens,* 26, 127, 231–35; for Hartford, see Stanley Johnson, "The Hartford Method for School Gardens: Vacation Times Where Work and Play Are Happily Combined," *The Craftsman* 12, 6 (September 1907): 647–58; for Poughkeepsie, see M. V. Fuller, "The Rejuvenation of Poughkeepsie," *American City* 4 (January 1911): 1–8.

85. Ellen Eddy Shaw, "The Playground Beautiful," *Garden Magazine* 13, 4 (May 1911): 242.

86. Joseph Lee, "Play as Landscape," *Charity and the Commons* 16, 14 (July 7, 1906): 427–94. Also see Lee, "Restoring Their Play Inheritance to City Children," *The Craftsman* 25, 6 (March 1914): 545–55; and Lee, *Constructive and Preventive Philanthropy* (New York: MacMillan and Company, 1902).

87. See Henry S. Curtis, "Nature in the Playgrounds," *The American City* 12, 2 (February 1915): 135–41.

88. Miller, *Children's Gardens for School and Home,* 66.

89. Jarvis, *Gardening in Elementary Schools,* 10.

90. Miller, *Children's Gardens for School and Home,* 99.

91. American Park and Outdoor Art Association (APOAA), The School Garden Papers of the Sixth Annual Meeting, Boston, August 1902, vol. 6, part 3 (Rochester, NY: American Park and Outdoor Art Association, 1902): 10.

92. Corbett, *The School Garden,* 6.

93. M. V. Fuller, "The Rejuvenation of Poughkeepsie," *The American City* 4 (January 1911): 1–8, quote p. 6.

94. *Garden Magazine* was first published in 1905 as a spin-off from the popular magazine *Country Life in America.* According to the description in its first edition, "The *Garden Magazine* is the logical working out of the growing interest in the garden not merely as a means of livelihood (though we expect to see more and more people turning to it as a life work), but as a delight and pursuit for busy people in the world who find a new fascination in things of the soil."

95. Hemenway, "School Gardens at the Hartford School," 33–34.

96. Lee Cleveland Corbett, "School Garden Work and the Department of Agriculture," *Charities* 11, 10 (September 5, 1903): 218–19. See also Corbett, *School Garden,* 7–12.

97. See Jarvis, *Gardening in Elementary Schools,* 48–50; Templin, *Information and Suggestions;* and Corbett, *The School Garden,* 7–12. As I discuss in chapter 3, the Home Gardening Association of Cleveland also provided seed packets, which many children's garden programs ordered.

98. Mrs. A. L. Livermore, *School Gardens: Report of the Fairview Garden School Association* (Yonkers, NY: Fairview Garden School Association, 1910).

99. Jarvis, *Gardening in Elementary Schools,* 28.

100. Ibid., 13.

101. Sipe, *Some Types of Children's Garden Work,* 7.

102. Kilpatrick, *Nature Education in the Cities.*

103. In his article "Little Machines in Their Gardens: A History of School Gardens in America, 1891–1920," *Landscape Journal* 16, 2 (Fall 1997): 161–73, Brian Trelstad argues that the demise of school gardens was the combined result of reduced funding after the war, lack of a central advocacy group, suburbanization, and the success of alternative solutions to some of the problems that gardens were supposed to solve, such as immigrant assimilation and child labor.

CHAPTER THREE. THE GOODNESS OF GARDENING

1. See William H. Wilson, *The City Beautiful Movement* (Baltimore: Johns Hopkins Press, 1989); Stanley Buder, *Visionaries and Planners* (New York: Oxford Press, 1990); and Jon Peterson, "The City Beautiful Movement: Forgotten Origins and Lost Meanings," *Journal of Urban History* 2 (August 1976): 415–34.

2. In *The Politics of Park Design* (Cambridge: MIT Press, 1982), Galen Cranz describes two phases that span this period: the pastoral park and the reform park.

3. See John Nolen, *The Industrial Village,* National Housing Association Publication 50 (New York: National Housing Association Publications, 1918); Graham Romeyn Taylor, *Satellite Cities* (New York: D. Appleton and Company,

1915); Margaret Crawford, *Building the Workingman's Paradise* (London: Verso, 1995); Gwendolyn Wright, *Building the Dream: A Social History of Housing in America* (Cambridge: MIT Press, 1981); and Kenneth T. Jackson, *Crabgrass Frontier: The Suburbanization of the United States* (New York: Oxford University Press, 1985).

4. See Edith Elmer Wood, *The Housing of the Unskilled Wage Earner* (New York: MacMillan Company, 1919); and Roy Lubove, *The Progressives and the Slums: Tenement Housing Reform in New York City: 1890–1917* (Pittsburgh: University of Pittsburgh Press, 1962). For an example of the popular promotion that Forest Hills and other model housing received, see Edward Hale Brush, "A Garden City for the Man of Moderate Means," *The Craftsman* 19, 5 (February 1911): 445–51.

5. See Susan Marie Wirka, "The City Social Movement," in *Planning the Twentieth-Century American City,* edited by Mary Corbin Sies and Christopher Silver (Baltimore: Johns Hopkins University Press, 1996), 55–76. The City Social or City Functional movement was largely concerned with sanitary reform and has been considered a starting point for later urban environmentalism. See Martin Melosi, ed., *Pollution and Reform in American Cities, 1879–1930* (Austin: University of Texas Press, 1980); and Robert Gottlieb, *Forcing the Spring: The Transformation of the American Environmental Movement* (Washington, DC: Island Press, 1993). As an illustration of the activism for sanitary reform during the 1880s–1930s in Chicago, see chapter 2 of David Naguib Pellow, *Garbage Wars: The Struggle for Environmental Justice in Chicago* (Cambridge: MIT Press, 2002).

6. Mary Simkhovitch, presentation at the First National Conference on City Planning, Washington, D.C., 1909; reprinted in *Proceedings of the First National Conference on City Planning* (Chicago: American Society of Planning Officials, 1969). For information on settlement houses, see Allan Davis, *Spearheads for Reform* (New York: Oxford University Press, 1976); and Judith Trulander, *Professionalism and Social Change* (New York: Columbia University Press, 1987).

7. Jessie M. Good, "The How of Improvement Work," *The Home Florist* 4, 1 (January 1901): 25.

8. Imogen B. Oakley, "The More Civic Work, the Less Need of Philanthropy," *The American City* 6, 6 (June 1912): 805–13, quote p. 805.

9. Warren H. Manning, "The History of Village Improvement in the United States," *The Craftsman* 5, 5 (February 1904): 423–32, quote p. 432. Charles Mulford Robinson, *The Improvement of Towns and Cities, or The Practical Basis of Civic Aesthetics,* 3rd rev. ed. (New York: G. P. Putnam and Sons, 1911).

10. Louise Klein Miller, "Civic Aspects of School Gardens," *Nature-Study Review* 8, 2 (February 1912): 74–76, quote p. 75.

11. Philadelphia Vacant Lots Cultivation Association (PVLCA), *Seventeenth Annual Report,* 1913, 6.

12. The mission of the National Plant, Fruit, and Flower Guild was to receive

donations of fruit, flowers, and potted plants to distribute to hospitals, tenements, kindergartens, and the poor, and to encourage poor people to care for plants in their homes. See Katherine Paul, "Results in New York City," *Garden Magazine* 19, 2 (March 1914): 102; see also Washington Branch of the National Plant, Fruit, and Flower Guild, *Report,* for the years 1896, 1898, 1902, and 1914.

13. Dick Crosby, "The School Garden Movement," *The School Garden Papers of the Sixth Annual Meeting,* vol. 6, part 3 (Rochester, NY: American Park and Outdoor Art Association, 1902): 9–16, quote p. 12.

14. Miller, "Civic Aspects of School Gardens," 76; capitalization in original text.

15. Alice J. Patterson, "Educational Values of Children's Gardens," *Nature-Study Review* 12, 3 (March 1916): 124–28, quote p. 127.

16. Patrick Geddes, "Introduction," in *School Gardening for Little Children,* by Lucy Latter (London: Swan Sonnenschein and Company, 1906), xiv.

17. Mary Rankin Cranston, "Converting Backyards into Gardens: The Happiness and Economy Found in Cultivating Flowers and Vegetables," *The Craftsman* 16, 1 (April 1909): 70–79, quote p. 70.

18. Mary Rankin Cranston, "The Garden as Civic Asset, and Some Simple Ways of Making It Beautiful," *The Craftsman* 6, 2 (May 1909): 205–10, quote p. 205.

19. E. L. Shuey, "Outdoor Art and Workingmen's Homes," *Second Annual Report of the American Park and Outdoor Art Association* (Rochester, NY: American Park and Outdoor Art Association, 1898): 112–23, quote p. 113.

20. See Arthur Comey, "Billerica Garden Suburb," *Landscape Architecture* 4, 4 (July 1914): 145–49; and Comey, "Neighborhood Centers," in *City Planning,* edited by John Nolen (New York: D. Appleton and Company, 1916).

21. This is illustrated in the papers presented by Lawrence Veiller, Raymond Unwin, and others that appear in the *Proceedings of the Third National Conference on City Planning,* Philadelphia, 1911 (Boston: University Press, 1911).

22. "Town and Village," *The American City* 3, 2 (August 1910): 90–92.

23. M. Louise Greene, *Among School Gardens* (New York: Russell Sage Foundation, 1910), 42.

24. The term *cooperative gardening* appears in the subtitle of Louise Klein Miller's 1904 book, *Children's Gardens for School and Home: A Manual of Cooperative Gardening.* In 1903, the editor of *Country Life in America* coined the term *extension gardening* to describe gardening done for the betterment of others, such as school gardens, vacant-lot gardens, and churchyard gardens. A 1915 editorial in *The Craftsman* called for a "Universal Garden Movement" that would unite various philanthropic garden programs that included agendas of beautification, social improvement, and economic opportunity. See "Gardens and the Unemployed," *The Craftsman* 27, 6 (March 1915): 708–10.

25. The American Park and Outdoor Arts Association (APOAA) was origi-

nally proposed by landscape architect Warren Manning, and its initial officers included respected designers such as Charles Mulford Robinson, Manning, and John Olmsted. In 1904 the National League of Civic Improvement Associations combined with the American Park and Outdoor Art Association to become the American Civic Association. All three of these organizations reflected the interactions between professional planning and local improvement societies, particularly the shifting roles of professionals belonging to the former and women belonging to the latter. See Bonj Szczyzgiel, "The City Beautiful Revisited: An Analysis of Nineteenth-Century Civic Improvement Efforts," *Journal of Urban History* 29, 2 (January 2003): 107–32.

26. Clinton Rogers Woodruff, " 'A More Beautiful America' as the Outward Manifestation of More Wholesome Communal Life," *Charities* 11, 5 (August 1, 1903): 101–105, quote p. 101.

27. Dick Crosby, *Children's Gardens: Prospectus of the Department,* American Civic Association, Department Leaflet no. 1 (Philadelphia: American Civic Association, February 1904), 1.

28. See Mary Ritter Beard, *Woman's Work in Municipalities* (New York: D. Appleton and Company, 1915; reprinted New York: Arno Press, 1972); and Sophonisba Breckinridge, *Women in the Twentieth Century* (New York: McGraw-Hill Book Company, 1933; reprinted New York: Arno Press, 1972). For more information on the gender distinctions between physical planning and social planning, see Suellen M. Hoy, "Municipal Housekeeping," in *Pollution and Reform in American Cities, 1870–1930,* edited by Martin V. Melosi (Austin: University of Texas Press, 1980); Suzanne M. Spencer-Wood, "Turn of the Century Women's Organizations, Urban Design, and the Origin of the American Playground Movement," *Landscape Journal* 13, 2 (fall 1994): 125–37; Wirka, "The City Social Movement," 55–76; and Wendy Kaminar, *Women Volunteering: The Pleasure, Pain, and Politics of Unpaid Work from 1830 to the Present* (Garden City, NY: Anchor Press, 1984).

29. Joseph Lee, *Constructive and Preventive Philanthropy* (New York: MacMillan Company, 1902), 89.

30. Richard Watrous, "The American Civic Association," *American City* 1, 2 (October 1909): 59–63, quote p. 62.

31. Francis King, "Significance of the 'Garden Clubs,' " *Garden Magazine* 17, 3 (April 1913): 186; Harlean James, "Civic Gardening Which Develops the City People," *The Craftsman* 25, 6 (March 1914): 574–84.

32. Mary Leland Butler, "A New Kind of School Garden," in Society of American Florists and Ornamental Horticulturists, *Report, Compilations and Suggestions on the Methods of Teaching Horticulture in Public Schools,* presented by Committee of the Society of American Florists and Ornamental Horticulturists at the convention in Dayton, Ohio, August 21, 1906: 10–12, quote p. 10.

33. Mrs. A. L. Livermore, *School Gardens: Report of the Fairview Garden*

School Association, Yonkers, NY (Yonkers: Fairview Garden School Association, May 1910), 20.

34. See Jessie M. Goode, "The Work of Civic Improvement," *The Home Florist* (October 1900): 11–16, quote p. 13.

35. George Townsend, "The Boys' Garden at the National Cash Register Company," *The School Garden Papers of the Sixth Annual Meeting,* vol. 6, part 3 (Rochester, NY: American Park and Outdoor Art Association, 1902): 27–31, quote p. 27.

36. National Cash Register Company, *Art, Nature and the Factory* (Dayton, OH: National Cash Register Company, 1904), n.p.

37. Home Gardening Association of Cleveland, *Annual Report* (Cleveland: Home Gardening Association, 1904), 29.

38. R. L. Templin, ed., *Information and Suggestions on School Gardens, Children's Home Gardens, Junior Clean-Up Work, and How to Make Your Home and Community a More Desirable Place to Live* (Cleveland: Children's Flower Mission, 1915), 35.

39. Leroy Boughner, "Vacant Lot Gardens in Minneapolis," address to the American Civic Association, Washington, D.C., December 13–15, 1911.

CHAPTER FOUR. PATRIOTIC VOLUNTEERISM

1. Charles Lathrop Pack, *The War Garden Victorious* (Philadelphia: J. B. Lippincott Company, 1919), 23.

2. Camp Dix, New Jersey, had a 400-acre garden to which 150 men were assigned, including alien enemies (Germans, Austrians, and Turks who could not go into active duty), conscientious objectors, and the physically unfit. Camp Grant, Illinois, had a 300-acre garden in 1918. Camp Devens, Massachusetts, had a 250-acre garden that was worked by war prisoners. The American Army Garden Service planted truck gardens in France to supply United States military troops. See Pack, *War Garden Victorious,* 46–52; Luther Burbank, "Over Here and Over There," *Garden Magazine* 28, 1 (August 1918): 15–16; and U.S. Food Administration, *Food Guide for War Service at Home* (New York: Charles Scribner's Sons, 1918), 55.

3. Bristol County Agricultural School, *Home Gardening Manual* (Fall River, MA: Alt-Win Co., 1918), 5.

4. Ibid., 3.

5. Excerpts from Woodrow Wilson's letter to the nation, dated April 15, 1917, were reprinted in *Garden Magazine* 25, 4 (May 1917): 220.

6. Henry Griscom Parsons, *City Gardens,* Columbia War Papers 1, 3 (New York: Division of Intelligence and Publicity of Columbia University, 1917), 3.

7. See Garrard Harris, "Gardening for Re-Educating Disabled Soldiers," *Garden Magazine* 28, 2 (September 1918): 45–46.

8. John Dewey, *Enlistment for the Farm,* Columbia War Papers 1, 1 (New York: Division of Intelligence and Publicity of Columbia University, 1917), 6.

9. The Food Administration existed from August 1917 through August 1919. Near its conclusion, it had over 8,000 full-time volunteers, 750,000 part-time volunteers, and approximately 3,000 paid staff. Refer to Maxcy Robson Dickson, *The Food Front in World War I* (Washington, DC: American Council on Public Affairs, 1944); William Clinton Mullendore, *History of the United States Food Administration, 1917–1919* (Palo Alto: Stanford University Press, 1941); and Charles R. Van Hise, *Conservation and Regulation in the United States during the World War* (Washington, DC: GPO, 1917).

10. Dickson, *The Food Front in World War I,* 182.

11. Ibid., 183.

12. Charles Lathrop Pack was a conservation advocate and controversial president of the American Forestry Association. See Alexandra Eyle, *Charles Lathrop Pack* (Syracuse: College of Environmental Science and Forestry Foundation, State University of New York, 1992). The National War Garden Commission included P. S. Risdale as executive secretary, along with the following commissioners:

Luther Burbank, horticulturist and member of the Academy of Sciences

P. P. Claxton, United States Commissioner of Education

Dr. Charles W. Eliot, president emeritus of Harvard and trustee of the Rockefeller Foundation

Dr. Irving Fisher, Yale professor of political economics

Frederick Goff, Cleveland Trust Company

John Hays Hammond, Massachusetts resident

Fairfax Harrison, Southern Railway Company president and chairman of the Railways War Board

Myron T. Herrick, former ambassador to France

Dr. John Grier Hibben, president of Princeton University

Emerson McMillin, New York City banker

A.W. Shaw, editor of *System Magazine*

Mrs. John Dickinson Sherman, conservation chairman of the General Federation of Women's Clubs

Carl Vrooman, assistant secretary of agriculture

John Baber White, timber baron and member of the United States Shipping Board

James Wilson, secretary of agriculture under Presidents McKinley, Roosevelt, and Taft

13. Ida Clyde Clarke, *American Women and the World War* (New York: D. Appleton and Company, 1918). In *The Woman's Committee, United States Council of National Defense* (Washington, DC: GPO, 1920), Emily Blair criticized the assumption that women were best reached through their clubs because it only perpetuated their limited political recognition and sidetracked the issue of granting women the right to vote. In addition to women's clubs, women's colleges, such as Mount Holyoke, Vassar, and Bryn Mawr, also actively promoted war gardening by providing classes in gardening and farming as well as garden sites.

14. The National Council of Defense, made up of the secretaries of war, navy, interior, agriculture, commerce, and labor, was established in 1916 to coordinate various resources for national security and to create new channels of cooperation among the various government departments. See Clarke, *American Women and the World War,* 19; and Blair, *The Woman's Committee, United States Council of National Defense.*

15. See Mayor's Committee of Women on National Defense, *The Activities of the Mayor's Committee on National Defense* (New York: n.p., 1918–1919): 56–58.

16. Miss Loines, letter to President Woodrow Wilson, quoted in Martha Nolan, *A Chronicle: The History of the Woman's National Farm and Garden Association, 1914–1984* (Fremont, OH: Lesher Printers, Inc., 1985), 21.

17. "An Announcement," *Bulletin of the Garden Club of America* 20 (May 1917): 3. For an overview of the Garden Club of America, see Marjorie Gibbons Battles and Catherine Holt Dickey, *Fifty Blooming Years* (Lake Forest, IL: Garden Club of America, 1963).

18. Modeled after England's Woman's Land Army, the Woman's Land Army of America was formed in 1917. See Woman's Land Army of America, *Women on the Land* (New York: Woman's Land Army, 1918); Woman's Land Army of America, *Report Wellesley College Training Camp and Experimental Station* (New York: Woman's Land Army, 1919); and Nolan, *A Chronicle.*

19. Dewey, *Enlistment for the Farm,* 5. The entire song excerpted in the section epigraph can be found in "The New Garden Song," United States School Garden Army (USSGA), *Fall Manual of the United States School Garden Army* (Washington, DC: GPO, 1918), 26. It was also reprinted in *Garden Magazine* 27, 5 (June 1918).

20. "A War Measure: Children in Farm Work and School Gardens," *American City* 16, 5 (May 1917): 503.

21. See USSGA, *An Outline of a Course of Study in Gardening* (Washington, DC: GPO, 1918); J. H. Francis, *The United States School Garden Army,* Department of Interior, Bureau of Education, Bulletin 26 (Washington, DC: GPO, 1919); and "The United States School Garden Army," *Nature-Study Review* 15, 3 (March 1919): 102–104. United States School Garden Army administrative officials included:

Franklin K. Lane, secretary of the interior

Philander P. Claxton, commissioner, Bureau of Education

John H. Francis, director; formerly superintendent of Columbus, Ohio public schools; later replaced by John L. Randall

Clarence M. Weed, regional director, northeastern states

Lester S. Ivies, regional director, central states

Frederick A. Merrill, regional director, southern states

Cyril A. Stebbins, regional director, western states; formerly of the University of California

John L. Randall, regional director, southeastern states

22. Quote from introduction by J. H. Francis, Director of the U.S. School Garden Army, in USSGA, *An Outline of a Course of Study,* 2.

23. Francis, *The United States School Garden Army,* 5.

24. Letter from Woodrow Wilson, dated February 25, 1918, reprinted in USSGA, *Fall Manual,* 3.

25. Letter from Herbert Hoover, dated December 23, 1918, reprinted in USSGA, *Spring Manual of the United States School Garden Army* (Washington, DC: GPO, 1919), 3.

26. Introductory letter from P. P. Claxton, Commissioner of Education, in USSGA, *Fall Manual,* 5.

27. O. H. Benson, "Accomplishments of Boys' and Girls' Clubs in Food Production and Conservation," *The World's Food* 74 (November 1917): 147–57.

28. Roland Guss, "Gardening and Nature-Study in Cincinnati Schools," *Nature-Study Review* 15, 3 (March 1919): 85–87; Guss, "Transportation of City Children to the Suburbs for Gardening," *Nature-Study Review* 15, 3 (March 1919): 87–88.

29. Ralph Stoddard, "Mobilizing Unused Land and Forces to Meet an Unprecedented Crisis," *The American City* 16, 5 (May 1917): 471–81.

30. Pack, *War Garden Victorious,* 103.

31. J. H. Prost, "The War Garden Activities in Chicago," *Parks and Recreation* 1, 4 (July 1918): 33–41, quote p. 33. For further information on the war garden campaign in Chicago, see T. J. Smergalski, "Vegetable Gardening in Chicago Parks," *Parks and Recreation* 1, 3 (April 1918): 24–28. For information on statewide efforts, see Frances Duncan, "Uncle Sam's Gardening," *Garden Magazine* 28, 3 (October 1918): 94; and Garden Club of America, "Summer Work on the Illinois Training Farm for Women," *Bulletin of the Garden Club of America* 26 (August 1918): n.p.

32. Smergalski, "Vegetable Gardening," 27.

33. *Garden Magazine* frequently included short highlights of company gardens. Companies listed in Pack, *War Garden Victorious:* New York Telephone;

Firestone Tire in Akron, Ohio; International Harvester; Oliver Chilled Plow; Du Pont de Nemours; American Rolling Mill; American Woolen; General Electric; United States Steel; American Optical; American Steel and Wire; J. I. Case Plow Works; Universal Portland Cement; Oliver Iron Mining; Ford Motor; Eastman Kodak; and Solvay Process.

34. Luther Burlingame, "Shop Gardening as a War Measure," *Industrial Management* 55, 3 (March 1918): 205–209.

35. Charles E. Hildreth, "Cooperative Shop Gardening," *Industrial Management* 55, 3 (March 1918): 204–205.

36. "Reports of the National War Garden Commission," *Garden Magazine* 28, 2 (September 1918): 58.

37. Quoted in Pack, *War Garden Victorious,* 55.

38. See W. C. Funk, *Value of a Small Plot of Ground to the Laboring Man,* United States Department of Agriculture Bulletin 602, March 5, 1918 (Washington, DC: GPO, 1918).

39. Railroad companies listed in Pack, *War Garden Victorious:* New York Central; Union Pacific; Northern Pacific; Missouri, Kansas, and Texas; Illinois Central; Atlantic Coast Line; Chicago, Rock Island and Pacific; Chicago, Burlington, and Quincy; Long Island; New York; New Haven and Hartford; Missouri Pacific; Erie; Boston and Albany; Delaware and Hudson; Chicago and Northwestern; Pere Marquette; Louisville and Nashville; Norfolk and Western; Seaboard Air Line; Chicago, Milwaukee, and St. Paul; and the Cleveland, Cincinnati, Chicago, and St. Louis.

40. "Time Tables and Garden Truck," *Garden Magazine* 28, 1 (August 1918): 24.

41. "Returns on the War Garden Campaign," *Garden Magazine* 28, 1 (August 1918): 20.

42. "Special Message from the Food Administrator to City and Town Officials," *American City* 17, 6 (December 1917): 493.

43. L. L. Sutton, "Conscripting Land and Gardeners," *American City* 17, 2 (August 1917): 172.

44. Edward I. Farrington, *The Backyard Garden: A Handbook for the Amateur, the Community and the School* (Chicago: Laird and Lee, 1918), 17.

45. Charlotte Perkins Gilman, "The Housekeeper and the Food Problem," *The World's Food,* Annals of the American Academy of Political and Social Science 74 (November 1917): 123–40.

46. O. R. Geyer, "Gardens Add $100,000,000 to Nation's Wealth," *Garden Magazine* 26, 3 (October 1917): 93–95. *Garden Magazine* ran a monthly feature, "Returns from the War Garden Campaign," that often included short descriptions of various cities' strategies and their successes in locating garden sites.

47. Charles Lathrop Pack, "Urban and Suburban Food Production," *The*

World's Food, Annals of the American Academy of Political and Social Science 74 (November 1917): 203–206.

48. Roscoe C. E. Brown, *Mobilizing the Country-Home Garden,* Columbia War Papers 1, 3 (New York: Division of Intelligence and Publicity of Columbia University, 1917), 6.

49. In the one article on food production that appeared in *Landscape Architecture* during the war, editor Charles Downing Lay encouraged landscape architects to advise clients on appropriate war garden sites on country estates. He emphasized the value of the designer's judgment in determining where a garden should go in order to avoid wasting energy and materials on inappropriate land. See Lay, "Food Production in War Time," *Landscape Architecture Magazine* 7, 3 (April 1918): 151–52.

50. W. R. Beattie, *The City Home Garden,* Farmer's Bulletin 1044 (Washington, DC: GPO, 1919).

51. Russell Sage Foundation, *War Gardens,* Bulletin of the Russell Sage Foundation Library 28 (April 1918).

52. U.S. Department of Agriculture, Bureau of Plant Industry, *The Small Vegetable Garden,* Farmers' Bulletin 818 (Washington, DC: GPO, 1917); H. M. Conolly, *The City and Suburban Vegetable Garden,* Farmers' Bulletin 936 (Washington, DC: GPO, 1918).

53. This garden was supervised by Grace Tabor, who also served as a field representative for the Woman's Committee of the Council for National Defense. See Tabor, "Making the Smallest Quantity Reach the Farthest," *Garden Magazine* 26, 6 (January 1918): 191–94.

54. F. F. Rockwell, "Getting a Jump on Garden Huns," *Garden Magazine* 27, 4 (May 1918): 187–88, quote p. 187.

55. Parsons, "City Gardens," 8.

56. Farrington, *Backyard Garden,* vii.

57. Susan Sipe Alburtis, "Editorial: War and the School Garden," *Nature-Study Review* 14, 3 (March 1918): 124–25, quote p. 124.

58. J. W. Lloyd, "The War Garden," *University of Illinois Bulletin* 15, 25 (February 1918).

59. Leonidas Willing Ramsey, "Make Your War Garden Attractive," *Garden Magazine* 26, 6 (January 1917): 200–201.

60. Frances Duncan, "Uncle Sam's Gardening," *Garden Magazine* 28, 2 (September 1918): 52.

61. Pack, *War Garden Victorious,* 98.

62. G. W. Hood, "The Small Garden Plot and the High Cost of Living," *Garden Magazine* 25, 3 (April 1917): 162–63; italics in original.

63. F. F. Rockwell, "Cashing-in the War Gardens," *Garden Magazine* 26, 1 (August 1917): 11–12.

64. Farrington, *Backyard Garden,* 11.

65. Frances Duncan, "Uncle Sam's Boom in Organized Gardening," *Garden Magazine* 28, 6 (January 1919): 176.

66. USSGA, *Fall Manual,* 27.

67. "America a Nation of Gardeners," *Garden Magazine* 29, 2 (March 1919): 76.

CHAPTER FIVE. AN ANTIDOTE FOR IDLENESS

1. See President's Emergency Committee for Employment, *The Godman Guild Community Garden: A Permanent Food-Garden Project Applied to Emergency Needs,* Community Plans and Actions Series no. 7 (Washington, DC: GPO, 1931, mimeograph); and U.S. Department of Commerce, *Subsistence Gardens: Some Brief Reports on Industrial, Community, and Municipal Projects Prepared from Reports Received from States and Local Communities,* prepared for the President's Organization on Unemployment Relief (Washington, DC: GPO, 1932).

2. See United States Federal Emergency Relief Administration (FERA), *The Emergency Work Relief Program of the FERA, April 1, 1934–July 1, 1935,* submitted by the Work Division, Federal Emergency Relief Administration, 60. In his report to the New York Governor's Commission on Unemployment Relief, Ernest J. Wolfe provided slightly different data. According to his state-by-state data, during the year 1934, 1,820,663 gardens were planted, comprising 398,593 acres, at a total cost of $5,561,554 and yielding a net return of $36,826,894. See Ernest J. Wolfe, *Industrial and Agricultural Work Relief Projects in the United States,* report submitted to the Governor's Commission on Unemployment Relief, New York, 1935, 18.

3. See Joanna Colcord, *Community Planning in Unemployment Emergencies: Recommendations Growing Out of Experience* (New York: Russell Sage Foundation, 1930); Leah Fedder, *Unemployment Relief in Periods of Depression: A Study of Measures Adopted in Certain Cities, 1857–1922* (New York: Arno Press, 1936); and Josephine Brown, *Public Relief 1929–1939* (New York: Octagon Books, 1940).

4. The evolution of work relief from earlier notions of the "work test" and "made work" to public-serving work relief is described in Joanna C. Colcord, William C. Koplovitz, and Russell H. Kurtz, *Emergency Work Relief* (New York: Russell Sage Foundation, 1932).

5. For instance, when the Federal Emergency Relief Administration (FERA) started funding work relief, the average wage was $27.70 per month, although this varied according to the prevailing rate for a given occupation and locality. See FERA, *Emergency Work Relief Program.*

6. Colcord, Koplovitz, and Kurtz, *Emergency Work Relief,* 7.

7. FERA, *Emergency Work Relief Program,* 58.

8. There is some confusion about the meaning of the term *community garden* as a result of these classifications. In most texts, *community gardens* referred to large tracts of land divided into individual garden plots for work-relief clients; however, some texts considered community gardens to be work-relief gardens where food was grown collectively. In a 1932 report prepared for the President's Organization on Unemployment Relief, five types of subsistence gardens were classified: private or backyard gardens, vacant-lot gardens, community plots divided into individual gardens, industrial gardens, and community gardens not divided into individual gardens (i.e., work-relief projects). See U.S. Department of Commerce, *Subsistence Gardens,* 2. In his 1935 report, Ernest Wolfe defines four categories, according to FERA's classifications, in which *community gardens* refers to tracts subdivided for individual work-relief clients to garden, as shown in table 5. See Wolfe, *Industrial and Agricultural Work Relief Projects,* 4–5. However, New York's agricultural advisor W. E. Georgia classifies community gardens as collectively gardened work-relief projects. See W. E. Georgia, *Subsistence Gardens in New York State in 1932,* report prepared for the New York Temporary Emergency Relief Administration (Ithaca, NY: Department of Vegetable Crops, New York State College of Agriculture, Cornell University, 1933).

9. Georgia, *Subsistence Gardens in New York State,* 10.

10. Joanna Colcord and Mary Johnston, *Community Programs for Subsistence Gardens* (New York: Russell Sage Foundation, 1933), 13.

11. U.S. Department of Commerce, *Subsistence Gardens,* 21.

12. "Food Gardens: Do They Offer a Remedy for Unemployment?" *Nature Magazine* 20, 1 (July 1932): 5.

13. The promotion of gardens by the Red Cross is briefly mentioned in Colcord and Johnston, *Community Programs for Subsistence Gardens,* 9.

14. Ibid., 15.

15. For information on rural industrial garden programs, see President's Organization on Unemployment Relief, "Subsistence Gardens in the Lumber Industry" (Washington, DC, May 3, 1932). The 1932 report, U.S. Department of Commerce, *Subsistence Gardens,* prepared for the President's Organization on Unemployment Relief, lists industrial gardens in Illinois, Indiana, Kentucky, Ohio, and West Virginia. See also President's Organization on Unemployment Relief, "Unemployment Relief: Industrial Firms Help Unemployed Grow Gardens," news release (March 26, 1932).

16. U.S. Department of Commerce, *Subsistence Gardens,* 5.

17. "Gardens for Unemployed Workers," *Monthly Labor Review* 35, 3 (September 1932): 495–97.

18. See President's Organization on Unemployment Relief, *Railroad Subsistence Gardens* (Washington, DC, June 4, 1932).

19. U.S. Department of Commerce, *Subsistence Gardens,* 4.

20. B. F. Goodrich, *Industrial Cooperative Gardening: The Story of a Cooperative Farm Plan* (Akron, OH: B. F. Goodrich Company, 1933), n.p.

21. Ibid.

22. Ibid.

23. For information on New York state's subsistence and work-relief garden programs, see Georgia, *Subsistence Gardens in New York State;* "Subsistence Gardens in Monroe County, N.Y.," *Monthly Labor Review* 45, 3 (September 1937): 623–25; Civic Committee on Unemployment (Rochester, NY), *Four Years Experience with Self-Help or Subsistence Gardens in Monroe County, New York,* June 1936; and excerpts pertaining to New York in Wolfe, *Industrial and Agricultural Work Relief Projects.* The New York Temporary Emergency Relief Administration produced multiple departmental letters with updates and advice for involved agencies and individuals, some of which are available at the National Agricultural Library in Greenbelt, Maryland.

24. See U.S. Department of Commerce, *Spruce Up Your Garden,* prepared for the President's Emergency Committee for Employment (Washington, DC: GPO, 1931).

25. President's Emergency Committee for Employment, *Home Gardens for Employment and Food,* Community Plans and Action Series no. 6 (Washington, DC: GPO, 1931), 1.

26. Tracking the federal relief organizational structure is difficult due to shifting names and short-lived agencies. Previous to the Federal Emergency Relief Administration, the Reconstruction Finance Corporation loaned federal funds to states for relief work. From November 1933 to March 1934, the Civil Works Administration oversaw work relief. The 1933 Federal Emergency Relief Act provided $500 million to states on a matching basis—whereby the state received one dollar for every three dollars expended by the state in the previous three months—and also targeted certain states in need, regardless of their ability to match. Through the Works Division of FERA, employment was provided to an average of two million men and women monthly. See Brown, *Public Relief.*

27. FERA, *Emergency Work Relief Program,* 2.

28. See Wolfe, *Industrial and Agricultural Work Relief Projects,* 1.

29. FERA, *Emergency Work Relief Program,* 56.

30. " 'Workaneat' Group Plans Garden Plot," *East Side Journal* 27 (March 17, 1938): 1. The *East Side Journal* was the official publication of the town of Kirkland, Washington.

31. Civic Committee on Unemployment (Rochester, NY), *Four Years Experience with Self-Help or Subsistence Gardens in Monroe County.*

32. See "Subsistence Gardens in Monroe County, N.Y."

33. See U.S. Department of Agriculture, Bureau of Agricultural Economics, *A Place on Earth: A Critical Appraisal of Subsistence Homesteads,* edited by Rus-

sell Lord and Paul H. Johnstone (Washington, DC: GPO, 1942). This report found that selection criteria resulted in a middle-class participant pool and concluded that the homesteads were not a solution to relief or industrial dependence but rather helped improve quality of life in conjunction with other economic opportunities.

34. Information on Washington state's Garden and Food Preservation Program was collected from archived materials at the University of Washington, in the Federal Emergency Relief Administration Collection held at the Washington State Archives, the Seattle Municipal Archive, and old newspaper articles kept at the Seattle Public Library. Early information was obtained through reports in the *Eastside Journal,* the weekly newspaper of Kirkland and communities along the eastern shores of Lake Washington. The Washington Emergency Relief Administration was the first centralized system of public relief in the state. It ran from April 1934 until 1935, when it was replaced by the State Department of Public Welfare. Each county had a welfare board to oversee work. Also under the board's jurisdiction were the Civilian Conservation Corps, Civil Works Administration, and other federally funded programs. See Genevieve Saxe, *Report of the Garden and Food Preservation Program in the State of Washington for the 1936 Season,* monograph no. 22 (Olympia, WA: Division of Special Programs of the State Department of Public Welfare, April 1936).

35. See Wolfe, *Industrial and Agricultural Work Relief Projects,* 17.

36. In a memorandum from Lewis Leonard dated May 28, 1934, the costs of gardens for 1935 are based on returns from a fifty-five-acre community garden near Seattle.

37. The phrase "no garden, no relief" was cited by Joanna Colcord and Mary Johnston in their 1933 report, *Community Programs for Subsistence Gardens,* for the Russell Sage Foundation. It refers to a policy in southern states to encourage sharecroppers to tend subsistence gardens rather than grow only cash crops. Also see Carl Rehder, "No Garden, No Relief," *Saturday Evening Post* (October 26, 1933): 31, 77–78.

38. See H. E. Young, "Employment Gardens and Their Relation to Public Welfare," source unknown (available at the National Agricultural Library). Excerpts from this source are included in U.S. Department of Commerce, *Subsistence Gardens,* 9–10.

39. Colcord and Johnston, *Community Programs for Subsistence Gardens,* 6.

40. FERA, *Emergency Work Relief Program,* 58.

41. H. E. Young, "Subsistence Garden Plan," Governor's Commission on Unemployment Relief, Indianapolis, 1933.

42. Colcord and Johnston, *Community Programs for Subsistence Gardens,* 33.

43. For information on Detroit Thrift Gardens, see U.S. Department of Commerce, *Subsistence Gardens,* 13–18.

44. Ibid., 15.
45. Colcord and Johnston, *Community Programs for Subsistence Gardens*, 7.
46. Wolfe, *Industrial and Agricultural Work Relief Projects*, 18.
47. FERA, *Emergency Work Relief Program*, 59.

CHAPTER SIX. VICTORY GARDENS OF WORLD WAR II

1. Prior to the United States' official entry into World War II, Americans had learned about Europe's food scarcity. Local groups and communities had started to promote gardens even before the official development of the victory garden campaign.

2. Introductory letter from M. L. Wilson to the secretary of agriculture, July 16, 1945. Attached to a report, *Progress Reports of State Directors of Cooperative Extension Work on the Victory Garden and Home Food Production Programs, June 15, 1945*, mimeograph.

3. This quote by Marvin Jones was provided as a "quotable quote" in U.S. Department of Agriculture, *Victory Garden Kit: Your Victory Garden Program—1945* (Washington, DC: GPO, 1945), n.p. The kit was composed of an assortment of materials, some reprinted from other sources and others mimeographed. The original source of the quote is not indicated.

4. U.S. Office of Program Coordination, Office of War Information, and Office of Price Administration, in Cooperation with the War Food Administration, *Food Fights for Freedom* (Washington, DC: GPO, 1943), 6. For a brief comparison of World War I and World War II food strategies, see Maxcy Robson Dickson, *The Food Front of World War I* (Washington, DC: American Council on Public Affairs, 1944), 181–85.

5. Offices of Program Coordination, War Information, and Price Administration, *Food Fights for Freedom*, 32.

6. The event was briefly summarized in U.S. Office of Civilian Defense, *Garden for Victory: Guide for Planning the Local Victory Garden Program* (Washington, DC: GPO, 1943).

7. U.S. Office of Civilian Defense, *Garden for Victory*, 3.

8. Refer to USDA, *A Coordinated Farm and Home Garden Program*, suggested by the Garden Committee of the U.S. Department of Agriculture and Coordinating Sub-Committee of the Federal Security Administration, mimeograph, 1941.

9. Refer to U.S. Department of Agriculture, *Handbook of Materials for the Victory Farm Volunteers* (1943, mimeograph); Fred Frutchey and Frank W. Lathrop, *Non-Farm Youth Work on Farms* (Extension Service, War Food Administration, February 1943, mimeograph); U.S. Department of Agriculture, *Farm Work for City Youth* (Program Aid 27, 1947); and U.S. Department of Agriculture,

Youth Can Help (Program Aid 28, March 1947), and "Using In-School Non-Farm Youth 16 Years Old or Over as Emergency Workers for Agriculture," *Education for Victory* 1, 7 (June 1, 1942): 15–16.

10. Arno H. Nehrling, "Victory Gardening in Boston," in National Victory Gardening Institute, *Gardening for Victory: A Digest of the Proceedings of the National Victory Garden Conference, 1943* (New York: National Victory Garden Institute, 1943), 24–26.

11. See USDA, *Victory Garden Kit,* n.p.

12. "A Testing Time for Garden Clubs," *Horticulture Magazine* (January 15, 1943); reprint of articles listed in Judith Nelson, "Gardening for Victory," *Horticulture Magazine* (July 1976): 25–40.

13. Information on San Francisco's victory garden program was gathered from articles in the *San Francisco Chronicle,* 1942–1948, and from archived materials kept at Golden Gate Park and the San Francisco Public Library.

14. The National Victory Garden Institute (NVGI) listed the following railroads as having victory garden programs: Union Pacific; Chicago and Eastern Illinois; Burlington Lines' Gulf, Mobile, and Ohio; New York Central; Santa Fe; Baltimore and Ohio; and the Illinois Central. See NVGI, *Industry Digs In* (New York: National Victory Garden Institute, 1944); and NVGI, *Manual for Company-Employee Gardens* (New York: National Victory Garden Institute, 1944).

15. William E. Lyons, "Running a Successful Company Garden," in NVGI, *Gardening for Victory,* 33–34.

16. NVGI, *News* (collection of mimeographed bulletins, 1945 to 1948, available at the National Agricultural Library, Greenbelt, Maryland).

17. L. A. Hawkins, *Have a Victory Garden* (Chicago: International Harvester, 1945).

18. Beech-Nut Packing Company, *Victory Garden Work Sheets* (Canajoharie, NY: Beech-Nut Packing Company, 1945). Also see General Mills, *Victory Guide for Officers of 4-H Clubs* (Minneapolis: General Mills, 1942).

19. As of 1943, the president of the National Victory Garden Institute was Paul Stark and the honorary chairman was M. L. Wilson, director, Extension Division, U.S. War Food Administration, U.S. Department of Agriculture. Vice presidents included:

Chester C. Davis, president, Federal Reserve Bank of St. Louis

Lester J. Norris, Illinois Food Director and Victory Garden Chairman

Richardson Wright, editor of *House and Garden*

John L. McCaffrey, vice president, International Harvester Company

David Burpee, president, W. Atlee Burpee Company

(Russell A. Firestone, Firestone Tire and Rubber Company, was added according to a 1945 report)

The institute's treasurer was James I. Clark, vice president of the Banker's Trust Company. Secretary-manager was Andrew S. Wing. The advisory committee included high-ranking representatives from Montgomery Ward, General Mills, General Foods, Standard Oil, B. F. Goodrich, United States Steel, Ford Motor Company, American Telephone and Telegraph, Studebaker, Chrysler, American Airlines, Goodyear Tire and Rubber Company, International Harvester, Union Carbide, and others.

20. Forty-four companies, representing manufacturers, railroad companies, power companies, and others, received National Victory Garden Institute plaques for all three years: 1943, 1944, and 1945.

21. After the war, the organization was renamed the National Gardening Institute. The institute justified its continued activism to promote gardening as integral to a higher standard of living for the nation, especially in connection with the estimated five million new homes proposed in postwar development. It intended to promote gardening for war veterans, garden therapy in veterans' hospitals, beautification projects, and youth programs for delinquents.

22. E. J. Condon, "Victory Gardens in Industry—A Third Front," in NVGI, *Gardening for Victory,* 10–12, quote p. 11.

23. "Victory Gardening through the Schools," *Education for Victory* 1, 1 (March 3, 1942): 16–17, quote p. 16.

24. For example, a victory garden handbook developed by the Pennsylvania Victory Garden Committee cited a test of seven thousand Pennsylvanians of diverse backgrounds, of whom one third were ill-fed and did not receive adequate nutrients. See Warren B. Mack, Helen Marshall Eliason, and Pauline Beery Mack, *Victory Gardens Handbook of the Victory Garden Committee, War Services, Pennsylvania State Council of Defense* (Pennsylvania State Council of Defense, 1944; reprinted from a publication of the Ellen H. Richards Institute, Pennsylvania State College). See Lydia J. Roberts, "Nutrition from the Victory Garden," in NVGI, *Gardening for Victory,* 40–42.

25. Frederick P. Moersch, "Health and Contentment in Gardening," in War Food Administration, *Report National Victory Garden Conference,* 49–51, quote p. 49; reprinted from *The Minnesota Horticulturist* (May 1943): 75–76.

26. Lester Norris, "Address of Welcome," in NVGI, *Gardening for Victory,* 3.

27. Mrs. Stephen G. Van Hoesen, "Gardening in an Army Hospital," in War Food Administration, *Report National Victory Garden Conference,* 22–25.

28. Judge Marvin Jones, "Victory Gardens in 1945," in War Food Administration, *Report National Victory Garden Conference,* 4–5, quote p. 5.

29. See G. Russell Steininger, *Victory Gardens for the Chicago Area* (Chicago: Chicago Daily News, 1943). Similarly, Washington, D.C.'s newspaper *The Star* published a garden book, *Our Victory Garden,* written by its garden editor, Wilber Youngman. California's growing conditions were covered in Rolly Langley, *War Gardens* (San Jose: Rosicrucian Press, 1943); and Ross Gast, *The Victory*

Garden: Vegetables in the California Garden (Los Angeles: Murray and Gee, 1943). The New York State College of Agriculture produced *Victory Gardens in Greater New York: Instructions for Vegetable Growing in Urban Areas* (New York: Greater New York Victory Garden Council, Civilian Defense Volunteer Office, 1943).

30. The National Committee of Victory Garden Harvest Shows was created in 1942 to raise money for Army Emergency and Navy Relief Programs, with *House and Garden* editor Richardson Wright as chairman. The committee's goal was to have twenty thousand flower, vegetable, and fruit shows, with the proceeds from admissions to be donated. See "Victory Garden Harvest Shows," *Parks and Recreation* 25, 11–12 (July–August 1942): 448–49.

31. *Paramount News* 3, 77 (May 24, 1944). This film clip, as well as others, is available on a reel at the National Archives Motion Picture Films Collection.

32. Quoted from a report by Mrs. Alexander J. Baron, Victory Garden Chairman for Allegheny County. See Mack, Eliason, and Mack, *Victory Gardens Handbook,* 3.

33. D. P. Trent, "Victory Garden Guide for Indian Families" (U.S. Indian Service, 1945, mimeograph, n.p.). The promotional pamphlets targeting Native Americans had a slightly different tone. In particular, they emphasized male participation, perhaps in reaction to the matriarchal structure of some tribes. For instance, the pamphlet stated, "Certainly the man or men in the family will want to do the plowing. No man will be willing for the women to do a lot of digging in hard ground which could much more easily and quickly be plowed."

34. Many publications provided direction for children's victory garden activities. See National Recreation Association, *Gardening: School, Home, Community* (New York: National Recreation Association, 1940); W. H. Gaumnitz, *School and War Gardens: Some Guides and Resources* (Washington, DC: GPO, 1943); and Cooperating Committee on School Lunches, *School Gardens for School Lunches,* circular no. 210, Federal Security Agency, U.S. Office of Education (Washington, DC: GPO, 1942). Articles on school garden programs were regularly included in the U.S. Department of Education's biweekly publication *Education for Victory,* which was sent out to schools and teachers around the nation. The 4-H clubs were very active in rural garden and farm promotion. Summary information on the activities of 4-H clubs and the Boy Scouts is provided in NVGI, *Gardening for Victory,* 20–22.

35. See "Los Angeles Schools Launch Food For Freedom Program," *Education for Victory* 1, 3 (April 1, 1942): 3–4.

36. *House and Garden* included several articles that promoted children's victory gardening: "A Victory Garden for Children" (June 1942): 21; "Your Child and His Garden" (April 1944): 72–78, 102; and "Youth Helpfuls: How Your Child Can Share the Home Work This Summer" (July 1943): 35–40.

37. USDA, *Victory Garden Kit,* n.p.

38. Quoted in *Gardengram* (the monthly newsletter of the National Victory

Garden Institute), n.d. An incomplete collection is available at the National Agricultural Library in Greenbelt, Maryland.

39. Fred G. Heuchling, "Victory Gardening in Chicago," in NVGI, *Gardening for Victory,* 17–20, quote p. 19.

40. District of Columbia Victory Garden Office, *Bulletin* 2 (April 10, 1945). Information on the D.C. victory garden programs is available at the Washingtoniana Division and Washington Star Collection, Martin Luther King Jr. Memorial Public Library, Washington, D.C.

41. See Heuchling, "Victory Gardening in Chicago," 17. Statewide, in 1945, one report indicated 522 victory garden chairmen throughout Illinois. Extension services reported extensive publication distribution, including 150,000 copies of "Garden Guide," 75,000 pamphlets on preserving fruits and vegetables, and 10,000 other garden-related publications. For information on Illinois and other states, see USDA, *Progress Report of the State Directors of Cooperative Extension Work on the Victory Garden and Home Food Preservation Programs, June 15, 1945.*

42. Several home economics experts spoke at the 1943 National Victory Garden Conference. See Mrs. Helen Kendal, "Home Food Preservation," Roberta Hershey, "Practical Home Canning," and J. M. Jackson, "For Success in Home Canning," in NVGI, *Gardening for Victory,* 43–49.

43. H. W. Hochbaum, *The 1943 Victory Garden Program* (Washington, DC: U.S. Office of Civilian Defense, 1943), 3.

44. E. L. D. Seymour, ed., *The New Garden Encyclopedia, Including Special Supplement for 1943 Victory Gardens* (New York: William H. Wise and Co., 1936, 1943), 1380.

45. Later changes in Victory Garden Fertilizer allowed for a 5–10–5 formula for the East Coast, 4–12–4 for the Midwest, and 6–10–4 for the West Coast. See "Time to Plan for Victory Gardens in 1944," *Horticulture* 21, 19 (November 1, 1943): 393–94.

46. Truman quoted in U.S. Department of Agriculture, *Garden and Conserve to Save What We've Won,* Victory Garden Program for 1946, pamphlet 4 (Washington, DC: GPO, 1946), 1.

47. Richardson Wright, "The Land and Our Survival," *House and Garden* 81 (June 1942): 5.

48. H. W. Hochbaum, "A Victory Garden Program for 1945 and Suggestions for a Post War Program," in War Food Administration, *Report National Victory Garden Conference,* 12–17, quote p. 14.

49. M. L. Wilson, "Gardening after Victory," in War Food Administration, *Report National Victory Garden Conference,* 9.

50. H. Tuescher, "Community Garden Park," *Parks and Recreation* 26, 8 (November–December 1943): 359–65, quotes pp. 359, 360.

51. "Rule Victory Gardens Out for Peacetime," *Times Herald,* September 29,

1945, part of the World War II, Victory Gardens 1942–1946, file kept in the Washingtoniana and Washington Star Collection, Martin Luther King Jr. Memorial Library, Washington, D.C. The victory garden program was initially discontinued in March 1946, then reinstated to spur an antifamine drive, and finally closed in June or July of that year. However, some gardens on land adjacent to Rock Creek Park, under the direction of the National Park Service, were allowed to continue. Some of these gardens are still in use today.

PART III. INTRODUCTION

1. Richard D. Parker, "City Gardeners Find Health, Recreation in Community Project," *Horticulture* (October 1952): 389.

2. Louise Bush-Brown was director of the School of Horticulture for Women at Ambler, Pennsylvania, from 1924 to 1952, editor of *Farmer's Digest Magazine,* and coauthor of a garden book with her husband, James Bush-Brown, a landscape architect. See Louise Bush-Brown, *Garden Blocks for Urban America* (New York: Charles Scribner's Sons, 1969). Additional information is from uncataloged archives at the Pennsylvania Horticultural Society Library.

3. See Charles A. Lewis, "Public Housing Gardens: Landscapes for the Soul," in *Landscape for Living: Yearbook of Agriculture, United States Department of Agriculture* (Washington, DC: GPO, 1972): 277–82; and Lewis, *Green Nature/Human Nature: The Meaning of Plants in Our Lives* (Urbana: University of Illinois Press, 1996).

4. In 1996, the author sent a questionnaire to the New York City Housing Authority to gather information on its programs. See Laura Lawson, *Creative Income Opportunities for Youth in the Public Landscape,* Report to National Endowment for the Arts, 1998.

5. Cleveland's school garden program was reinstated in 1983 through the work of the Cooperative Extension, the Master Gardener program, and the National Gardening Association. See Mary Coe, *Growing with Community Gardens* (Taftsville, VT: Countryman Press, 1978); Suzanne Gerety, "Knox Park in Cleveland," *Journal of Community Gardening* 5, 3 (Fall 1996): 8–9; and Pamela Kirschbaum, "Gardens in the School Yard," *Community Greening Review* 9 (1999): 2–13.

6. Ellen Eddy Shaw wrote a regular column on school and children's gardens for *Garden Magazine* from 1909 through World War I. After Shaw's retirement, her assistant, Frances Miner, directed the program until 1974. See Frances Miner, "The Children's Gardens in the Brooklyn Botanical Garden," reprint from the *Journal of the Royal Horticultural Society;* and Doris Stone, "Brooklyn Botanic History," *Journal of Community Gardening* 5, 3 (Fall 1986): 6.

CHAPTER SEVEN. THE COMMUNITY GARDEN MOVEMENT OF THE 1970S AND 1980S

1. For a historical account and pictorial view of Boston's communal gardens, see Sam Bass Warner, *To Dwell Is to Garden* (Boston: Northeastern University Press, 1987).

2. Charlotte Kahn, interview by author, Boston, April 13, 1999.

3. Gardens for All's initial focus on community gardening shifted to more general support of gardening at home, at schools and companies, and in community plots. Through funding from the Troy-Built Manufacturing Company, Gardens for All published books and a bimonthly magazine (established 1979) and sponsored the National Gardening Survey, conducted by the Gallup Organization. In 1979, the organization was renamed the National Gardening Association (NGA). Loss of its main funding source in 1984 led the NGA to establish a for-profit enterprise to sell garden supplies.

4. This fact is cited in Cecil Blackwell, "Why Folks Garden and What They Face," *USDA: Gardening for Food and Fun: Yearbook of Agriculture, 1977* (Washington, DC: GPO, 1977), 2.

5. For a review of environmentalism, see Robert Gottlieb, *Forcing the Spring: The Transformation of the American Environmental Movement* (Washington, DC: Island Press, 1993). Urban ecologists have continued to promote community gardens and other forms of urban agriculture. See Ann Whiston Spirn, *The Granite Garden: Urban Nature and Human Design* (New York: Basic Books, 1984); Nancy and John Todd, *Bioshelters, Ocean Arks, City Farming: The Ecological Basis for Design* (San Francisco: Sierra Club Books, 1984); and Michael Hough, *City Form and Natural Process: Towards a New Urban Vernacular* (New York: Van Nostrand Reinhold, 1984).

6. Gallup Poll, National Garden Survey (Princeton, NJ: Gallup, 1976). Survey conducted for Gardens for All; see Gardens for All, *News about Community Gardening—1976* (January 15, 1976): 1.

7. Helga and William Olkowski, *The City People's Book of Raising Food* (Emmaus, PA: Rodale Press, 1975). Also see Helga Olkowski, Bill Olkowski, Tom Javits, and the Farallones Institute Staff, *The Integral Urban House: Self-Reliant Living in the City* (San Francisco: Sierra Club Books, 1979). A variety of sources have provided regular publications on self-reliant living and gardening, including the Institute for Local Self-Reliance in Washington, D.C.; Tilth in the state of Washington; Ecology Action in Palo Alto, California; Mother Earth News; and Rodale Press. Publications on organic and intensive gardening include Robert Rodale, *The Basic Book of Organic Gardening* (Emmaus, PA: Rodale Press, 1971); and John Jeavons, *How to Grow More Vegetables Than You Ever Thought Possible on Less Land Than You Can Imagine* (Berkeley: Ten Speed Press, 1974). Sugges-

tions for diet were provided in Frances Moore Lappé, *Diet for a Small Planet,* rev. ed. (New York: Ballantine Books, 1975).

8. Rachel Kaplan, "Some Psychological Benefits of Gardening," *Environment and Behavior* 5, 2 (June 1973): 143–62. Later studies that document human response to plants include Rachel Kaplan, "The Role of Nature in the Urban Context," and Roger Ulrich, "Aesthetic and Affective Responses to the Natural Environment," both in *Behavior and the Natural Environment,* edited by Irwin Altman and Joachim Wohlwill (New York: Plenum Press, 1983).

9. Charles Lewis, "Healing in the Urban Environment: A Person/Plant Viewpoint," *Journal of the American Planning Association* 45 (July 1979): 330–38, quote p. 334.

10. Gardens for All regularly sponsored Gallup polls to collect data on national gardening trends. These reports were available for a fee and are currently difficult to find except as secondhand references in other materials. For instance, some highlights of the 1979 gardening survey are included in U.S. Office of Consumer Affairs, *People Power: What Communities Are Doing to Counter Inflation,* 1980, p. 66. Highlights from the 1982 poll are cited in Mark Francis, Lisa Cashdan, and Lynn Paxson, *Community Open Spaces: Greening Neighborhoods Through Community Action and Land Conservation* (Washington, DC: Island Press, 1984), 19. The data are rarely more than basic percentages used to prove or rationalize support for gardening. The NGA occasionally summarized the findings in its newsletters.

11. See John Mollenkopf, *The Contested City* (Princeton: Princeton University Press, 1983); Robert Halpern, *Rebuilding the Inner City: A History of Neighborhood Initiatives to Address Poverty in the United States* (New York: Columbia University Press, 1995).

12. The 1970s witnessed a renewed "back to nature" movement of urbanites to farming communities, rural towns, communes, and so on. Between 1970 and 1976, rural areas and small towns grew by 4.3 million. Though only about 7 percent of the population was actually employed in agriculture in the 1970s, many people looked nostalgically to an agrarian lifestyle as an alternative to the city. See "Living on a Few Acres," in U.S. Department of Agriculture, *Yearbook of Agriculture* (Washington, DC: GPO, 1978).

13. A 1982 study by the Neighborhood Open Space Coalition in New York found that most open-space projects (the main type being community gardens) were clustered in neighborhoods that had experienced social and environmental disruption in the 1960s. See Tom Fox, Ian Koeppel, and Susan Kellam, *Struggle for Space: The Greening of New York City, 1970–1984* (New York: Neighborhood Space Coalition, 1985).

14. Gardens for All, *Gardens for All: A Guide to a Greener, Happier Community* (Charlotte, VT: Gardens for All, 1973), 1.

15. Jamie Jobb, *The Complete Book of Community Gardening* (New York: William Morrow and Company, 1979), 56.

16. Catherine Lerza, "Poisoned Gardens," *The Elements* 11 (September 1975): 1–3. *The Elements* was a monthly newsletter published by the Transnational Institute, a program of the Institute for Policy Studies in Washington, D.C. Rebuttals included one by Gill Friend of the Institute for Local Self-Reliance, "Poisoned Cities and Urban Gardens," *The Elements* (January 1976): 8–11. For a summary of ways to deal with lead in soil, see Ellen Cohen, "A Look at City Soil," *Organic Gardening* 33, 4 (April 1986): 102–104, 106–109.

17. Charles Lewis, "The Harvest Is More Than Vegetables or Flowers," *Brooklyn Botanical Garden Record/Plants and Gardens* 35, 1 (Spring 1979): 14–15.

18. Gardens for All, *Gardens for All;* Mary Coe, *Growing with Community Gardening* (Taftsville, VT: Countryman Press, 1978); Boston Urban Gardeners, *A Handbook of Community Gardening* (New York: Charles Scribner's Sons, 1982). Also see Larry Sommers, *The Community Garden Book* (Burlington, VT: Gardens for All/National Association for Gardening, 1984); Jobb, *Complete Book of Community Gardening;* Susan Drake, *Recreational Community Gardening,* Technical Assistance Bulletin no. 4, U.S. Department of Interior, Bureau of Outdoor Recreation, Lake Central Region (Ann Arbor: U.S. Department of Interior, Bureau of Outdoor Recreation, 1976).

19. By 1994, San Jose had fifteen city-run gardens totaling twenty-seven acres with 850 plot holders. See John Dotter, "On Undeveloped Park Land," *Brooklyn Botanical Garden Record/Plants and Gardens* 35, 1 (Spring 1979): 58–59; John Dotter, "Cultivating People-Plant Relationships in Community and Cultural Heritage Gardens: San Jose, California," *Journal of Home and Consumer Horticulture* 1, 2/3 (1994): 153–70; and Steve Radosevich, "Permanent Sites in San Jose," *Journal of Community Gardening* 1, 2 (May 1982): 8.

20. Jobb, *The Complete Book of Community Gardening,* 55–65, lists a range of funding sources available in 1978, including the Bureau of Outdoor Recreation's Land and Water Conservation Fund; Department of Health, Education, and Welfare funds for seniors; Comprehensive Employment and Training Act (CETA) funds; the Community Service Administration's Community Food and Nutrition Program (CFNP); the Community Development Block Grant Program (CDBG); and ACTION minigrants.

21. Tracking the development of the national USDA Urban Garden Program is difficult because of missing records at the federal level. Some original letters and memos available at the Bancroft Library, University of California, Berkeley, refer mostly to the Los Angeles program. An interim report for the 4-H club written by Allison Brown provides some background. See David Malakoff, "Final Harvest?" *Community Greening Review* 4 (1994): 4–12. In addition, I conducted interviews with people who had been involved in the early stages of the program—including Libby Goldstein (Philadelphia, February 11, 1999) and

Terry Mushovic (Philadelphia, September 6, 1999), who had been directors of the Pennsylvania State Urban Garden Program, and Allison Brown (Washington, DC, February 7, 1999).

22. Quoted in Allison Brown, "Extension Urban Gardening: The Sixteen Cities Experience, Intern Report," 1970.

23. According to research conducted by Carolina A. Ojeda-Kimbrough, in 1995 there were at least nine major organizations promoting community gardens in Los Angeles: Common Ground, Pacific Asian Consortium in Employment, Urban Resources Partnership, LA Harvest, Los Angeles Food Security Network, Interfaith Hunger Coalition/Americorps, Natural Resources Conservation Service, Los Angeles Conservation Corps, and LA Works. Ojeda-Kimbrough, "Investing in Gardens: A Critical Assessment of Community Gardening Programs in Los Angeles," M.A. thesis, University of California, Los Angeles, 1996.

24. C. S. Koehler, "Summary of 1976 Survey of the Home/Community Vegetable Gardening and Other Aspects of Plant Science Public Service," October 1, 1976. Report is available at Bancroft Library, University of California, Berkeley.

25. Tessa Huxley, "Recycled Materials—Building Community Gardens without Lots of Money," *Journal of Community Greening* 2, 1 (Spring 1983): 10.

26. Sally McCabe continues to work in community gardening as the outreach coordinator for Philadelphia Green and an active member of the American Community Gardening Association. Interview by author, February 11, 1999.

27. Charlotte Kahn, interview by author, Boston, April 13, 1999.

28. Terry Mushovic, Philadelphia, interview by author, September 6, 1999. Terry Mushovic was a teacher and social worker before getting a degree in horticulture. She started as a staff member of the Penn State Urban Garden Program in 1977 and later served as director for twelve years. She is currently executive director of the Neighborhood Garden Association, a land trust in Philadelphia.

29. Letter from J. Blaine Bonham, director of Philadelphia Green, to Diane Torrens, coordinator, Mayor Byrne Neighborhood Farm and Garden Program, City of Chicago, August 21, 1979. Letter in American Community Garden files kept at Philadelphia Green, Pennsylvania Horticultural Society. For current information, see website: www.communitygarden.org.

30. As of 1985, the Life Lab Science Program was located in fifty urban and rural schools. By 1988, Life Lab had served over one thousand schools across the country and had two award-winning curricular resources, *Life Lab Science for K-5* and *The Growing Classroom.*

31. Lynne Oscone with Eve Pranis, *National Gardening Association Guide to Kids' Gardening* (New York: John Wiley and Sons, 1983).

32. See Kevin Shank, "The Bronx's GLIE Farms: A Successful Urban Agriculture Business," *Journal of Community Gardening* 1, 4 (December 1982): 18–19.

33. See Julie Stone, "BUGs Sow Landscape Skills," *Journal of Community Gardening* 7, 1 (1988): 8–9.

34. See Mark Francis, Lisa Cashdan, and Lynn Paxson, *Community Open Spaces* (Washington, DC: Island Press, 1984); Randolph T. Hester, *Planning Neighborhood Space,* 2nd ed. (New York: Van Nostrand Reinhold Company, 1975).

35. See Fox, Koeppel, and Kellam, *Struggle for Space.*

36. Andrew Stone, "New York City Garden Leasing Program: Green Guerillas," *Journal of Community Gardening* 6, 3 (Fall 1987): 11.

CHAPTER EIGHT. COMMUNITY GREENING

1. See Marc Breslav, "The Common Ground of Green Words: One Author's Search for a Definition of Community Greening," *Community Greening Review* 1, 1 (1991): 4–9.

2. Jeanette Abi-Nader, Kendall Dunnigan, and Kristen Markely, *Growing Communities Curriculum* (Philadelphia: American Community Gardening Association, 2001).

3. Both surveys were nonrandom surveys mailed to ACGA-affiliated garden programs. The 1996 survey was conducted by the ACGA through the assistance of Suzanne Monroe-Santos, who completed the research as part of her Masters of Science thesis at the University of California, Davis. Both monographs have errors in tables and summaries, so the data included in this book may be affected. I have tried to provide the best numbers; however, there may be some discrepancies with other researchers' data. See American Community Gardening Association, *Findings from the National Community Gardening Survey,* ACGA monograph (1992); American Community Gardening Association, *National Community Gardening Survey,* ACGA monograph (1998).

4. In 1990, responses came from Asheville, Baltimore, Boston (four organizations), Burlington, Cincinnati (three organizations), Cleveland (two organizations), Davis, Dayton, Denver (three organizations), Detroit, Honolulu, Indianapolis, Lansing, Louisville (two organizations), Madison, Minneapolis/St. Paul (two organizations), New York City (six organizations), Pittsburgh, Providence, St. Louis, San Francisco, Seattle (two organizations), Trenton, and Troy. The 1996 survey included responses from Albany, Austin, Boston, Cheshire (Connecticut), Cincinnati, Columbia (Missouri), Davis, Dayton, Denver, Duluth, Durham, Grand Rapids, Houston, Idaho Falls, Indianapolis, Lansing, Lubbock, Madison, Manhattan (Kansas), Minneapolis, Newark, New Orleans, New York, Philadelphia, Pittsburgh, Portland, San Francisco, Santa Barbara, Sarasota (Florida), Seattle, Somerville (Massachusetts), Spokane, Springfield (Ohio), Trenton, Troy, Tucson, the District of Columbia, and Wilmington. In several

cases, several organizations within one city collaborated to reflect their combined work. Follow-up letters were sent to clarify information as necessary.

5. Terry Mushovic, interview by author, Philadelphia, September 6, 1999.

6. For examples of how community gardens can be included in planning and zoning, see Lenny Librizzi, "Comprehensive Plans, Zoning Regulations, Open Space Policies and Goals Concerning Community Gardens and Open Green Space from the Cities of Seattle, Berkeley, Boston, and Chicago." Presented at the GreenThumb Grow Together Workshop, March 20, 1999, and available online through the ACGA website: www.communitygarden.org.

7. See Garden Futures, *Rooted in Our Neighborhoods: A Sustainable Future for Boston's Community Gardens* (Jamaica Plains, MA: Garden Futures, 1997); loose-leaf report available from Garden Futures, 11 Green St., Jamaica Plains, MA, 02130. As of 2004, Garden Futures includes nine organizations.

8. Lenny Librizzi, interview by author, New York, April 15, 1999. As of 1999, Librizzi had been a staff member of the Council on the Environment in New York for thirteen years. His title is Assistant Director of Open Space Greening.

9. Abi-Nader, Dunnigan, and Markely, *Growing Communities Curriculum,* 30.

10. The Trust for Public Land was founded in 1972 as a nonprofit organization devoted to protecting and conserving land for recreation and quality of life. For information on the Trust for Public Land's early work to preserve gardens, see the May 1982 edition of the *ACGA Journal of Community Gardening.* Also see Larry Sommers, *The Community Garden Book* (Burlington, VT: Gardens for All/National Association for Gardening, 1984), 43. For more recent information, see its website: www.tpl.org.

11. Various Boston organizations have acquired community garden land: the Boston Natural Areas Fund, created in 1977, owns thirty community gardens in Roxbury, Dorchester, Jamaica Plain, and East Boston; Boston Urban Gardeners, founded in 1977, currently owns nine properties and plans to acquire two more; the South End Lower Roxbury Open Space Land Trust, founded in 1991, owns eight community gardens and pocket parks; and Dorchester Gardenlands Preserve and Development Corporation, founded in 1977, owns nine properties. See Garden Futures, *Rooted in Our Neighborhoods.*

12. This history of the Seattle P-Patch is based on newspaper articles in the *Seattle Times* and *Post Intelligencer,* P-Patch literature and website information, and site visits.

13. Seattle's P-Patch has shifted offices over its lifetime. It was initially housed in Seattle's Department of Public Resources, then the Department of Housing and Human Services. It is currently in the Department of Neighborhoods.

14. Information on the San Francisco League of Urban Gardeners (SLUG) is culled from SLUG literature, several interviews with staff, and reports as of 2004. See Rosemary Menninger, *Community Gardens in California* (Sacramento: Of-

fice of Planning and Research and Office of Appropriate Technology, 1977); and Kristina Elmstrom, "Community Garden Master Plan," a report submitted to the Open Space Committee, September 30, 1986, prepared by the San Francisco League of Urban Gardeners. Recent information collected from newspaper articles available online (www.sfgate.com) and phone interview with SLUG director Cletis Young by author, July 21, 2004.

15. Cory Calandra, interview by author, San Francisco, December 7, 2001.

16. Information on Philadelphia Green is collected from its literature, interviews, reports, and files kept at the library of the Pennsylvania Horticultural Society. Many documents and articles are available on its website: www.pennsylvaniahorticulturalsociety.com/home/index.html.

17. Patricia Schrieber, "The Philadelphia Experience: A City of Neighborhoods," in *Proceedings of the Second National Urban Forestry Conference* (Washington, DC: Forestry Association, 1982), 86–88.

18. Eight Greene Countrie Townes were implemented:

West Hagert, the first Greene Countrie Towne, started in 1982 on eight blocks.

Point Breeze, started in 1982–83 to address scattered sites in a two-hundred-block neighborhood. Forty gardens were planted and forty-five blocks lined with trees and planters. The greening project led to other community development efforts in youth employment, housing, and development of a neighborhood center.

West Shore, started in 1985 on ten blocks.

Francisville, started in 1987 and resulting in forty greening projects.

Champlost Homes, Oak Lane, started in the late 1980s at a public housing project, focusing on planting front yards and common areas.

Susquehanna, started in 1990 with three main focal points in the 150-block region, including the well-known four-acre Glenwood Green Acres vegetable garden.

Strawberry Mansion, started in 1992 on 125 blocks.

Norris Square, started in 1993, the first Latino Greene Countrie Towne.

19. See Pennsylvania Horticultural Society, *Green Countrie Towne Development Guide: A Guide to Community Development through Horticulture* (Philadelphia: Pennsylvania Horticultural Society, 1988), 1.

20. J. Blaine Bonham, interview by author, Philadelphia, September 6, 1999.

21. Ibid.

22. Hope Wohl Associates, "Feasibility Analysis of For-Profit Urban Agricultural Businesses in Philadelphia as Vacant Land Re-Use Options," report to the Pennsylvania Horticultural Society, 2000. See J. Blaine Bonham, Gerri Spilka, and Darl Rastorfer, *Old Cities/Green Cities: Communities Transform Un-*

managed Land, Planning Advisory Service Report no. 506/507 (Washington, DC: American Planning Association, 2002).

23. Terry Mushovic, interview by author, Philadelphia, September 6, 1999.

24. Besides the Open Space Greening Program, the Council on the Environment of New York City runs the Greenmarket farmers' markets, a waste prevention program, and environmental education programs. Information on New York's community gardens is based on literature, brochures, reports of various organizations, newspaper articles, and the author's interviews with longtime activists, including Lenny Librizzi and Tessa Huxley, in April 1999.

25. See Tom Fox, Ian Koeppel, and Susan Kellam, *Struggle for Space: The Greening of New York City, 1970–1984* (New York: Neighborhood Space Coalition, 1985); and Mark Francis, Lisa Cashdan, and Lynn Paxson, *Community Open Spaces* (Washington, DC: Island Press, 1984).

26. Even with new lease arrangements, activists were concerned about securing sites. In his 1987 critique of New York's leasing program, Andrew Stone noted that most gardens on city property still had only one-year leases that could be revoked with one month's notice, and that some of the most successful community gardens were not eligible for five-year leases. He also noted the drop in approvals for garden leases in general. He reported that the number of GreenThumb leases had dropped; in 1985 GreenThumb submitted 242 applications, of which 161 (67 percent) were approved by the Division of Real Property, while in 1986 GreenThumb submitted 202, of which 81 (40 percent) were approved. See Andrew Stone, "New York City Leasing Program: Green Guerillas," *Journal of Community Gardening* 6, 3 (Fall 1987): 11; Jane Weissman, "New York City Garden Leasing Program: the City Perspective," *Journal of Community Gardening* 6, 3 (Fall 1987): 10–11.

27. Michael Grunwald, "Mayor Giuliani Holds a Garden Sale," *Washington Post,* May 12, 1999: A1.

28. For accounts of this transaction, see Richard M. Stapleton, "Bringing Peace to the Garden of Tranquility," *Land and People* 11, 2 (Fall 1999): 2–7.

29. Green Guerillas, "Gardeners Fight for Livable Communities and Win," Green Guerillas Vitis Vine Special Report, n.d.

30. Tessa Huxley, interview by author, New York, April 15, 1999.

CHAPTER NINE. A LOOK AT GARDENS TODAY

1. Mark Francis, "Some Different Meanings Attached to a City Park and Community Gardens," *Landscape Journal* 6, 2 (Fall 1987): 101–12.

2. Various studies have sought a range of information about gardeners. A 1987 study evaluated 144 gardeners involved in the Penn State Urban Garden

Program. By comparing a stratified random sample of garden sites in Philadelphia with a control group, this study attempted to distinguish ethnic differences among gardeners as well as changes in nutrition and social patterns for gardeners in contrast to nongardeners. See Dorothy Blair, Carol Giesecke, and Sandra Sherman, "A Dietary, Social, and Economic Evaluation of the Philadelphia Urban Garden Project," *Journal of Nutritional Education* 23, 4 (1991): 161–67. County agricultural agent Ishwarbhai Patel conducted a study of 178 gardeners in Newark to determine the socioeconomic impact of gardens. See Ishwarbhai Patel, "Gardening's Socioeconomic Impacts: Community Gardening in an Urban Setting," *Journal of Extension* 29 (Winter 1991): 7–8. In 1992, the American Community Garden Association and Kansas State University conducted a nonrandom survey of the demographics and perspectives of 361 gardeners representing 36 gardens. See Richard Mattson, Jeanne Merkle, Bashir Hassan, and Tina Waliczek, "The Benefits of Community Gardening: Survey Suggests Gardens Contribute Economic and Quality of Life Benefits," *Community Greening Review* 4 (1994): 13–15; and Bashir Nur Hassan, "Educational Models for Community Garden Programs in the United States and Their Potential Application for Sub-Saharan Africa," Ph.D. diss., Department of Horticulture, Forestry, and Recreation, Kansas State University, 1995. Another study focuses on Latino gardeners in Los Angeles and New York City: see Gloria Beatriz Aquino Ramirez, "Social and Nutritional Benefits of Community Gardens for Hispanic-Americans in New York City and Los Angeles," M.A. thesis, Department of Horticulture, Forestry, and Recreation, Kansas State University, 1991.

3. See Luis Aponte-Pares, "Casitas: Place and Culture," *Places* 11, 1 (1996): 54–61; and Joseph Sciorra, " 'I Feel Like I Am in My Country': Puerto Rican Casitas in New York City," *The Drama Review* 34, 4 (Winter 1990): 156–68.

4. Information on the Wattles Farm and Neighborhood Garden was collected from the "Wattles Farm Gardeners' Manual" and from personal interviews with gardeners and visits to the site in the summers of 2001 and 2002.

5. See Andrew Fisher and Robert Gottlieb, "Community Food Security: Policies for a More Sustainable Food System in Context of the 1995 Farm Bill and Beyond" (Los Angeles: Lewis Center for Regional Policy Studies, University of California, Los Angeles, 1995); and Robert Gottlieb, "Community Food Security: A Basic Strategy for Community Health," University of California Wellness Lecture Series, 1996 (n.p.). The *Community Food Security News* is a quarterly publication of the Community Food Security Coalition.

6. See Patel, "Gardening's Socioeconomic Impacts."

7. The Urban Garden of the Los Angeles Regional Food Bank is located on 41st Street between Long Beach Avenue and Alameda Avenue. Information on the garden was collected through a Food Bank press packet that includes articles from *The New York Times, Los Angeles Times,* and *Tu Mundo.* Another useful resource was Carolina A. Ojeda-Kimbrough, "Investing in Gardens: A Critical As-

sessment of Community Gardens in Los Angeles," M.A. thesis, University of California, Los Angeles, 1996. Site visits in the summers of 2001 and 2002 completed this research.

8. This estimate of just over $8,000 was provided by Doris Block, executive director of the Los Angeles Food Bank, in "Harvest of Hope," *Los Angeles Times* (August 19, 1993), B1.

9. See Roberts Foundation, *The New Social Entrepreneur: The Successes, Challenges, and Lessons of Non-Profit Enterprise* (San Francisco: Roberts Foundation, 1996). Two recent studies have explored the economic viability and training agendas of entrepreneurial garden projects. See Gail Feenstra and David Campbell's 1999 study, "Entrepreneurial Community Gardening," University of California Agricultural and Natural Resource Publication 21587; and Laura Lawson and Marcia McNally, "Rethinking Direct Marketing Approaches Appropriate to Low/Moderate Income Communities and Urban Market Gardens," report to Sustainable Agricultural Research and Education Program, University of California, Davis, 1999.

10. See Laura Lawson, "Creative Income Opportunities for Youth in the Public Landscape," report to the National Endowment for the Arts, 1998; Lawson and McNally, "Rethinking Direct Marketing."

11. "Mission Statement and Guiding Principles of the King Edible School Yard," n.d. Information on the progress of the Edible Schoolyard is available online: www.edibleschoolyard.org/garden.html.

12. See Learning in the Real World, *The Edible Schoolyard* (Berkeley: Center for Ecoliteracy, 1999).

CONCLUSION. SUSTAINING A CITY BOUNTIFUL

1. Robert C. Baron, ed., *The Garden and Farm Books of Thomas Jefferson* (Golden, CO: Fulcrum, 1987), 19.

2. See, for instance, the chapter on city farming in Michael Hough, *City Form and Natural Process* (New York: Routledge, 1984).

3. Frederick W. Speirs, Samuel McCune Lindsay, and Franklin B. Kirkbride, "Vacant Lot Cultivation," *Charities Review* 8, 1 (March 1898): 74–107, quote p. 75.

4. Richardson Wright, "Victory Garden Harvest Shows," *House and Garden* (August 1942): 65.

5. Tessa Huxley, interview by author, New York, April 15, 1999.

INDEX

Page references in italics indicate illustrations and tables.

ACGA. *See* American Community Gardening Association
advocacy. *See* promotional efforts
African Americans, 36, 80, *81*, 314n69
agrarianism, 59, 309n21. *See also* nature, reconnection with
agricultural training: school gardens and, 58–60; vacant-lot associations and, 30, 42; war gardens and, 138–40. *See also* job-training gardens
AICP. *See* New York Association for Improving the Condition of the Poor
Airport Farm (Seattle), 162, *163*
Akron (OH): B. F. Goodrich cooperative farm, 153–55. *See also* Firestone Tire and Rubber Company
Alabama, Depression-era garden programs in, *156*
Albertis, Susan Sipe, 69, 78, 139–40
Alemany Public Housing Project (San Francisco), 251, *252*, 253
allotment garden tradition, 23–24, 48–49. *See also* vacant-lot cultivation associations
American City, 135, 138
American Civic Association. *See* American Park and Outdoor Art Association
American Community Gardening Association (ACGA), 12, 215, 231–33, 238–39; annual conferences of, 232–33; community garden contest of, 232, *233;* support by, 244–45; surveys of (1990, 1996), 239–42
American Park and Outdoor Art Association (APOAA), 99, 100–2, 318n25
American Women's Voluntary Services, 179
Among School Gardens (Greene), 52, 77, 89, *101*. *See also* Greene, M. Louise
APOAA. *See* American Park and Outdoor Art Association
Arizona, Depression-era garden programs in, *156*
Arkansas, Depression-era garden programs in, *156*
Atlanta (GA): USDA Cooperative Extension Urban Garden Program and, 226, *227;* war gardens in, *141*

Babcock, Ernest, 59
backyard gardening. *See* home gardens

Baldwin, W. A., 75
Baltimore (MD), USDA Cooperative Extension Urban Garden Program and, 226, *227*
Beattie, W. R., 138
beautification. *See* civic improvement
Beech-Nut Packing Company, 184
benefits of urban gardens: to community, 217; company gardens and, 193; Depression-era gardens and, 151–52, 168; institutional programs and, 46, 48; land use benefits (1990s), 243; vacant-lot programs (1890s) and, 27–31. *See also* civic improvement; health benefits of gardens
Berkeley (CA): BYA Community Garden Patch in, *9, 10,* 274–81, *276–80;* Edible Schoolyard program in, 274–81, *276–80,* 282–85, *283–85;* school garden tradition in, 284–85. *See also* University of California, Berkeley
Berkeley Youth Alternatives (BYA) Community Garden Patch, *9, 10,* 274–81, *276–80*
B. F. Goodrich cooperative farm, 153–55, *154*
Billerica Garden Suburb (Boston), 99
Blair, Emily, 322n13
Bolster, Fred, 74, *79*
Bonham, J. Blaine, 255–56
Boston (MA): Boston Common, 2, 188; community gardens in, 213, *214;* Fenway Victory Garden, 207; 1990s gardens in, 243, 245; postwar programs and, 207–8; Revival program in, 229; school gardens in, 60–62; USDA Cooperative Extension Urban Garden Program and, *227;* vacant-lot associations in, 27, *28*
Boston Urban Gardeners (BUG), 213–14, 224, 231, 235
Breslav, Marc, 238–39
Bridgeport (CT), USDA Cooperative Extension Urban Garden Program and, *227*
Brooklyn (NY): Boys' and Girls' Club in, 210–11; vacant-lot cultivation in, *28*
Brooklyn Botanical Garden, 210
Brown and Sharpe Manufacturing Company (Providence, RI), 134
Brucato, John, 180
Buffalo (NY), 27, *28*
Buffalo, Rochester, and Pittsburgh Railroad, 134–35
BUG. *See* Boston Urban Gardeners
Burbank, Luther, 121
Bureau of Education. *See* U.S. Bureau of Education
Burlington (IA), war gardens in, *141*
Burlington (VT), 214
Burpee, David, 198
Burpee Company. *See* W. Atlee Burpee Company
Butler, Mary Leland, 76, 104
BYA Garden Patch. *See* Berkeley Youth Alternatives Community Garden Patch

Calandra, Cory, 253
Caldwell, Otis, 57–58
California, 216, 233; cooperative extension service program in, 229; current programs in, 264–85; Depression-era garden program in, *156;* Life Lab Science Program in, 234; school gardens in, 69–74, 281–85; victory gardens in, 180–81, *182–83, 187, 189,* 190–91, *191, 192, 194, 198, 199. See also* Berkeley; Los Angeles; San Francisco; University of California, Berkeley
Camp Dix (NJ), 320n2
Canada, 53–54
Carson, Rachel, 216
Carver, George Washington, 75
Casady, Mark, 267
Cashdan, Lisa, 236
casita gardens, 266
Chapman, Bertha, 81–82
Charities, 90, 100
charity: gardening as form of, 23, 27, 29, 31–33; relief efforts and, 145–46. *See also* philanthropic programs; vacant-lot cultivation associations; work-relief gardens

Cheyenne Community Botanical Garden (CO), *233*
Chicago (IL): American Community Gardening Association conference and, 231–32; Chicago Parental School, 81; Housing Authority programs and, 210; 1990s gardens in, 245; USDA Cooperative Extension Urban Garden Program and, 226, *227;* vacant-lot cultivation and, *28;* victory gardens in, 196; war gardens in, 130–33, *134*
Chico State Normal School (CA), 69
child development, 52, 57, 58
Children's Flower Mission. *See* Home Garden Association of Cleveland
children's gardens, 234; community garden movement and, 234; Fairview Garden, 104; National Cash Register Company Boys' Garden, 62, 104–6, *106,* 152; rationales for, 4–5; victory gardens and, 190–93. *See also* school garden movement; school gardens after World War I; U.S. School Garden Army
Children's Gardens for School and Home (Miller), *56,* 85, *108*
Christy, Liz, 258
Cincinnati (OH), 124, 128; vacant-lot cultivation in, *28*
City Beautiful movement, 94–95, 96
City Functional Movement (City Social Movement), 95
Civic Center of Washington, 68
civic improvement, 22, 93–110, 215; American Park and Outdoor Art Association and, 99, 100–2; Cleveland Home Garden Association and, 106–9, *108;* National Cash Register Company efforts and, 62, 104–6, *106,* 152; 1900s movement and, 94–97; 1990s activism and, 218–20; Philadelphia Neighborhood Garden Association and, 209; urban gardens as projects for, 97–100; vacant-lot garden programs and, 97–98, 104, 109–10; victory gardens and, 196–98; women's organizations and, 95, 96, 102–3; Fairview Garden and, 76, 86, 90, 103–4, *105. See also* community activism; community development; community greening programs
Clapp, Henry Lincoln, 60
Claxton, Philander P., 51, 65–66, 121, 126
Cleveland (OH), 226; Home Garden Association of, 106–9, *108;* Junior Civic Improvement League pledge card, 107, 109; 1930s depression and, 150, *151;* school gardens in, 66, 210; USDA Cooperative Extension Urban Garden Program and, *227*
Clinton Community Garden (NYC), 259–60, *261*
Coe, Mary, 223–24
Colcord, Joanna, 151–52, 164
Colorado, Depression-era garden program in, *156*
Comey, Arthur, 99
Committee on Congestion of the Population, 20
community activism, 206; 1970s programs and, 218–20, 230–31; 1990s programs and, 259–63; sustainability and, 296–97
community development: New York City Housing Authority Tenant Garden Program and, 209–10; 1990s gardens and, 243; Philadelphia Green and, 256–57; urban gardening's sustainability and, 293–97
community garden movement (1970s–'80s), 205–11, 213–15, 220; community activism and, 218–20, 230–31; community open space and, 236–37; company gardens and, 233–34; expansion of, 233–35; Fenway Community Garden, 207–8; gardening organizations and, 224–25; garden management and, 221–23; government support and, 225–29; motivations behind, 215–18; need for local support, 229–30; New York City Housing Authority Tenant Garden Program, 209–10; Philadelphia

community garden movement *(continued)*
Neighborhood Garden Association, 208–9; school gardens and, 210–11; urban revitalization and, 218–20, 230–31. *See also* American Community Gardening Association
community gardens, *6;* acreage of, by state (1934), *156–57; community garden* as term, 327n8; current programs, 264, 265–70; revitalization of, 219; in 1990s, 240, *241;* subsistence gardens, 148, *149;* victory gardens, 201–2; war garden campaign, 137–38. *See also* community garden movement; community greening programs; community open space
community greening programs (1990 onward), 238–63; examples of, 245–63; extension of capacity of, 242–45; Philadelphia Green, 253–57; Seattle's P-Patch, 245, 246–48, *247;* SLUG, 248–53; types of, 240, *241. See also* community development; community garden movement; community gardens; community open space
community open space: 1970s programs, 235–36; 1990s programs, 243, 257, 258. *See also* parks
Community Open Space (Francis, Cashdan, and Paxson), 236
community-supported agriculture, 248, 273
company gardens, *156–57,* 185, 206; benefits of, 193; in Depression era, 152, 153–55; in 1970s, 206, 233–34; postwar promotion of, 201–2; unemployment and, 155; victory gardens and, 181–85, *199;* war garden campaign and, 133–35. *See also* B. F. Goodrich cooperative farm; National Cash Register Company Boys' Garden
Condon, E. J., 185
Connecticut, Depression-era garden program in, *156*
Consolidated Coal Company, 152
Constructive and Preventive Philanthropy (Lee), 102
contests: ACGA community garden contest, 232, *233;* for children's gardens, 86; civic improvement movement and, 106, 107, 109; MacArthur medal program, 192–93; New York City Housing Authority Tenant Garden Program and, 217, 221
Continental Machines (Minneapolis), 182
cooperative extension service programs, 200–1, 215, 225–29, *227. See also* U.S. Department of Agriculture
cooperative farms (1890s), 43–44. *See also* work-relief gardens
Corbett, L. C., 60, 68, 86
cost effectiveness of gardens: of Depression-era relief gardens, 150, 158–60; of school gardens, 87, 270; turn-of-the-century charity and, 31–32; of vacant-lot cultivation associations, 25–26, 44; of victory gardens, 179
Country Life Commission, 19, 58
Country Life in America (magazine), 97
The Craftsman (magazine), 19, 81, 96, 99, 103
Cranston, Mary Rankin, 99
Crosby, Dick, 85–86, 98, 102
Cultivating Communities program (Seattle), 247–48
Curtis, Henry, 84

Dallas (TX), war gardens in, *141*
Davis, B. M., 69
Dayton (OH), 129; vacant-lot cultivation in, *28. See also* Edgemont Solar Garden; National Cash Register Company Boys' Garden
Dean, Arthur, 74
Delaware, Depression-era garden program in, *156*
demonstration gardens: education and, 132–33, 249; in mid-1970s, *222,* 226; Neighborhood Garden Association and, 208–9; victory gardens and, 181, 188, 196; war gardens, 130–33, *131*
Denison Engineering Company (Columbus, OH), 184

Denver (CO): USDA Cooperative Extension Urban Garden Program and, *227;* vacant-lot cultivation in, *28;* war gardens in, *141*
Department of Agriculture. *See* U.S. Department of Agriculture
Department of Labor. *See* U.S. Department of Labor
Depression-era garden campaign (1930s), 114–15, 144–46; benefits of, 151–52, 168; decline of, 168–69; federal support for, 158–60, 161, 168–69; idle land used by, 164–65; local initiatives, 144–45; participation in, 162–64; as relief strategy, 146–49, 151–52, 155–58, 160–62; rules/procedures of, 165–67; state involvement in, 155–58, 160–62. *See also* subsistence gardens; work-relief gardens
Des Moines (IA), 136–37
Detroit (MI), 33; Bureau of Governmental Research, 67; Depression-era gardening in, 166–67, *167;* 1890s vacant-lot associations in, 21, 23, 24–26, *28,* 33–34, 40; USDA Cooperative Extension Urban Garden Program and, 226, *227*
Dewey, John, 120, 125
DeWitt Clinton Farm School (NYC), 4–5, 54, 62–63, *64,* 84, 86, 102
Dickerson, F. B., 30
District of Columbia. *See* Washington, D.C.
Dix, James, 36, 39
Duluth (MN), vacant-lot cultivation in, *28*
Dumond, Frank V., 122, *123*
Duncan, Frances, 140, 142
Durante, Jimmy, 188

Eastin, Delaine, 281
East Orange (NJ), vacant-lot cultivation in, *28*
East Seventh Street School (Los Angeles), 69, *71*
East Side Journal (Seattle), 145
ecological living, 216. *See also* environmental concerns
Ecology Action (Palo Alto, CA), 216
economic depressions. *See* Depression-era garden campaign; vacant-lot cultivation associations
Edgemont Solar Garden (Dayton, OH), 232
Edible Schoolyard (Berkeley, CA), 282–85, *283–85*
education, 7–8, 277, 295–96; later twentieth-century programs and, 228, 249; reform of, 56–58; war gardens and, 132–33, 138–40. *See also* agricultural training; job-training gardens; school garden movement; school gardens after World War I
Education for Victory (Office of Education publication), 177, 190–91
1890s urban garden movements. *See* civic improvement; school garden movement; vacant-lot cultivation associations
The Elements (newsletter), 221
Eliot, Charles, 121
Enclosure Acts (England), 23–24
English allotment system, 23–24
Enright, Maginal Wright, 122
entrepreneurial gardens: community-supported agriculture and, 248; current garden programs and, 272–81; 1970s community garden movement and, 235; SLUG and, 251; urban gardening's sustainability and, 296; vacant-lot associations and, 21, 25–26, 31–32. *See also* job-training gardens
environmental concerns: 1970s community gardens and, 216–17, 221; 1990s urban programs and, 249, 250, 253, 258, 262; reconnection with nature, 5–7, *6,* 22, 54–56, 60; turn-of-the-century programs and, 18, 19–20, 95; urban gardening's sustainability, 297. *See also* health benefits of gardens
environmental determinism, 20–22, 94
Environmental Justice Program (SLUG), 250
Europe, urban gardens in, 23–24, 48–49, 53

extension gardening. *See* civic improvement

Fairview Garden (Yonkers, NY), 76, 86, 90, 103–4, *105*
Farrington, Edward, 135, 139, 142
federal support: for Depression-era gardens, 158–60, 161, 168–69; for national service programs, 242; for national war garden campaigns, 120–25; for 1970s community gardening, 225–29, *227;* for victory garden program, 170, 174–78. *See also entries under* U.S.
Fenway Community Garden (Boston), 207–8
FERA. *See* U.S. Federal Emergency Relief Administration
fertilizers, during World War II, 198–99
Firestone Tire and Rubber Company (Akron, OH), 152, 182, 184
Flagg, James Montgomery, 122
Florida, Depression-era garden program in, *156*
flowers: New York City Housing Authority programs and, 209; Philadelphia Flower Show, 257; in victory gardens, 197. *See also* civic improvement
Food Administration. *See* U.S. Food Administration
Food Fights for Freedom campaign, 172–73, *173,* 190
food security: current garden programs and, 270–72; differences between World Wars I and II, 171–72; 1970s gardening and, 215; subsistence gardens, 148–49; urban gardening's sustainability and, 270–72, 294; vacant-lot cultivation and, 42–43; war garden campaign and, 118–20, 171–72
Ford, Mrs. Henry, 180
Ford Motor Company, 152
Forest Hills Gardens (Long Island, NY), 95
Fort Mason Community Garden (San Francisco), *233*
Foster, Merriam and Company, 134
Fox, George, 51
Francis, Mark, 236
Froebel, Frederick, 53
From the Roots Up (AGCA mentorship program), 244
funding: for company gardens, 153; for current garden programs, 271–72, 283; for Depression-era gardens, 145, 150, *151,* 159, 168–69; federal, 145, 159; for New York City community gardens, 258, 259; for 1970s community garden movement, 224–26, 228, 229; for Philadelphia Green, 257; for school gardens, 63–67; for SLUG, 248, 251, 253; for vacant-lot cultivation associations, 31–32; for victory gardens, 179; for war gardens, 128–30. *See also* entrepreneurial gardens

Galloway, Beverly, 68
Gardena Agricultural High School (Los Angeles), 69, 191
Garden and Food Preservation Program (WA), 161–62
Garden City (U.C. Berkeley), 71–73, *73*
Garden Club of America, 124–25, 180
garden clubs: Philadelphia Neighborhood Garden Association, 208; victory gardens and, 179–80. *See also* women's organizations
garden design, 140, 166, 196–98, 220–22, 275–76; garden plans, *61, 70, 87, 88, 199, 269, 278*
Gardener, Cornelius, 27, 33
Garden for the Environment (SLUG), 249, 253
Garden Futures (Boston), 243
Gardengram (newsletter), 185
gardening as hobby: civic improvement movement and, 99–100; 1970s gardens and, 215–18; urban gardening's sustainability and, 295; victory gardens and, 200–1; war garden campaign and, 141–42
Gardening in Elementary City Schools (Bureau of Education), 65–66

Gardening Opportunities (Greenville, KY), *233*
Garden Magazine, 69, 77, 97, 316n94; children's garden contest run by, 86; war garden campaign and, 138, 139, 140, 142; women's groups and, 103
Gardens for All (organization), 214, 220, 234, 237, 336n3; *Gardens for All* (1973 book), 220, 223
garden supervisors: at community gardens, 230–31; in Depression era, 165–66; at school gardens, 66, 76–78; at vacant-lot cultivation associations, 33, 42
Geddes, Patrick, 98
General Electric, 182
General Federation of Women's Clubs, 96, 234
George Putnam Grammar School (Boston), 60–61
Georgia, 1930s garden programs in, *156*
Georgia, W. E., 148
Gilman, Charlotte Perkins, 135–36
Giuliani, Rudolph, 260–63
Glad Bag Company, 232, *233*
GLIE Farms (Bronx, NY), 235
Goddess of Victory poster (World War I), 122, *123*
Godman Guild Community Garden (Columbus, OH), 144, 149
Golden Gate Park victory gardens (San Francisco), 181, *183*
Good, Jessie M., 96
Goodrich cooperative farm. *See* B. F. Goodrich cooperative farm
Goodrich Settlement House (Cleveland, OH), 107
Goodyear Tire and Rubber Company, 182
government agencies, 138; in Depression era, 114–15, 145; Depression relief efforts and, 146; and Food Fights for Freedom campaign, 172–73, 190; post–World War I, 114; and school gardens, 58–59. *See also* U.S. Department of Agriculture, cooperative extension service of; *entries under* U.S.
Grand Rapids (MI), war gardens in, *141*
Greenbelt (MD), 160
Greene, M. Louise, 52, 65, 76, 78, 79, 89, 100
Greene Countrie Towne program (Philadelphia Green), 255–56, 342n18
Green Guerillas (NYC), 214, 257–58, 262–63
Growing Communities Curriculum (AGCA handbook), 244–45
Growing with Community Gardening (Coe), 223–24
Guss, Roland, 74

Hall, Bolton, 26, 46, 306n55
Hampton Normal and Agricultural Institute (VA), 80
Hampton Victory Garden (Hampton, NH), *233*
A Handbook of Community Gardening (Boston Urban Gardeners), 224
Hartford School of Horticulture (CT), 77, 89–90
Hartman, Edward, 99–100
health benefits of gardens, 215; 1890s allotment gardens and, 29–30; 1970s gardening and, 216–17; urban health concerns, 18, 20, 216; victory gardens and, 185–88, 201
Hemenway, H. D., 59–60, 89–90, 313n63
Heuchling, Fred, 195
Hewlett-Packard (Palo Alto, CA), 233
Higgins, Thomas J., 84
Hochbaum, H. W., 197, 200
Hofstader, Richard, 20
Hollywood High School (CA), 69
Home Garden Association of Cleveland, 106–9, *108*
home gardens, 85; acreage of, by state (1934), *156–57;* civic improvement and, 99, 104; Depression-era gardens and, 158; postwar expansion of, 205; School Garden Army and, 125; school gardens and, 82–83; victory gardens and, 192; war garden campaign and, 118, 135, 137
Hood, G. W., 141–42

Hoover, Herbert, 120; Depression-era gardens and, 158; Food Administration and, 126
Hopkins, Harry, 157, 159
horticultural therapy, 241; school gardens and, 80; victory gardens and, 186, 187. *See also* institutional gardens
Horticulture (magazine), on victory gardens, 179, 197
House and Garden (magazine), 200; on victory gardens, 188, 192, 197
Houston, USDA Cooperative Extension Urban Garden Program and, 226, *227*
Huxley, Tessa, 263, 296

Idaho, 1930s garden programs in, *156*
idle land: Depression-era gardens' use of, 164–65; 1970s community gardens and, 219, 220–24; war garden campaign and, 136–38, *137. See also* vacant-lot cultivation associations; vacant-lot garden programs
Illinois: 1930s garden programs in, *156;* victory gardens in, 334n41; war gardens in, 130–33, *132, 134*
The Improvement of Towns and Cities (Robinson), 96
Indiana, 1930s garden programs in, *156*
Indianapolis (IN), USDA Cooperative Extension Urban Garden Program and, *227;* vacant-lot cultivation in, *28;* war gardens in, *141*
Industrial Cooperative Gardening (1953 Goodrich report), *154,* 155
industrial gardens. *See* company gardens
industrialization, 17–21
institutional gardens, 46, 48. *See also* horticultural therapy
instruction. *See* resource materials; technical support
International Farm League, 64, 65, 310n33
International Harvester Company, 184, 196
Iowa, 1930s garden programs in, *156*

Jacksonville (FL), USDA Cooperative Extension Urban Garden Program and, *227*
Jarvis, C. D., 65–66, 74, 83, 84, 85, 90–91
Jeavons, John, 216
Jefferson, Thomas, 289
Jewell, James, 55
Jobb, Jamie, 220
job-training gardens, 8; BUG Landscape Skills Training Program, 235; in current garden programs, 272–81; in 1970s community garden movement, 235; in 1990s programs, 240, *241;* SLUG program, 249–51, 251; in turn-of-the-century programs, 22, 30; urban gardening's sustainability and, 296; vacant-lot cultivation associations and, 42. *See also* agricultural training; entrepreneurial gardens
Johnson, Lady Bird, 208–9
Johnston, Mary, 151–52, 164
Jones, Marvin, 171, 187–88
Jonesboro (TN), vacant-lot cultivation in, *28*
Journal of Community Gardening, 237
The Junior Agriculturist (newsletter), 72–73

Kahn, Charlotte, 213–14, 231
Kansas: 1930s garden programs in, *156;* vacant-lot cultivation in Kansas City, *28*
Kaplan, Rachel, 217
Kelgaard, J. W., 30, 33
Kentucky, 1930s garden programs in, *156*
Kilpatrick, Van Evrie, 65, 83, 91
King, Francis, 103
King, Mel, 213
Kline, Kevin, 260
Knox, Margaret, 55
Koch, Edward, 260

Lacrosse (WI), vacant-lot cultivation in, *28*
Lakeside (NJ), community garden in, 150
land acquisition: acreage of garden programs, by state (1934), *156–57;* Depression-era programs and, *156–57,* 164–65; New York community gardens and, 259; school gardens and, 82–85; site permanence and, 236–37;

urban gardening's sustainability and, 299–300; vacant-lot cultivation associations and, 37–41; victory gardens and, 196; war gardens and, 136–38. *See also* temporariness of urban gardens; vacant-lot garden programs
Land Reutilization Authority (MO), 219
Landscape Skills Training Program (BUG), 235
land trusts, 242, 245, 262, 299. *See also* Neighborhood Garden Association/ A Philadelphia Land Trust
land use planning, 12–13, 243, 300
Larkey, Marie Aloysius, 69
Lay, Charles Downing, 325n49
leadership: community garden activists and, 230–31; in community garden movement, 206–7; urban gardening's sustainability and, 297–99. *See also* organizational structure
Lee, Joseph, 84, 102
Lend-Lease program (World War II), 171
Lettuce-Link (Seattle), 246
Lewis, Charles, 209, 217, 221
Livermore, Henrietta, 90
Lloyd, J. W., 140
local initiatives: community garden movement and, 229; Depression-era gardens and, 144–45, 149, 160–62; federal support and, 158–60; 1970s movements and, 215–16; relief agencies and, 146–47; victory gardens and, 178–79; war garden campaign and, 128–30; in Washington state, 160–62
Los Angeles (CA), *7;* Food Bank Urban Garden Program, 270–72, *274;* Gardens and Farms Program, 267; LA Common Ground program, 226, 228; school gardens in, 69–71, 190–91, *192;* USDA Cooperative Extension Urban Garden Program and, *227;* victory gardens in, 190, *191, 192, 194;* war gardens in, *141;* Wattles Farm and Neighborhood Garden, 267–70, *268, 269, 271;* Watts Growing Project, 7, 266
Louisiana, 1930s garden programs in, *156*
Louisville (KY), war gardens in, *141*

MacArthur, Douglas, 192
Maine, 1930s garden programs in, *156*
Manhattan Community Garden (KS), *233*
Manning, Warren, 96, 99, 319n25
Maryland, 1930s garden programs in, *156*
Massachusetts, 86, 134, 178, 312n52; 1930s garden programs in, *156. See also* Boston
Massachusetts Horticultural Society, 61
Mather Field Station Hospital victory garden (San Francisco), *187*
McCabe, Sally, 230
McInerney, Colleen, 181
McLaren, John, 81
McNutt, Paul V., 174–75
Memphis (TN), 226; USDA Cooperative Extension Urban Garden Program and, *227*
Metropolitan Life Company (San Francisco), *199*
Michigan: 1930s garden programs in, *156;* school gardens in, 79
Midler, Bette, 262
Miller, Louise Klein, 66, 85, 97
Milwaukee (WI), USDA Cooperative Extension Urban Garden Program and, 226, *227*
Minneapolis (MN), 109–10, 240, *241;* vacant-lot cultivation in, *28;* war gardens in, *141*
Minnesota, 1930s garden programs in, *156*
Mississippi, 1930s garden programs in, *156*
Missouri: Missouri Land Reutilization Authority, 219; 1930s garden programs in, *156*
Monroe County (NY), 160
Monroe-Santos, Suzanne, 340n3
Montana, 1930s garden programs in, *156*
moral development, 59–60
Mortin, Franklin, 102
Mount Auburn Food Park (Cincinnati, OH), *233*
Muncie (IN), work-relief program in, 163–64

municipal programs: acreage of garden programs, by state (1934), *156–57;* Depression-era gardens and, 150; gardening organizations and, 224–25; subsistence gardens and, 148, *149. See also entries under specific cities*
Mushovic, Terry, 231, 242–43, 339n28

Nashville (TN), 129–30
National Advisory Garden Committee, 175
national campaigns. *See* federal support; *entries under* U.S.
National Cash Register Company Boys' Garden (Dayton, OH), 62, 104–6, *106,* 152
National Child Labor Committee, 125
National Committee of Victory Garden Harvest Shows, 333n30
National Council of Defense, 322n14
National Defense Gardening Conference (1941), 174–75
National Education Association Department of Garden Education, 91–92
National Emergency Food Garden Commission. *See* National War Garden Commission
National Federation of Women's Clubs, 103
National Gardening Association, 233–34, 282, 336n3
National League for Woman's Service, 125
National League of Improvement Associations, 96
National Plant, Flower, and Fruit Guild, 68, 98, 103
National Victory Garden Conferences, 175, 187–88, 195, 198, 200
National Victory Garden Institute (NVGI), 181–85, 184–85, 192–93, 196, 331n14, 331n19
National War Garden Commission, 114, 120, 121–22, 135, 141, 321n12
Native Americans, 80, 190, 333n33
nature, reconnection with, 5–7, *6,* 22, 54–56, 60, 289–90, 337n12
Nature-Study Review (journal), 75, 77, 91, 98, 308n14
Nebraska, 1930s garden programs in, *156*
Neighborhood Garden Association/ A Philadelphia Land Trust, 208–9, 254, 257
neighborhood gardens. *See* community garden movement; community gardens
Neighborhood Open Space Coalition (NYC), 236, 258
Nevada, 1930s garden programs in, *156*
Newark (NJ), 226, 240, 241; squatters' gardens near, *169;* USDA Cooperative Extension Urban Garden Program and, *227;* war gardens in, *141*
New Deal programs, 115, 160, 168. *See also* Depression-era garden campaign
The New Garden Encyclopedia (Seymour, ed.), 197
New Hampshire, 1930s garden programs in, *156*
New Jersey, 1930s garden programs in, *156*
New Kensington Project (Philadelphia Green), 257
New Mexico, 1930s garden programs in, *156*
New Orleans (LA): USDA Cooperative Extension Urban Garden Program and, 226, *227;* war gardens in, *141*
Newport Garden Club (RI), 124–25
New York Association for Improving the Condition of the Poor (AICP), 19, 26, 27, *28, 35,* 39, 43–44, 44
New York City, 33, 237; Brooklyn Botanical Garden, 210; Bryant Park war garden, 130, *131;* community gardens in, *6,* 229, 236, 257–63, *260, 261,* 299; Council on the Environment of New York, 343n24; Department of Housing Preservation and Development (HPD), 260; DeWitt Clinton Farm School, 4–5, 54, 62–63, *64,* 84, 86, 102; Green Guerillas, 224; Housing Authority Tenant Garden Program, 209–10, 217, 221, 240; model developments in, 95; number of gardens in (1996), 240, *241;* Operation GreenThumb, 258–59, 260; postwar

programs in, 209–10; school gardens in, 54, 62–63, *64,* 76, 84, 86, 90; urban garden organizing in, 257–63; USDA Cooperative Extension Urban Garden Program and, 226, *227;* vacant-lot gardens and, 26, *28,* 214; victory gardens in, *195. See also* Brooklyn
New York Municipal Art Society, 98
New York School Garden Association, 84, 310n32
New York state: Depression-era gardens in, 155–58; 1930s garden programs in, *156;* Reformatory for Women, 46, 48; Supreme Court of, 262
nonprofit organizations: community gardens and, 224–25; New York City community gardens and, 262; 1990s greening and, 243; P-Patch, 246–47; SLUG, 248. *See also* charity; philanthropic programs
Norris, Lester, 186
North Carolina, 1930s garden programs in, *156*
North Dakota: 1930s garden programs in, *157;* victory gardens in, 178–79
Nuru, Mohammed, 249
NVGI. *See* National Victory Garden Institute

Oakley, Imogen, 96
Ohio, 1930s garden programs in, *157*
Ojeda-Kimbrough, Carolina A., 339n23
Oklahoma: 1930s garden programs in, *157;* war gardens in Oklahoma City, *141*
Olkowski, Helga, 216
Olkowski, William, 216
Olmsted, Frederick Law, Jr., 52
Omaha (NE), vacant-lot cultivation in, *28*
Open Space Greening Program (NYC), 258
Operation GreenThumb (NYC), 258–59, 260
Oregon, 1930s garden programs in, *157*
organizational structure: of community gardens, 224–25; community leadership (1970s), 206; in Depression era, 146, 159; of mid-century national programs, 114, 115; of school gardens, 63–67; of vacant-lot cultivation associations, 33–34; of victory garden campaigns, 178–79; of war gardens, 121, 124, 127–30

Pack, Charles Lathrop, 10, 121, 137, 150. *See also* National War Garden Commission
Panama-Pacific International Exposition (San Francisco, 1915), 86, *88*
Parker, Richard, 207
parks: civic improvement and, 100–2; demonstration gardens and, *131,* 188; Fenway Community Garden, 207–8; school gardens in, 62–63, *64,* 84; victory gardens and, 190, 196, 201; war garden campaign and, 131–32, 133. *See also* community open space
Parsons, Fannie Griscom, 54, 62–63, 124, 139. *See also* DeWitt Clinton Farm School
Parsons, Henry Griscom, 59, 120
participation: in community gardens, 220–25; in current programs, 266; in Depression-era gardens, 162–64; in 1990s gardens, 242; in school gardens, 78–82; social mix and, 8, *10,* 320n2; sustainability and, 12, 300; in vacant-lot cultivation associations, 34–37, 45; in victory gardens, 188–90; in war gardens, 135–36. *See also* community activism; leadership
Patterson, Alice, 98
Patterson, F. J., 105–6
Patterson, John H., 104–5
Paxson, Lynn, 236
Paxton Potato Syndicate, 134
Penn, William, 255
Penn State Urban Garden Program, *222,* 231, 254, 343n2
Pennsylvania, 1930s garden programs in, *157*

Pennsylvania Horticultural Society, 254, 255–56, 256–57; Philadelphia Green and, 234, 245, 253–57
Philadelphia (PA), *9*, 33, 226; locations of Philadelphia Vacant Lots Cultivation Association Gardens, *40, 41;* Neighborhood Garden Association, 208–9; 1990s gardens in, 240, *241,* 245; Penn State Urban Garden Program, 226; Philadelphia Green, 234, 245, 253–57; school gardens in, 51, 66–67; Southwark/Queen Village Community Garden, *222, 223;* urban garden organizing in, 253–57; USDA Cooperative Extension Urban Garden Program and, *227;* vacant-lot cultivation in, *28. See also* Philadelphia Vacant Lots Cultivation Association
Philadelphia Vacant Lots Cultivation Association, 26, 27, 36–37, 39–41, 67, 97, 253, 311n40; continued growth of, 44–49, *47*
philanthropic programs: Depression-era gardens and, 146, 149–50; school gardens and, 64–65; vacant-lot cultivation associations and, 27–31, 305n10. *See also* charity; nonprofit organizations; reform efforts
Phoenix (AZ), USDA Cooperative Extension Urban Garden Program and, *227*
Pingree, Hazen, 23, 24–26, 34, 44
Pittsburgh (PA), 188–90, 240, *241,* 242; vacant-lot cultivation in, *28;* war gardens in, *141*
planned communities, 95, 96
plant selection, 140, 197–98, 221. *See also* garden design
playgrounds, 12–13, 84
Pleasant Village Community Garden (NYC), *233*
pledges, 107, 109, 122, 128, 166–67
plot size: in Depression-era gardens, 165; in vacant-lot associations (1890s), 38; in victory gardens, 181, 196–97; war garden campaign and, 140
Portland (OR), 240, *241*
posters: for victory gardens, *176,* 188; for war garden campaign, 122, *123, 125, 127,* 131, *132*
Potato Patch Farms (Detroit), 24–26, 44
Poughkeepsie (NY), school gardens in, 84, 86, *87*
Powell, George T., 19
Powell, R. F., 33, 42
P-Patch program (Seattle), 214, *233,* 245, 246–48, *247*
President's Emergency Committee for Employment, 158
President's Organization on Unemployment Relief, 151, 152
produce, use of: Depression-era gardens and, 144, 145, 155, 162, 164, 166–67; school gardens and, 89–91; vacant-lot cultivation associations (1890s) and, 42–43; victory gardens and, 197; war gardens and, 140. *See also* entrepreneurial gardens; food security
progressivism. *See* reform efforts
promotional efforts: for Depression-era gardens, 150–52, 158; by 1970s community garden movement, 233; for school gardens, 52, 63–67, 281; for state work-relief gardens, 162; urban gardening's sustainability and, 300–1; by vacant-lot cultivation associations (1890s), 33–34; for victory gardens, 184, 186, 188–90, *191,* 193, 196–98; by war garden campaign, 138–40; by women's organizations, 102–3, 104. *See also* contests; posters
Providence (RI), vacant-lot cultivation in, *28*
psychological health, 186, 217
public housing gardens, 240, *241,* 248; New York City Housing Authority Tenant Garden Program, 209–10, 217, 221, 240. *See also* Alemany Public Housing Project
Public Landscape Project (Philadelphia Green project), 256
public-private collaborations, 66–67, 149–50, 178, 213, 243, 258–59, 297–99

Purdue University Horticulture Department (IN), 163
Putnam School (Boston), 51

Rafter, Elizabeth, 68
railroad garden projects: Depression-era gardens and, 152–53; victory gardens and, 184; war garden campaign and, 134–35
Ramsey, Leonidas, 140
Reading (PA), vacant-lot cultivation in, *28*
recreation. *See* gardening as hobby; parks
Red Cross, 150, 155
reform efforts, 17–22; civic improvement and, 95–97, 99; 1970s vacant-lot gardens and, 219–20; school garden movement and, 54–56; systemic factors and, 20–21, 292–93. *See also* philanthropic programs; self-help; voluntary efforts
relief efforts. *See* Depression-era garden campaign; job-training gardens; vacant-lot cultivation associations; work-relief gardens
resource materials: American Community Gardening Association and, 232–33; 1970s community gardens and, 223–24; victory gardens and, 177, 188–90, 196–97. *See also* technical support
Rhode Island, 1930s garden programs in, *157*
Richmond, Frederick, 225, 226
Riis, Jacob, 4, 22, 31, 52, 55–57
Robinson, Charles Mulford, 96
Rochester (NY): vacant-lot cultivation in, *28;* war gardens in, *141*
Rockwell, F. F., 139, 142
Roosevelt, Franklin Delano, 159
Roosevelt, Theodore, 19
rules and procedures: of current garden programs, 267–68; of Depression-era gardens, 165–67; of 1890s vacant-lot cultivation associations, 34; of 1970s community gardens, 222–23. *See also* pledges
rural conditions, 18–19, 201
Russell Sage Foundation, 95, 138; reports by, 148–49, 151–52, 164, 167

St. Louis (MO), 219, 226; USDA Cooperative Extension Urban Garden Program and, *227;* vacant-lot cultivation in, *28*
St. Mary's Urban Farm (San Francisco), 251, *252,* 253
Salt Lake City (UT), war gardens in, *141*
San Francisco (CA): earthquake in, and tent school gardens, 81–82; SLUG, 5, 245, 248–53, *250;* urban garden organizing in, 248–53; victory garden campaign in, 180–81, *182, 183, 198, 199;* war garden campaign in, 117–18
San Francisco Garden Advisory Committee, 180
San Francisco League of Urban Gardeners (SLUG), 5, 245, 248–53, *250*
San Jose (CA), 224
Santa Fe (NM), 2
school curricula and gardens, 74–76
The School Garden (USDA bulletin), 78
School Garden Army. *See* U.S. School Garden Army
School Garden Association of America (SGAA), 64–65, 86, 91, 310n32, 313n63
school garden movement (1890s–1920), 21; advocacy for, 63–67; in California, 69–74; children reached by, 78–82; civic improvement and, 52, 98, 102; correlative studies and, 75–76; cost of running, 87–89; crop selection for, 89–90; curricula for, 74–76; decline of support for, 91–92; design and, 85–87; discipline and, 76; garden location, 82–85; international precedents for, 53–54; as national movement, 64–65; nature-based experiences and, 54–56, 60; public schools and, 65–67; social reform and, 54–56; teacher training and, 76–78; urban conditions and, 54–56; use of crops from, 90; war

school garden movement *(continued)* garden campaign and, 118, 133; in Washington, D.C., 67–69
School Gardens (USDA bulletin), 78
school gardens after World War I: in Cleveland, *101;* current garden programs and, 281–85; in 1970s and 1980s, 234; postwar programs, 210–11; School Garden Army, 128; teacher training and, 211; victory gardens and, 177, 190–93. *See also* Edible Schoolyard
School Gardens for California Schools (Davis), 69
Schwab, Charles, *195*
Schwab, Erasmus, 53
Scranton (PA), vacant-lot cultivation in, *28;* war gardens in, *141*
Seattle (WA): Airport Farm, 162, *163;* community gardens in, 214, 246–48; P-Patch program, 245, 246–48, *247;* urban garden organizing in, 245, 246–48; USDA Cooperative Extension Urban Garden Program and, *227;* vacant-lot associations (1890s) in, 27
Seaver, B. E., 153
security concerns: of current garden programs, 271; of Depression-era gardens, 165; of National Cash Register Company Boys' Garden, 105; of 1970s community gardens, 221; of victory gardens, 193
seeds, 107, *108,* 166, 198
Seeger, Pete, 260
self-help, 8, 219, 290–92; garden management and, 221–23; urban gardening's sustainability and, 290–91; urban reform and, 20–22; vacant-lot cultivation associations and, 22, 29, 31, 33–34
Seymour, E. L. D., 197
SGAA. *See* School Garden Association of America
Shaw, Ellen Eddy, 55, 77, 84, 210
Shuey, Edwin, 99
Simkhovitch, Mary, 95
Sipe, Susan. *See* Albertis, Susan Sipe
slacker land. *See* idle land
SLUG. *See* San Francisco League of Urban Gardeners
Smergalski, T. J., 133
Smith, Merle, 69
Smith, Neil, 282
soil contamination, 221
Some Types of Children's Garden Work (USDA bulletin), 78
South Carolina, 1930s garden programs in, *157*
South Dakota, 1930s garden programs in, *157*
South End Lower Roxbury Open Space Land Trust (Boston), *214*
South End Settlement House (Boston), 62
Southern Workman (magazine), 80
Spitzer, Eliot, 262
squatters' gardens, 168, *169*
state programs: Depression-era gardens and, 145, 155–58, *156–57;* Federal Emergency Relief Act and, 159, 161; victory gardens and, 178–79
Stebbins, Cyril, 59, 72, 73
Stone, Andrew, 343n26
Stover, Charles B., 84
subsistence gardens, 115, 148–49, 151, 162; categories of, 327n8; federal support for, 160, *161;* financial justification for, 168; types of, 148, *149;* in Washington state, 160–62
Subsistence Homestead Program, *161*
suburban development, 18, 218
Suggestions for Garden Work in California Schools (U.C. Berkeley), 72
sustainability of urban gardening, 12, 287–88; community development and, 293–97; future outlook on, 301–2; gardens' purposes and, 288–93; key strategies for, 297–301
Sutton, L. L., 135
Syracuse (NY), vacant-lot cultivation in, *28*

Talley, Marion, *191*
teaching gardens, 210–11. *See also* demon-

stration gardens; education; school garden movement; school gardens after World War I
technical support: for community garden movement, 216, 224; for current garden programs, 272; Depression-era advisors, 163–64, 165–66; for gardening organizations, 224; garden organization training programs, 244; for postwar gardening, 201; School Garden Army and, 127–28; for school gardens, 77–78; SLUG and, 249; urban toxins and, 221; for vacant-lot cultivation associations, 33, 42; war garden instruction, 138–40. *See also* resource materials
temporariness of urban gardens, 11–12; borrowed land and, 13; community garden movement and, 236; Fenway Community Garden and, 207; garden protection, 259–60; land trusts and, 245; license agreements and, 260; New York City community gardens and, 259–63; 1990s gardens and, 242; rental and leasing programs and, 259–63; urban gardening's sustainability and, 299–300; vacant-lot associations (1890s) and, 38–41, 50
Temporary Emergency Relief Administration, New York (TERA), 157
Tennessee, 1930s garden programs in, *157*
Tether, Chester, 80
Texaco, 182
Texas, 1930s garden programs in, *157*
Thrift Gardens Program (Detroit), 166–67, *167*
Toledo (OH), vacant-lot cultivation in, *28*
Topeka (KS), vacant-lot cultivation in, *28*
TPL. *See* Trust for Public Land
Tracy School (Lynn, MA), 86
Trelstad, Brian, 316n103
Trenton (NJ), 240, *241*
Truman, Harry, 200
Trust for Public Land (TPL), 245, 262
Tuescher, H., 201–2
Tuskegee Normal and Industrial Institute (AL), 80
unemployment: during 1930s depression, 147–48; SLUG and, 249–50; turn-of-the-century programs and, 21, 43–44. *See also* company gardens; Depression-era garden campaign; vacant-lot cultivation associations
U.S. Bureau of Education, 21, 75, 91; Office of School and Home Gardening, 52, 65–66, 83, 91, 126; school gardens and, 85; teacher training and, 77. *See also* U.S. School Garden Army
U.S. Congress, 69
U.S. Council of National Defense, 114
U.S. Department of Agriculture, 138, 270; bulletins from, 78; cooperative extension service of, 200–1, 215, 225–29, *227;* gardening as a hobby and, 215; *Land Policy Review,* 193; Office of Experimental Stations, 78; promotion of rural gardening by, 201; school gardens and, 52; teacher training and, 78; Urban Gardening Program, 201, 206, 225–29, *227;* victory gardens and, 174, 175, 177, 189, 192, 196; war garden campaign and, 118–19, 134, 171; Washington, D.C., school gardens and, 68–69
U.S. Department of Labor, 160
U.S. Federal Emergency Relief Administration (FERA), 145, 159–60, *163,* 164, 168, 328n26
U.S. Food Administration, 114, 120–21, 122
U.S. Office of Civilian Defense, 175, 177
U.S. Office of Defense, Health, and Welfare Services, 175
U.S. Office of Education, 175, 177–78
U.S. School Garden Army (USSGA), 114, 118, 120, 125–28, *127, 129,* 133, 142, 322n21
United States Steel Company, 152
universal gardening movement. *See* civic improvement
University of California, Berkeley, 71–74, *79,* 312n48
University of California, Davis, 73–74
urban development patterns. *See* temporariness of urban gardens

urban garden programs: forms of, 3–4, 17, 148, *149,* 193–96, 240, *241;* future of, 12–14, 301–2 (*see also* sustainability of urban gardening); in late twentieth century, 205–7 (*see also* community garden movement; community greening programs); organizing for, in sample cities, 245–63; origins of, 1–3; rationales for, 4–12, 14, 21, 52, 99, 242–45, 258, 264, 288–93; at turn of the century, 17–22 (*see also* civic improvement; school garden movement; vacant-lot cultivation associations); World War I through World War II, 113–15 (*see also* Depression-era garden campaign; victory garden campaign; war garden campaign). *See also* benefits of urban gardens
Urban Herbals (SLUG brand), 251, 253
urban planning: New Deal and, 166, *167;* 1990s gardens and, 243; planned communities, 95, 96; in Seattle, 247
urban renewal, 218–19. *See also* temporariness of urban gardens
USDA. *See* U.S. Department of Agriculture
USSCG. *See* U.S. School Garden Army
Utah, 1930s garden programs in, *157*

vacant-lot cultivation associations (1890s), *28;* cooperative farms and, 43–44; cost effectiveness of, 25–26, 31–32; depression programs (1890s) and, 23–44; "Detroit Experiment," 24–26; as economical charity, 31–33; as ethical form of relief, 27–31; evaluation of, 49–50; formation of, 33–34; growth of, 26–27, *28;* human benefits of, 27–31; international influence of, 48–49; land acquisition and, 37–41; participants in, 34–37; plot size and, 38; post-Depression continuation of, 44–49; self-help ethic and, 26, 29, 31, 33–34; temporariness and, 24, 39, 50; training programs and, 30, 42
vacant-lot garden programs: acreage of, by state (1934), *156–57;* civic improvement and, 97–98, 104, 109–10; Fairview Garden, 76, 86, 90, 103–4, *105;* 1970s projects and, 213–15; Philadelphia Green, 256–57; rationales for, 21; school gardens and, 83–84; urban revitalization and, 219–20; victory gardens and, 195–96; war garden campaign and, 136–38. *See also* community garden movement
vandalism, 105, 221
Van Hoesen, Mrs. Stephen, 186
Vermont, 1930s garden programs in, *157*
Verrees, J. Paul, 122
victory garden campaign (World War II), 11–12, 115, 170–71; children and, 190–93; commercial programs and, 181–85; design and, 196–98; federal support for, 170, 174–78; Food Fights for Freedom campaign, 172–73, 190; garden types, 193–96; health benefits and, 185–88, 201; origins of, 170, 174–78; postwar gardening and, 200–2; promotional efforts for, 184, 186, 188–90, *191,* 193, 196–98; in San Francisco, 180–81, *182, 183, 198, 199;* state and local planning and, 178–79; supplies rationing and, 198–99; victory garden manuals, 184; war garden campaign and, 140, 171–72; women's organizations and, 179–80
Victory Garden Leader's Handbook (USDA), 197
Virginia, 1930s garden programs in, *157*
voluntary efforts, 114; civic improvement and, 95–97; school gardens and, 21, 64–65; vacant-lot associations (1890s) and, 33. *See also* philanthropic programs; self-help; women's organizations

W. Atlee Burpee Company, 166, 198
Wagner, Judith, 213
War Food Administration (World War II), 171, 177, 200

war garden campaign (World War I), 118; commercial programs and, 133–35; crop value of, *141;* effects of, 140–43; food security and, 118–20; funding of, 127, 129; gardening instruction in, 138–40; household participation in, 135–36; idle land used by, 136–38; in Illinois, 130–33, *134;* local efforts, 128–30; national organization of, 120–25; posters for, 122, *123, 125, 127,* 131, *132;* School Garden Army, 125–28, *127, 129*
War Garden Guyed (National War Garden Commission), 12[illegible] *[illegible]36*
war gardens. *See* vict[illegible] garden campaign (World War I[illegible] war garden campaign (World War[illegible]
War Garden Vi[illegible]ious (Pack), 10, *141*
Washington, D.C., 67–69, 102; Depression-era garden program in, *156;* vacant-lot cultivation in, *28;* war gardens in, *141*
Washington state, relief gardens in, *157,* 160–62, 329n34
Washington State College and Agricultural Extension, 162
Waters, Alice, 282
Watrous, Richard, 103
Wattles Farm and Neighborhood Garden (Los Angeles), 267–70, *268, 269, 271*
Westside Community Garden (NYC), 259, *260*
West Virginia, 1930s garden programs in, *157*
Whitcomb and Blaisdell Machine Tool Company (Worcester, MA), 134
Whiting Corporation, 182
Whitten, James, 225, 226, 228
Whittier School, Hampton Institute (VA), 80, 89
Wickard, Claude R., 174–75
Williams, Niculia, 275
Wilmington (DE), vacant-lot cultivation in, *28*
Wilson, M. L., 171, 185, 200–1
Wilson, Woodrow, 52, 119, 120, 126, 127
window-box gardens, 98
Wisconsin, 1930s garden programs in, *157*
Wolfe, Ernest J., 326n2, 327n8
Woman's Committee of the National Defense Council, 120, 122–25
Woman's Land Army, 125
Woman's National Farm and Garden Association, 103, 124, 180
women's organizations, 103; civic improvement and, 95, 96, 102–3; 1970s community garden movement and, 234; urban garden promotion and, 102; victory gardens and, 179–80; war garden campaign and, 122–25, 135. *See also specific organizations;* garden clubs
Woodruff, Clinton, 100
Woodward, Joanne, 260
Worcester (MA): Good Citizens' Factory, 312n52; war gardens in, *141*
Workaneat Group (Seattle), 160
work-relief gardens, 115, 146; federal support for, 160; general issues for relief programs and, 147; Muncie Plan and, 163; subsistence gardens and, 148–49; in Washington state, 160–62
Works Progress Administration (WPA), 160
World War I. *See* war garden campaign
World War II. *See* victory garden campaign
WPA. *See* Works Progress Administration
Wright, Richardson, 200, 290–91
Wyoming, 1930s garden programs in, *157*

Yearbook of Agriculture (Department of Agriculture; 1973), 215
Young, Cletis, 253
Young, H. E., 166
Youth Garden Internship Program (SLUG), 250–51

zoning, 243

Text: 11.25/13.5 Adobe Garamond
Display: Adobe Garamond
Compositor: Binghamton Valley Composition, LLC
Printer and Binder: Maple-Vail Manufacturing Group

CPSIA information can be obtained
at www.ICGtesting.com
Printed in the USA
LVHW011531040521
686466LV00001B/101

9 780520 243439